A River Flows Through It

A River Flows Through It: A Comparative Study of Transboundary Water Disputes and Cooperation in Asia explores water disputes in Asia and addresses the question of how states sharing a river system can be incentivized to cooperate.

Water scarcity is a major environmental, societal, and economic problem around the world. Increasing demand for water as a result of rapid economic development, high population growth and density has depleted the world's water resources, leading to floods, droughts, environmental disasters, and societal displacement. Shared river basins are therefore often a source of tension and conflict between states. In regions where relations between countries have historically been conflictual, scarce river water resources have exacerbated tensions and have even sparked wars. Yet, more often than not, states sharing a river basin are able to come to some form of agreement, whether they are far-reaching ones such as water-sharing agreements or those that are more limited such as the sharing of hydrological data. Why do riparian states cooperate, especially when power asymmetries between upstream and downstream countries are characteristic of transboundary river basins? How do non-state actors affect the management of international rivers? What are the conditions that facilitate or hinder cooperation? This book wrestles with these questions by exploring water disputes and cooperation in the major river systems in Asia, and by comparing them with cases in Africa, Europe, and the United States.

This book will be of great value to scholars, students, and policymakers interested in transboundary water disputes and cooperation, hydro-diplomacy, and river activism. It was originally published as special issues of *Water International*.

Selina Ho researches and writes on Chinese politics and foreign policy, focusing on water disputes and infrastructural politics. She has published widely on China's water disputes, and is the author of two books, *Thirsty Cities: Social Contracts and Public Goods Provision in China and India* (Cambridge University Press, 2019) and *Rivers of Iron: Railroads and Chinese Power in Southeast Asia* (co-author, University of California Press, 2020). She is Assistant Professor of International Affairs at the Lee Kuan Yew School of Public Policy, National University of Singapore.

Routledge Special Issues on Water Policy and Governance

Edited by:
Cecilia Tortajada (IJWRD) – Institute of Water Policy, Lee Kuan Yew School of Public Policy, NUS, Singapore
James Nickum (WI) – International Water Resources Association, France

Most of the world's water problems, and their solutions, are directly related to policies and governance, both specific to water and in general. Two of the world's leading journals in this area, the *International Journal of Water Resources Development* and *Water International* (the official journal of the International Water Resources Association), contribute to this special issues series, aimed at disseminating new knowledge on the policy and governance of water resources to a very broad and diverse readership all over the world. The series should be of direct interest to all policy makers, professionals and lay readers concerned with obtaining the latest perspectives on addressing the world's many water issues.

Urban Resilience to Droughts and Floods
The Role of Policies and Governance
Edited by Cecilia Tortajada, James Horne and Larry Harrington

Politics and Policies for Water Resources Management in India
Edited by M. Dinesh Kumar

Rural–Urban Water Struggles
Urbanizing Hydrosocial Territories and Evolving Connections, Discourses and Identities
Edited by Lena Holmes, Rutgerd Boelens, Leila M. Harris and Gert Jan Veldwisch

Legal Perspectives on Bridging Science and Policy
Edited by Mara Tignino, Raya Marina Stephan, Renée Martin-Nagle and Owen McIntyre

Virtual Water
Implications for Agriculture and Trade
Edited by Chittaranjan Ray, David McInnes and Matthew R. Sanderson

Flood Resilience of Private Properties
Edited by Thomas Hartmann, Willemijn van Doorn-Hoekveld, Helena F.M.W. van Rijswick and Tejo Spit

A River Flows Through It
A Comparative Study of Transboundary Water Disputes and Cooperation in Asia
Edited by Selina Ho

For more information about this series, please visit: https://www.routledge.com/series/WATER

A River Flows Through It

A Comparative Study of Transboundary Water
Disputes and Cooperation in Asia

Edited by
Selina Ho

Routledge
Taylor & Francis Group
LONDON AND NEW YORK

IWRA

First published 2021
by Routledge
2 Park Square, Milton Park, Abingdon, Oxon, OX14 4RN

and by Routledge
52 Vanderbilt Avenue, New York, NY 10017

Routledge is an imprint of the Taylor & Francis Group, an informa business

British Library Cataloguing-in-Publication Data
A catalogue record for this book is available from the British Library

ISBN13: 978-0-367-63676-0

Typeset in Minion Pro
by codeMantra

Publisher's Note
The publisher accepts responsibility for any inconsistencies that may have arisen during the conversion of this book from journal articles to book chapters, namely the inclusion of journal terminology.

Disclaimer
Every effort has been made to contact copyright holders for their permission to reprint material in this book. The publishers would be grateful to hear from any copyright holder who is not here acknowledged and will undertake to rectify any errors or omissions in future editions of this book.

Contents

PART II
Lessons from Africa, Europe, and the United States

Citation Information

The following chapters were originally published in the *Water International*, volume 42, issue 2 (Feb 2017) and volume 43, issue 5 (July 2018). When citing this material, please use the original page numbering for each article, as follows:

Chapter 1

International water conflict and cooperation: challenges and opportunities
Jacob D. Petersen-Perlman, Jennifer C. Veilleux and Aaron T. Wolf
Water International, volume 42, issue 2 (Feb 2017) pp. 105–120

Chapter 2

China's transboundary river policies towards Kazakhstan: issue-linkages and incentives for cooperation
Selina Ho
Water International, volume 42, issue 2 (Feb 2017) pp. 142–162

Chapter 3

China's 'old' and 'new' Mekong River politics: the Lancang-Mekong Cooperation from a comparative benefit-sharing perspective
Sebastian Biba
Water International, volume 43, issue 5 (July 2018) pp. 622–641

Chapter 4

Assessing the Indus Waters Treaty from a comparative perspective
Neda Zawahri and David Michel
Water International, volume 43, issue 5 (July 2018) pp. 696–712

Chapter 5

The Heilongjiang (Amur) River in Sino-Russian relations: from conflict towards cooperation
Wan Wang and Xing Li
Water International, volume 43, issue 5 (July 2018) pp. 665–695

Chapter 6

River activism, policy entrepreneurship and transboundary water disputes in Asia
Pichamon Yeophantong
Water International, volume 43, issue 5 (July 2018) pp. 163–186

Chapter 7
Dam diplomacy? China's new neighbourhood policy and Chinese dam-building companies
Carla P. Freeman
Water International, volume 43, issue 5 (July 2018) pp. 187–206

Chapter 8
Multi-track diplomacy: current and potential future cooperation over the Brahmaputra River Basin
Yumiko Yasuda, Douglas Hill, Dipankar Aich, Patrick Huntjens and Ashok Swain
Water International, volume 43, issue 5 (July 2018) pp. 642–664

Chapter 9
Infrastructure development and the economics of cooperation in the Eastern Nile
Marc Jeuland, Xun Wu and Dale Whittington
Water International, volume 42, issue 2 (Feb 2017) pp. 121–141

Chapter 10
The remarkable restoration of the Rhine: plural rationalities in regional water politics
Marco Verweij
Water International, volume 42, issue 2 (Feb 2017) pp. 207–221

Chapter 11
The dilemma of autonomy: decentralization and water politics at the subnational level
Scott M. Moore
Water International, volume 42, issue 2 (Feb 2017) pp. 222–239

For any permission-related enquiries please visit:
http://www.tandfonline.com/page/help/permissions

Contributors

Dipankar Aich PlanAdapt, Berlin, Germany.

Sebastian Biba Institute for Political Science, Goethe University Frankfurt, Frankfurt am Main, Germany.

Carla P. Freeman School of Advanced International Studies, Johns Hopkins SAIS, Washington, DC, USA.

Douglas Hill School of Geography, University of Otago, New Zealand.

Selina Ho Lee Kuan Yew School of Public Policy, National University of Singapore.

Patrick Huntjens Inholland University of Applied Sciences.

Marc Jeuland Sanford School of Public Policy and Duke Global Health Institute, Duke University, Durham, USA. Institute of Water Policy, Lee Kwan Yew School of Public Policy, National University of Singapore, Singapore.

Xing Li Department of Government, Beijing Normal University, China.

David Michel Stockholm International Peace Research Institute, Sweden.

Scott M. Moore Penn Global China Program, University of Pennsylvania, USA.

Jacob D. Petersen-Perlman Department of Geography, Planning & Environment, East Carolina University, USA.

Ashok Swain Department of Peace and Conflict Research, Uppsala University, Sweden.

Jennifer C. Veilleux Institute for Water and the Evironment, School of Earth, Arts, & Sciences, Florida International University, Miami, USA.

Marco Verweij Department of Social Sciences and Humanities, Department of Political Science, Jacobs University, Bremen, Germany.

Wan Wang Chinese Academy of Macroeconomic Research, Beijing.

Dale Whittington Departments of Environmental Sciences & Engineering and City & Regional Planning, University of North Carolina, Chapel Hill, USA. Lee Kwan Yew School of Public Policy, National University of Singapore, Manchester Business School, Manchester, UK.

Aaron T. Wolf College of Earth, Ocean, and Atmospheric Sciences, Oregon State University, Corvallis, USA.

Xun Wu Division of Public Policy, Division of Social Science and Division of Environment and Sustainability, Hong Kong University of Science and Technology, China.

Yumiko Yasuda Global Water Partnership, Sweden.

Pichamon Yeophantong School of Humanities and Social Sciences, University of New South Wales Canberra at the Australian Defence Force Academy, Canberra, Australia.

Neda Zawahri Department of Political Science, Cleveland State University, Cleveland, Ohio, USA.

Introduction: Comparing transboundary river cooperation in Asia

Selina Ho

A River Flows Through It consists of a collection of essays that appeared in a special issue (42:2, 2017) and special section (43:5, 2018) of *Water International*. The essays for the special issue were carefully selected from papers written and presented by a team of leading scholars at a workshop organized by the Centre on Asia Globalisation, Lee Kuan Yew School of Public Policy, from May 21 to 23, 2015. The authors in the special section were later invited to provide valuable insights into international rivers in Asia that were not covered by the special issue. A total of 21 authors contributed to the eleven chapters in this volume.

The essays seek to answer two related questions: under what conditions do riparian states cooperate in water disputes, and how do various actors/stakeholders facilitate or hinder cooperation? It is widely acknowledged that both cooperation and conflict exist on a spectrum of relations among riparian states. Conflict among riparians, however, tends to grab the attention of policy-makers, scholars, and the media, more than cooperation. The aim of this collection of essays is to refocus the discussion on the issue of riparian cooperation, by drawing from a range of international relations theories to explain the conditions that promote cooperation, including rationalist, institutionalist, constructivist, realist, and international political economy perspectives. The focus on cooperation is not to say that cooperation is always good (Zeitoun and Mirumachi, 2008); the quality of cooperation matters, that is, how genuine is the agreement to cooperate, do they address underlying or foundational issues, do they deal with the roots of the conflict, and do they entrench the interests of the more powerful actor at the expense of the weaker? The aim of this collection is to examine the factors that contribute to or hinder cooperation, not to address the issue of whether cooperation is a normative good.

Specifically, this volume looks at the roles of various stakeholders or actors in promoting or hindering cooperation in transboundary river basins around the world. These stakeholders include states, both at the national and sub-national levels, and non-state actors, such as international organizations, non-governmental organizations, civil society, private and state-owned enterprises, and even individuals. By focusing on the various actors involved in international river basins, we seek to understand the motivations of different actors and groups, the strategies they use to promote cooperation, and the extent of their impact on the management of international rivers. This collection uses a comparative approach to understand these issues. The focus is on the major transboundary rivers in Asia, and lessons are drawn from Africa, Europe, and the United States.

The problems of cooperation vary across regions, due to differences in geography, climate, resource endowment, and economic, political and societal systems. Not only are the findings important theoretically and empirically, they also have significant policy implications; by identifying the conditions that facilitate cooperation among riparian stakeholders, the articles will help policy-makers formulate policies that encourage cooperation, reduce conflict, and thus, help promote peace and stability among countries that share rivers. They show that managing transboundary river basins is an exercise in foreign policy-making and diplomacy that goes beyond the technicalities of river basin management.

Competition and Cooperation in Transboundary River Basins

In international politics, the competition for water is a transnational non-traditional security[1] issue that has serious repercussions for regional stability. Water is a primordial concern that touches on issues of survival and basic sustenance. Rivers are particularly contentious, as they meander across political boundaries, and hence, are subject to competing interests. Most nation states see themselves as sole proprietors of the portion of the river that flows through their territory; rivers are regarded as a national resource that states have sovereign rights to utilize as they deem appropriate for their self-interests. Water diversions and dam constructions by one riparian, because they could potentially reduce water resources available to another riparian, often provoke great emotions among those negatively affected by such activities. Tensions and conflict often ensue. A potential ultimate outcome could be war. Cutting off the water supply of a country, for instance, can be regarded as a *casus belli*.

Rivers are also contentious because they are a key resource closely knitted into the history of human development. Water resources can be used to alleviate abject poverty. River economies sustain millions of people who live by rivers, and who depend on rivers for their livelihoods. This is particularly true in Asia, where poorer sections of the population tend to congregate around rivers, but less so in Europe. Rivers have multiple uses and are a renewable energy resource. They are a source of drinking water, and are important for navigation, transportation, flood control, irrigation, and hydropower generation. As a result, there are often conflicts over how rivers should be used. For instance, building dams to generate hydropower affects flood levels, fisheries, and navigation. Diverting water from one area to another pits one group of users against another. In regions where relations among riparians are historically conflictive and where water resources are scarce, for example in the Jordan River area, the management of water resources is even more contentious. This is because water disputes are seldom about water itself, but are an extension of broader relations between states in the Middle East.

Wars are, however, rare occurrences. Water disputes are more accurately defined as a form of low-intensity conflict, that is, a political-economic-military confrontation between contending states or groups below conventional forms of war, and above routine, peaceful competition (US Army and US Air Force, 1991). Conflict short of armed violence is a regular occurrence between parties, whether within or between nation states, sharing water resources. For instance, in the history of Malaysia and Singapore, Malaysia's occasional threat to "turn off the tap" was a constant source of insecurity for water-scarce Singapore, which until recently was dependent on Malaysia for most of its water supply. Such episodic threats were a source of tension between the two neighboring states.

Harold Lasswell (1936) in his book, *Politics: Who Gets What, When and How*, states that "the study of politics is the study of influence and the influential", and "the influential are those who get the most of what there is to get" (p. 295). In other words, politics is about power manifested in the ability to influence the allocation and distribution of re-sources. The politics surrounding the allocation of water resources clearly illustrates Lass-well's thesis. The power distribution among riparian states influences the allocation of scare water resources (Ho, 2016; Ho, 2018). Power in transboundary river basins has two aspects: the geographical position of the riparian, and the economic and military prowess of the riparian. Upstream riparians usually have the upper hand vis-à-vis downstream riparians; they are able to export negative externalities to lower riparians while suffer-ing few consequences for their action. In some cases, downstream riparians can be more powerful than upstream riparians if they have control over a disproportionate share of the river. The power distribution among states, independent of their geographical position on the river they share, also impacts the management and allocation of water resources. States with a larger economy, military, and population are more likely to dominate river systems. These states are known as hydro-hegemons (Zeitoun and Warner, 2006).

When hydro-hegemons are also upstream countries, there are few restraints on their behavior. This is particularly true in the case of China; not only do the most important rivers in Asia have their water source in the Tibetan plateau, China is also the largest and most powerful actor in Asia. Realism predicts that in such a situation, cooperation can only come about when the hydro-hegemon imposes cooperation. By this reasoning, it seems that when hydro-hegemons are downstream countries, cooperation is more likely to result. Yet, other studies have shown that conflict is more probable when a downstream country is highly dependent on river resources, is stronger compared to upstream coun-tries, and believes that it has the military capacity to rectify the allocation of water re-sources to its advantage (Homer-Dixon, 1994).

Given the difficulties in upstream-downstream relationships, collective action or joint management of shared river basins around the world has been difficult. Studies by John Waterbury on the Nile River Basin illustrate the problem of collective action among 11 riparian states (Waterbury, 2002). Waterbury, in his keynote speech for the workshop in Singapore from which this collection of essays is drawn from, spelt out the problems of commitment, compliance and defection in collective action. Interestingly, he concludes that first, voluntary compliance is probably the only way to sustain collective action, and second, positive unilateralism in terms of "to do at home what you would hope all your neighbors would do", in the absence of commitment, can set the stage for collective action (Waterbury, 2015). However, there are significant obstacles for states to arrive at voluntary compliance and positive unilateralism at the inter-state level, since states themselves often find it difficult to exercise best practices, such as water recycling, efficient and rational use, and water pollution control, at home. Collective action is also difficult when large num-bers of domestic actors with differing interests are involved in decision-making. Local governments, businesses, non-governmental groups, civil society, and private individuals can influence the management of rivers in multiple ways. Scholars have also pursued the ideas of benefit-sharing (Saddoff and Grey, 2002) and reciprocity (Devlaeminck, 2018) in solving collective action problems, but there are significant obstacles to implementation.

However, despite the difficulties of collective action and headline-grabbing cries of "water wars", armed violence is rare in transboundary river basins – "there has never been a single war fought over water" (Wolf, 1998, p. 257). In the limited incidents of wars between states where water is involved, the underlying or primary cause was not water. In these incidents of armed conflict, hostile relations already exist, and water either act as a catalyst to the conflict or becomes embroiled in the conflict. Indeed, studies have shown that there are high incidences of cooperation among riparians. An Oregon State University-UN study found that the instances of cooperation among riparian states outnumbered the number of conflicts; since 1948, there have been 37 incidents of acute violent conflicts over water (30 of these between Israel and its neighbors), while during the same period, about 295 international water agreements were negotiated and signed (Wolf, 2002).

Of course, history is not necessarily the best guide for the future. Changes in our economic and physical conditions produce new circumstances that could lead to increased water conflict and even war. The review chapter by Jacob D. Petersen-Perlman, Jennifer C. Vellieux and Aaron T. Wolf, "International Water Conflict and Cooperation: challenges and opportunities", argues that as transboundary river basins undergo both rapid physical and economic changes, the field of water conflict and cooperation needs to be re-examined in light of these new realities. Water conflicts could increase as populations grow and climate change continues to manifest. Human security and water security become increasingly intertwined, as a result. The authors propose solutions for reducing water conflicts, and underscore the importance of institutional capacity and third party involvement for resolving conflicts and promoting cooperation. They also pointed to future directions and techniques for addressing the new problems that have arisen in transboundary water resources management.

Greater awareness of the possibility of increased conflict has led to both state and non-state actors taking action to lower the risks of acute conflict. The high costs of war and the sensitivity surrounding water resource issues often prompt states to engage in strategies to prevent the outbreak of armed hostilities. For instance, states may try to desecuritize water as an issue with their neighbors and employ inclusive rhetoric, such as promising not to harm the interests of others, to diffuse some of these tensions (Ho et al., 2019). They also join international conventions and rely on international law to ensure the peaceful settlement of disputes. States also engage in institution-building and joint development programs at the river basin level. No matter how imperfect water institutions and treaties can be, they nevertheless help routinize issues, lower the risks of crises, facilitate issue-linkages, and provide platforms for reducing differences, thus helping to promote stability at both the river basin and regional levels.

Civil society and non-state actors are also playing an active role to deal with water shortage, water pollution and dam-building problems. Many of these groups have cross-border networks and links and hence, are helpful in encouraging cooperation across national borders. Some work closely with national governments to address issues such as water pollution, while others create areas of contestation to encourage change in water management. By skilfully framing water-related issues and constructing narratives around these issues, non-state actors have become effective as a source of pressure on national governments and in encouraging a process of dialogue among various stakeholders that helps improve transboundary water governance.

Incentives and Strategies for Change and Cooperation

Some questions this special issue seeks to address include:

1. Why and how do riparians cooperate? What are the strengths and weaknesses of these cooperative efforts?

2. How does the power distribution among riparian states affect the allocation of water resources? Why would hydro-hegemons cooperate? What strategies could weaker downstream states use to induce cooperation from stronger upstream states?
3. How do non-state actors affect the management of international rivers? What are their strategies and tools for ensuring that they are able to influence national policies?

Following Petersen-Perlman et al.'s chapter, the next four chapters focus on the role of national governments in shaping transboundary river cooperation. Selina Ho's chapter, "China's Transboundary River Policies towards Kazakhstan: issue-linkages and incentives for cooperation", examines power asymmetries in riparian relations and assesses the conditions under which the Chinese government is likely to cooperate with its smaller neighbors in managing transboundary rivers. She argues that China cooperates with some of its riparian neighbors more than others. Using the Basins at Risk scale, she shows that China cooperates with Kazakhstan on the Ili and Irtysh rivers relatively more than it does with the Indochina states of the Mekong, and cooperates least with India on the Brahmaputra. Cooperation takes the form of institution-building and joint development. Although problems in managing shared waters persist and China does not always consider the impact of its actions on Kazakhstan, China has nevertheless taken the unprecedented step of beginning negotiations on a water-sharing agreement with Kazakhstan. Why does China engage in a higher level of institutionalized cooperation with Kazakhstan, a weaker downstream neighbor? Ho demonstrates that issue-linkages between water and issues "beyond" the river basin offer a plausible explanation for China-Kazakhstan transboundary river cooperation. A reciprocal relationship exists between China and Kazakhstan; in return for Kazakhstan's cooperation on a variety of issues vital to its security, economic and strategic interests, China is incentivized to cooperate and accommodates Kazakhstan's concerns over their shared rivers. China-Kazakhstan transboundary water cooperation offers important lessons for weaker downstream states seeking cooperation from stronger upstream states.

The chapter by Sebastian Biba on the Mekong River Basin continues to examine China's relations with its smaller riparian neighbors. By using Sadoff and Grey's benefit-sharing framework, Biba provides useful insights into how China's Lancang-Mekong Cooperation (LMC) is progressing. The LMC was officially launched by China in 2016, and is the only Mekong multilateral initiative, apart from the Greater Mekong Subregion, that comprises all six riparians of the Mekong. Biba's article compares China's policies towards the Mekong before and after the creation of the LMC, by using the benefit-sharing framework. While the LMC signals a new Chinese approach and greater willingness to embrace water resources management, it nevertheless suffers from significant deficiencies at least at this point in time. A key deficiency is that China has continued to emphasize the economic benefits to be derived from the river and beyond while neglecting to generate ecological benefits to ensure sustainability. Nevertheless, Biba sees it as a positive that China has agreed to put the topic of water resource management on the table.

Neda Zawahri and David Michel's chapter on the Indus Water Treaty (IWT) also addresses the theme of cooperation between two parties of unequal power. The IWT is significant because it survived wars between India and Pakistan. Nevertheless the IWT is subject to many criticisms. For instance, it has been criticized for being unable to deal with increasing demand, climate change, water quality issues, mismanagement of water supply, and fragmented governance, among others. The main charge is that the IWT in its current form is unable to deal with the mounting challenges that could lead to future violent water conflicts between India and Pakistan. By comparing the IWT with other river treaties, Zawahri and Michel's study found that while many of the criticisms leveled against the IWT is legitimate, the IWT is not significantly different or lacking compared with other treaties that govern river basins in the Middle East or other parts of Asia. Instead of revising or revisiting the IWT, India and Pakistan can consider using memorandums of understanding or other accords outside the IWT to deal with climate change and water quality issues. This will avoid reopening questions of rights and allocations in the IWT, which will likely drag negotiations into a quagmire.

Wan Wang and Xing Li's chapter examines the extent of cooperation and conflict in the management of the Heilongjiang or Amur River between China and Russia. Historically and presently, Wang and Li found that cooperation between China and Russia on the Heilongjiang tends to follow the broad contours of relations between the two countries; there is more cooperation when overall relations between the two are good. Using the Basins at Risk scale, the authors found that during the time period, 1951–2016, there are more cooperative incidents than conflictual ones between China and Russia on the Heilongjiang. Of the 221 water-related incidents that were uncovered, 89.14% were cooperative, 10.41% were conflictive, and the remaining 0.45% were neutral. While China and Russia presently have a cooperative relationship with respect to the Heilongjiang, the authors also point out some of the problems related to current cooperation, such as a lack of a comprehensive plan for developing and protecting the river, and a weak legal basis for cooperation, among others. China and Russia also have different priorities when it comes to the river, with China focusing on development and Russia focusing on protection.

In contrast to the state-centric views of the preceding chapters, Pichamon Yeophantong's "River Activism, Policy Entrepreneurship and Transboundary Water Disputes in Asia" takes a bottom-up approach to understanding the issue of cooperation. Based on three case studies, namely the Mekong, the Nu-Salween, and the Brahmaputra, Yeophantong focuses on the role of river activists across different countries and contexts in Asia. Drawing from a policy entrepreneurship framework, Yeophantong investigates how river activists strategically frame and narrate rivers as areas of contestation. She argues that contestation in turn becomes desirable for policy change, which in the long run, is potentially constructive for transboundary water cooperation. By defining hydropower development as a "problem" worthy of public concern with attendant effects on legitimacy and national interests, policy entrepreneurship arising from contestation can trigger procedural changes that enhance the reach and effectiveness of transboundary water governance through the inclusion of a wider array of stakeholders. Yeophantong's study also shows how non-state actors work together across borders to advocate for different aspects of their cause. For instance, in the case of Laos's Xayaburi Dam in the Mekong River, Thai activists focused on the legal aspects of the issue, while Cambodian and Vietnamese activities emphasized the dam's threat to regional stability.

Carla Freeman's "Dam Diplomacy?: China's new neighborhood policy and Chinese dam-building companies" plugs a critical gap in the existing literature on the interactions between the Chinese government and the business sector in dam buildings, and the consequent implications for China's relations with its neighbors. Under President Xi Jinping, China has announced a "new neighborhood policy", which emphasizes good neighborliness and deepening of economic benefits towards the region. Freeman's paper investigates whether China's renewed emphasis on managing relations with countries along its periphery is reflected in changes to its role in hydropower development affecting neighboring countries. Freeman focuses on the principal-agent problem, specifically how the Chinese government is managing its dam-building corporations as foreign policy actors within its neighborhood. In her paper, Freeman details the activities of Chinese dam companies in Asia, and preliminarily concludes that despite exhortations from Chinese officials to improve their environmental practices, there is little evidence that Chinese dam building companies have fundamentally altered the way they build dams. Their primary motivation remains profit-seeking. Much of Freeman's paper focuses on the Myitsone dam controversy and its impact on Chinese policies. Her paper shows how despite the intentions of policy-makers, agents on the ground, in this case, the business sector, can be a stumbling block to greater cooperation at the river basin level. It also dispels the myth that Chinese companies operating abroad necessarily represent the strategic interests of the Chinese government.

Like Biba, Yumiko Yasuda *et al.* use the idea of benefit sharing but from the perspective of multi-track water diplomacy with respect to the Brahmaputra River. The authors advocate an all-inclusive approach, encompassing multiple actors using both formal and informal (customary) institutions of cooperation. They proposed the idea of a Zone of Possible Effective Cooperation (ZOPEC), which combines future action situations that facilitate mutual gains through cooperation. ZOPEC takes a political economy approach that looks at how multiple layers of cooperation through trade routes via waterways, basin development, hydroelectricity generation, agriculture land, and aquatic resources could lead to benefit-sharing among different stakeholders.

One of the strengths of this book is the comparative approach it took to understanding water disputes and cooperation not only in Asia, but also in other regions, specifically Africa, Europe, and the United States. This cross-regional approach provides insights into the kinds of problems different regions face in managing transboundary rivers as well as the strategies that are deployed. The lessons drawn from the last three chapters could thus be useful for resolving some of Asia's water disputes bearing in mind differences in geography, resources, terrain, population densities, and political and social conflicts. The experience of the Grand Ethiopian Renaissance Dam tells us how large dams could impact on incentives for cooperation and shift power dynamics; lessons drawn from this case could be useful for considering dam-building activities in the Mekong region. Water pollution is highlighted in the chapter on the Rhine. Asia's rivers also suffer from water pollution. The chapter thus offers lessons on how multiple stakeholders can come together in a virtuous cycle to reduce water pollution. The last chapter on the Colorado River reminds us that sub-national politics is an integral part of international river disputes, an important fact which policy-makers and scholars of international politics should be mindful of. The salient points of these three chapters are summarized below.

Marc Jeuland, Xun Wu and Dale Whittington's "Infrastructure Development and the Economics of Cooperation: the case of the Eastern Nile" uses an international political economy approach to consider riparian cooperation. Their article demonstrates how large infrastructure projects can alter the dynamics of cooperation between riparian countries in transboundary river basins. The authors employ a hydroeconomic optimization model to analyze the effects of the Grand Ethiopian Renaissance Dam (GERD) on the distribution and magnitude of benefits in the Eastern Nile. They show that large infra-structures can shift power asymmetries at the basin level in complex ways, modifying incentives to participate in river basin institutions, as well as opportunities for cooper-ation. For example, new cooperative institutions and approaches to benefit sharing may become viable as a result of infrastructure development. At the same time, shifting power dynamics may strain existing and previously stable governance regimes. Infrastructure development therefore has important implications for cooperation, which perhaps ex-plains why large projects are often contentious even when they are developed to include compensation, benefit-sharing mechanisms, or provisions to limit adverse downstream impacts. This in turn may help to explain why some riparians and stakeholders appear willing to obstruct projects rather than allowing such efforts to move forward, even when they could deliver substantial basin-wide economic benefits.

Marco Verweij's chapter, "The Remarkable Restoration of the Rhine Explained: plural rationalities in water politics", examines how the complex interactions between state and non-state actors can lead to cooperation. In Europe, the problems encountering trans-boundary rivers like the Rhine and the Danube are different from Asia and the Middle East. The impact of dams and water diversions is less salient, while the primary focus is on cooperation to deal with water pollution. According to Verweij, the cleaning up of the Rhine has been characterised by a number of puzzling developments. These in-clude chemical companies reducing their toxic effluents by more than legally required, and riparian governments quarrelling internationally over environmental measures that each of them are undertaking domestically. To address these puzzles, Verweij explains how states, at the national and municipal levels, and non-state actors, such as international institutions, chemical multinationals, water treatment companies, scientists, environmen-tal NGOs, farmers' associations, and the media came together to clean up the Rhine. His paper argues that the plural rationality theory explains how a virtuous cycle consisting of creative, though not always harmonious, interactions between stakeholders with different interests and alternative policy perspectives, has resulted in the cleaning up of the Rhine.

The last chapter by Scott Moore, "The Dilemma of Autonomy: decentralization and water politics at the subnational level", gives a different take on transboundary river cooperation from the other articles by focusing on the sub-national politics of international rivers, specifically the Colorado River in the United States. Moore's chapter considers how conflict dynamics at the international level differ from those at the sub-national level. It develops a framework for understanding the role of sub-national states in water politics in decentralized federal systems. First, the process of decentralization has increased the role of sub-national states in water resource management. Second, the quest for autonomy sometimes leads sub-national officials to prefer loose forms of cooperation. Third, the in-teractions among sub-national states, central governments, and non-governmental actors create the constituency for collaboration in shared river basins. This framework helps explain the long-term shift from weak institution-building to more institutionalized and cooperative relationships between the riparian states of the Colorado River.

Acknowledgements

The editor would like to thank the Lee Kuan Yew School of Public Policy for sponsoring the "Water Politics and Regional Stability" workshop in Singapore in May 2015. She is also grateful to the team of research assistants and event organizers for ensuring the smooth-running of the work-shop. This collection would not have been possible without the valuable comments and feedback provided by the discussants of the workshop, reviewers and Editor-in-Chief of *Water International*. Special thanks to Jing Bo-jiun for his assistance in helping to put together the special issue. She is also grateful to the IWRA and Taylor and Francis for agreeing to publish this as part of the Routledge book series.

Selina Ho

Lee Kuan Yew School of Public Policy,

National University of Singapore

Note

1. Non-traditional security issues are non-military threats to state and human security. They are usu-ally transnational, arise quickly, and can be mitigated even if not prevented. Definition is used by the Consortium of Non-Traditional Security Studies in Asia. http://www.risis-ntsasia.org

References

Devlaeminck, D. (2018). Revisiting the substantive rules of the law of international watercourses: an analysis through the lens of reciprocity and the interests of China. *Water Policy*, 20(2), 323–335. https://doi.org/10.2166/wp.2017.069

Ho, S. (2016). Big Brother, Little Brothers: Comparing China's and India's transboundary river policies. *Water Policy*, 18(S1), 32–49. https://doi.org/10.2166/wp.2016.103

Ho, S. (2018). Power asymmetry and the China-India water dispute. In T.V. Paul (Ed.), *The China-India Rivalry in the Globalization Era*. Washington, DC: Georgetown University Press.

Ho, S, Qian, N. & Yan, Y. (2019). The role of ideas in the China-India Water dispute. *Chinese Journal of International Politics*, 12(2), 263–294. https://doi.org/10.1093/cjip/poz005

Homer-Dixon, T. (1994). Environmental scarcity and violent conflicts: Evidence from cases. *International Security*, 19(1), 5–40. DOI: 10.2307/2539147

Laswell, H. (1936). *Politics: Who gets what, when, how*. New York, NY: McGraw-Hill Book Co.

Sadoff, C., & Grey, D. (2002). Beyond the river: The benefits of cooperation on international rivers. *Water Policy*, 4(5), 389–403.

US Army and US Air Force. (1991). *FM 100–20/AFP 3-20: military operations in low intensity conflict*. Washington, DC: Department of the Army. Retrieved from https://www.globalsecurity.org/military/library/policy/army/fm/100-20/

Waterbury, J. (2002). *The Nile basin: National determinants of collective action*. New Haven, CT: Yale University Press.

Waterbury, J. (2015). *Puzzles of commitment, compliance and defection in collective action*. Keynote Speech at the Workshop on Water Politics and Regional Stability, May 2015. Singapore.

Wolf, A. (1998). Conflict and cooperation along international waterways. *Water Policy*, 1(2), 251–265. https://doi.org/10.1016/S1366-7017(98)00019-1

Wolf, A. (2002). Atlas of international fresh water agreements. Nairobi, Kenya: UNEP.

Zeitoun, M. & Mirumachi. (2008). Transboundary water interaction I: Reconsidering conflict and cooperation., 8(4), 297–316.

Zeitoun, M., & Warner, J. (2006). Hydro-hegemony – a framework for analysis of trans-boundary water conflicts. *Water Policy*, 8, 435–460. https://doi.org/10.2166/wp.2006.054

Taking Stoke: International Water Conflict and Cooperation: challenges and opportunities

Jacob D. Petersen-Perlman, Jennifer C. Veilleux and Aaron T. Wolf

ABSTRACT

Though awareness of the nature of water conflict and cooperation has improved over time, the likelihood of water conflicts could increase as populations continue to grow and climate change continues to manifest. This article details the nature of water conflict and water cooperation. We discuss how water conflicts can be resolved, how water can be seen as a vehicle for change between states, and future directions that can be taken in transboundary water conflict research.

Introduction

All societies are dependent upon readily available freshwater for domestic needs, cultural practices, food production, livelihoods, power generation, industry and/or navigation. However, water resources are subject to change over space and time due to precipitation and temperature cycles, which are becoming increasingly unpredictable due to the effects of climate change. Responding to this variability, users have altered water resources through various engineering efforts, changing the availability, quantity, or quality of water resources for other users. Such alterations can potentially create conflict. Therefore, the management of shared water resources requires innovative and flexible approaches to ensure cooperation between various communities of users.

The complex challenges brought forth by transboundary waters are prevalent across the world. Since first assessing transboundary basins for the International Water Resources Association's Committee for International Collaboration in 1999 (Wolf, Natharius, Danielson, Ward, & Pender, 1999), Oregon State University's Transboundary Freshwater Dispute Database (TFDD) has most recently identified 286 surface water basins that cross international boundaries. In addition, the International Groundwater Resources Assessment Centre (2015) has identified 592 transboundary aquifers. The surface water basins alone cover nearly half of the earth's land area and are also home to 40% of the world's population (TFDD, 2016). Countless more watersheds cross subnational jurisdictions. Rising water demand in these basins, coupled with increasing variability, leads to higher potential for transboundary conflict. Whether conflict increases or decreases, however, depends on several factors that will be discussed in this article.

As transboundary river basins are experiencing rapid changes through both physical and economic pathways around the world, the field of water conflict and cooperation deserves a re-examination based on these new realities. This article begins with an overview of how water conflicts manifest, including which factors may lead to water conflicts and why water conflicts happen (and do not happen) in certain locations. Following this, it discusses how human and water security may be impacted by global change and how threats to human and water security may lead to water conflict. We next discuss how water conflicts may be resolved, first by arguing that parties must overcome the potential risks of cooperation before cooperation may occur. We also argue that power dynamics in transboundary basins need to be reconsidered regarding water cooperation and conflict. We stress that building institutional capacity is the strongest method to prevent and resolve water conflicts, despite its imperfections. After, we discuss the evolution of attempts to install water principles and build institutional capacity in the international arena. We then elaborate on the role of third parties in resolving, managing, and exacerbating water conflicts. The article concludes with suggestions on how new challenges for practitioners managing transboundary waters can be met.

Water conflict

Given the complexity created by different economies, ecosystems, climates, politics and cultures within watersheds, transboundary water management can be considered a type of conflict management and/or conflict prevention. Therefore, successful transboundary water resources management must consider the dimensions of potential conflict. For the context of this article, 'transboundary water conflict' is defined as verbal, economic, or militarily hostile actions between stakeholders over internationally shared water resources. Within this group of actions, 'violent transboundary water conflict' is reserved to describe militarily hostile actions.

Research has shown that a precedent of coordination between stakeholders, through the establishment of institutional capacity in the form of agreements, treaties or informal working relationships, can help reduce the likelihood of conflict (Wolf, Stahl, & Macomber, 2003; Yoffe, Wolf, & Giordano, 2003; Yoffe et al., 2004). Once institutional capacity is established between parties it has been proven to be resilient over time, even as conflict was being waged over other issues (Wolf, 1999).

The complexities within watersheds make managing water a complicated task, but there are opportunities for both conflict and cooperation within these complexities. Competing demands within any given jurisdiction increase the difficulty of finding consensus among a river basin's users, unless there are diplomatic, economic or other institutional precedents. And the chances of mismanagement due to misunderstanding, mistrust or lack of information increase as a watershed crosses more administrative and conceptual boundaries. The boundaries in question can be actual or conceptual spaces, economic sectors, sovereign nation-states, contested areas, ethnic or language regions, other legal jurisdictions, climate zones, mountain ranges, infrastructure, or socially constructed concepts of the environment, space or history.

When competing interests clash and one stakeholder perceives wrongdoing by another stakeholder in a shared basin, transboundary water conflicts are likely to occur through economically and verbally hostile actions. Verbal actions can raise tensions when, for example, upstream and downstream interests clash. One instance of verbal hostilities occurred in the La Plata basin in 1993, when Argentina's foreign minister stated that Argentina would not close a canal that diverted flows from the Pilcomayo River (shared between Argentina and Paraguay) towards its territory, which could impact irrigation within Paraguay (TFDD, 2016).

Economically hostile actions that can lead to transboundary water conflicts include upstream actions that are perceived as creating negative impacts downstream. An example of this is when upstream Namibia proposed building a pipe that would pass through the Caprivi Strip to supply drinking water to its capital city, which caused a dispute with downstream Botswana due to its concerns about protecting the Okavango Delta and its ecotourism industry (Wolf, Kramer, Carius, & Dabelko, 2005). Economically hostile actions regarding transboundary water management have also manifested in tensions in the Amu Darya basin between Uzbekistan and Tajikistan due to downstream concerns over how upstream development would affect summer irrigation and winter hydropower production needs (Wegerich, 2008). Reports in Tajikistan suggested that the destruction of a rail bridge near the Uzbekistan–Afghanistan border, crippling shipments of goods, was due to Uzbekistan's opposition to the Tajikistani Rogun Dam project (Kucera, 2011). As of 2015, the countries are still at odds over Tajikistan's proposed Rogun Dam and how it would impact the state-run production of cotton in Uzbekistan, the third-largest exporter of the resource in the world (Putz, 2015).

Though there are several examples of verbally and/or economically hostile actions regarding shared waters, the risk of conflict over water resources, particularly violent conflict, has been debated throughout the years (e.g. Gleditsch, Furlong, Hegre, Lacina, & Owen, 2006; Gleick, 1993; Homer-Dixon, 1994; Lowi, 1995; Toset, Gleditsch, & Hegre, 2000; Wolf, 2000, 2007). Wolf (2000) addressed this argument, stating that while many scholars have identified water as a historical cause and by extrapolation a future cause of warfare, the term 'water war' has been inconsistently defined. Using a definition of 'water dispute' where water was identified as the explicit cause of military action, De Stefano, Edwards, De Silva, and Wolf (2010) found 38 'acute' disputes (those involving water-related violence) between 1948 and 2008; of those, 31 were between Israel and one or more of its neighbours, with none of the violent events occurring after 1970. Yet most of the cases identified by De Stefano et al. were either (1) political tensions or instability rather than true acts of war, or (2) involved using water as a tool, target, or victim of armed conflict. It is important to note that even though water-related violence still exists (Cooley & Gleick, 2011), and the perceived and actual risk of future conflict over water exists, interactions over water resources to date are largely cooperative. Wolf, Yoffe, and Giordano (2003)'s 'Basins at Risk' study, which catalogued over 1800 events involving water conflict and cooperation between nations from 1948 to 2000, found that cooperative events outnumbered conflictive events by over two to one. There is also an extensive history of formalized water cooperation; over 650 treaties related to water have been signed since 1820 (TFDD, 2016). War is usually not a realistic 'best alternative to a negotiated agreement'; this is increasingly accepted as true in international settings. Violent conflict, in the form of shots fired or troops mobilized, has been present in the international setting, but water resources have never been the sole cause of an all-out war. One analyst familiar with both strategic issues and water resources wrote, "Why go to war over water? For the price of 1 week's fighting, you could build five desalination plants. No loss of life, no international pressure, and a reliable supply you don't have to defend in hostile territory" (Wolf, 1995).

We should note that work by the London Water Research Group, based largely on Zeitoun and Mirumachi (2008) and Zeitoun and Warner (2006), makes the important contribution that transboundary water relations are more complex than individual interactions, and are often both conflictive and cooperative at the same time. Moreover, they point out that not all conflict is bad, as conflict is often the method for disputes to be addressed, and not all cooperation is good, as power imbalances are often solidified in agreements.

Water security and human security

Water conflicts could become more likely between users as their water security becomes threatened. 'Water security' is a concept that captures the threats to sustainable and safe water uses from natural and manmade pressures on water resources, through either water's presence, as in floods or inundation, or its absence, as in drought or contamination (Grey & Sadoff, 2007). The concept has been defined as a population's capacity to ensure sustainable access to sufficient quantities of water at an acceptable quality for human, economic and environmental well-being (United Nations-Water, 2013). Water security assessments of transboundary waters are rare, but given the various potential threats to sustainable water, they will become more common. Water security focuses on concerns regarding changes due to climate change, water development and armed conflict.

'Human security' is loosely defined as freedom from fear and freedom from want (Eldering, 2010). However, the accepted concept of human security can have different applications, depending on the discipline, and as such has been used interchangeably with water security – as water is the basis for human civilization in a broad sense, and human life in a specific sense. The relationships of an individual or community to a water resource as pertains to livelihoods, health, identity, culture or transportation can all have direct relationships with that individual or community's vulnerability, risk or stability. Assessing human security as it relates to water is especially important in situations of water scarcity, active conflict and natural disasters, including effects of climate change or economic development.

In situations involving water resources development that either modifies or eliminates water use by other basin stakeholders, the human security question becomes quite important. Dam development is one example of this sort of change. When a nation or a private corporation receives permission from a nation to develop a shared water resource, the quality, quantity and access to that resource may change for existing users. Even in cases of relocation, success rates for the river community's security are called into question. Occasionally, people with deep roots near a river are asked to become farmers or city-dwellers, or compensation packages may not be sustainable. Measuring human security calls for both quantitative information, such as economics and mortality, and qualitative information, such as identity and culture. Changes to shared water resources on multiple scales can threaten the stability of human security and result in reverberating problems throughout a region (Veilleux, 2014).

Resolving water conflicts

It is not shortage or lack of water that leads to conflict (Yoffe et al., 2003) but how water is governed and managed. To regulate water use and enable sustainable and equitable management in areas stricken with water shortages, stronger policies need to be put in place. Yet water management institutions, especially in developing countries, often lack the human, technical and financial resources to develop and implement comprehensive management plans that can properly accomplish the installation of sufficient governing mechanisms.

Parties have to weigh whether the opportunities that may come from entering into a cooperative agreement will outweigh the risk of not cooperating. Some of the categories of risk perceived by decision makers include the following, identified by Subramanian, Brown, and Wolf (2012):

Capacity and knowledge. This is when parties fear that they will be at a disadvantage at the negotiating table. The risk manifests in two major ways: (1) parties have a perception of less negotiating capacity than others; or (2) parties have a perception that they do not have accurate information about the shared watercourse.

Accountability and voice. Decision makers fear that other basin countries, third parties, or the regional institution may not deliver benefits. Parties perceive that it is highly probable that the proposed institutional arrangement would not result in the flow of benefits, and are concerned that their party's interests would not be adequately considered in joint decision-making processes.

Sovereignty and autonomy. This risk involves sensing the danger that a sovereign's authority may be intruded upon in decision-making processes. It addresses both the desire to have control of its development goals, resources and infrastructure, and the right to make independent decisions.

Equity and access. Parties are acutely aware of ensuring fairness in any agreement, whether it involves specified water quantities and/or qualities, benefit flows, or project costs. Parties also want to ensure their entitlement to use the watercourse, which could mean the right to continue with historic uses, gaining access to a river that runs through (or originates in) its territory, and/or attaining benefits in proportion to its relative size in (or contribution to) the basin.

Stability and support. The final risk is an important one for all parties, but particularly for those that have diversified and powerful stakeholders. Parties consider the implementability of an agreement based on whether key stakeholders support or oppose the agreement and the positive or negative public image of the decision maker.

In the process of overcoming these risks, parties can start working together towards resolving present conflicts and preventing future ones. To create successful transboundary water management, Blomquist and Ingram (2003) suggested building institutional capital, achieving equity and fairness, and meeting needs that are harmonious with both parties' cultural values.

An example of building institutional capacity from a recent water dispute is the Nile River negotiations between Egypt, Ethiopia and Sudan over the Grand Ethiopian

Renaissance Dam. To date, the Nile River basin countries have respected the 1959 Nile Waters Treaty, a document formed and amended before many of the Nile Basin countries were independent from colonial rule. The treaty allots 100% of Nile water resources to Egypt and Sudan.

Egypt, Ethiopia and Sudan have engaged in a series of negotiations starting in 2013 that most recently resulted in a Declaration of Principles in March 2015. New negotiations can be representative of political changes in Egypt and their relationship with upstream, more economically and politically stable countries, such as Ethiopia, as well as economic improvements that the upstream Nile basin countries are experiencing (Veilleux, 2015). The very act of the ongoing negotiations demonstrates a willingness of historically rival countries to come to the table over shared Nile River water resources. The alternative, conflict, was contained in rhetoric, particularly by the press and politicians, but the choice, cooperation, was demonstrated in action (Veilleux, 2015).

The importance of institutional capacity

Throughout, this article has referenced how building institutional capacity has been observed to be a successful strategy in resolving and preventing water conflicts. Building institutional capacity, through signing agreements (treaties) and creating river basin organizations, is described as a tool which can reduce the likelihood of water conflict (Wolf, Stahl, & Macomber, 2003; Yoffe et al., 2004, 2003). McCaffrey (2003, p. 157) wrote, "Treaties stabilize [the relations of states sharing a river] giving them a certain level of certainty and predictability that is often not present otherwise." To overcome the risks of cooperation mentioned in the previous section, institutions responsible for managing a watercourse's resources must be strong enough to balance competing interests of allocation and use.

Countries are more likely to sign treaties when they are in conflict with others over management, are dependent on the water resource, and/or have less control over the resource (Espey & Towfique, 2004). Riparians who have countervailing economic and political power and share 'Western civilization' characteristics are also much more likely to sign treaties (Song & Whittington, 2004).

Institutions may also need to manage human-induced water scarcity. Because of the possible contentiousness surrounding these decisions related to managing water scarcity, institutions (and their structural makeup) can themselves become central settings for disputes. International water conflicts may happen when there is no institution that delineates each nation's rights and responsibilities with regard to the shared body of water, nor any agreements or implicit cooperative arrangements.

The mere presence of institutional capacity does not imply its effectiveness; in an analysis of 153 water-related agreements in Africa identified by Lautze and Giordano (2005), only 108 were considered substantive, and many of these either were never implemented in practice or are no longer enforced. For these agreements to be considered substantive, they need to have characteristics that are operative in preventing conflicts. Ambiguity (intentional or not) in agreements may prove to increase the agreement's resilience towards conflict by allowing each side to present the treaty differently at home to defuse domestic opposition and/or providing leeway to adjust allocations during crises (Fischhendler, 2008). Other characteristics that make institutional capacity effective for preventing water conflict include:

- an adaptable management structure (including flexibility, allowing for public input, changing basin priorities, and new information and monitoring technologies)
- clear and flexible allocating criteria among riparians
- equitable distribution of benefits
- detailed conflict-resolution mechanisms (Giordano & Wolf, 2003).

Cooperative water management institutions that have these characteristics are much better at anticipating conflict and solving long-simmering disputes, especially when the adaptable management structure allows stakeholders to be included in decision-making processes and given the necessary data, trained staff, and financing for the parties to work together as equals. These institutions may reduce the potential for conflict in a number of ways. First, these institutions may act to provide a forum where joint negotiations may take place, allowing each party's interests to be considered during decision-making processes. This forum can allow multiple perspectives and interests to be heard, in turn revealing new management options and creating win-win solutions. Institutions may also provide opportunities for parties to collaborate and engage in joint fact-finding, hopefully leading to decisions that are much more likely to be accepted by stakeholders and stakeholder groups in both parties, even if total consensus may not be reached. Once in place, effective institutions are tremendously resilient over time, particularly if they incorporate characteristics such as effective monitoring, a clear allocation framework, and adequate enforcement. An often-cited example is the Indus Waters Treaty, which is still in place after over 60 years despite two wars between India and Pakistan. While the treaty does have flexible allocating criteria, the allocating criteria are clear, third-party referrals for resolution allow some management flexibility, and the treaty contains conflict-resolution mechanisms that have been deemed effective, including the Permanent Indus Commission, which must meet at least once every year (Sarfraz, 2013). Institutions without these characteristics can lead to future conflict, as when Iran and Iraq confronted a militarized water dispute in 1981 due to the lack of enforcement mechanisms in their 1975 treaty over joint control of the Shatt al-Arab waterway (Mitchell & Zawahri, 2015; TFDD, 2016).

Political power dynamics in transboundary basins

Though installing institutions in the form of treaties and river basin organizations has the potential to create a more equitable power dynamic between riparians, sometimes these institutions can reinforce or even strengthen disparities. Power inequities are defined by relationships between basin riparian countries and, often, important actors outside the region. It has been argued that unequal resources, usually of a financial or political nature, result in real-world inequities. These inequities can be present within institutions, weakening their effectiveness (Conca, Wu, & Mei, 2006).

Riparian parties can exploit institutions in numerous ways. For example, treaties are not easily enforceable and can be structured in a manner that reflects (or worsens) existing inequities between parties, and can lead to a lack of participation by other riparians. This exploitation by more powerful riparian parties forms the basis of the theory of hydro-hegemony, which postulates that the most powerful country in the basin, the hydro-hegemon, can create its preferred mechanisms of transboundary water management due to its relative power within the watershed (Zeitoun & Warner, 2006). In some situations, hegemons have sufficient structural power to coerce asymmetric cooperation (Weinthal, 2002).

Hydro-hegemons have been identified by the presence of advantages that these riparians may have over others, including their relative power, riparian position, and technological potential to exploit the resource (Zeitoun & Warner, 2006). However, it can be argued that the presence of these advantages for riparians has led to assumptions that the so-called non-hegemons are unable to achieve their own positive outcomes. Countries may use interlinkages between water and non-water issues, internal and external expectations of riparian behaviour, and consideration of whether the water-related issue at hand is crucial to each party's survival or whether the party has the luxury to survive the outcome of the resolution (Petersen-Perlman & Fischhendler, forthcoming). Certain dynamics where the non-hegemons have been able to achieve positive outcomes have been prevalent in such transboundary basins as the Mekong, La Plata and Nile. Take for instance Paraguay's refusal to change the frequency of electricity produced by its generators at the Brazilian-Paraguayan Itaipu Dam, despite the hegemon, Brazil, intensively pressuring Paraguay to do so and offering to pay for the conversion (Nickson, 1982). Other examples include: Syria (the non-hegemon) securing water from Turkey (the hegemon) by allowing the Kurdistan Workers' Party (PKK) to base themselves within Syria's borders (MacQuarrie, 2004); Afghanistan (the non-hegemon) unilaterally capturing water resources in the Harirud basin shared with Iran, the hegemon (Thomas & Warner, 2015); Ethiopia (the non-hegemon) building the Grand Ethiopian Renaissance Dam over the downstream objections of Egypt (Oestigaard, 2012); and upstream riparian countries in the Nile pursuing ratification of a Cooperative Framework Agreement, again over the downstream objections of Egypt, the hegemon, and Sudan (Heuler, 2013).

International water law

It is still arguably more difficult for non-hegemons to achieve all desired outcomes. Resolving international disputes through legal means can prove difficult in certain cases due to some jurisdictions having to rely upon poorly defined water law or customary water law, few enforcement mechanisms, and a dispute settlement system (the International Court of Justice) where the disputing parties themselves have to decide on jurisdiction and frames of reference before a case can be heard. This results in very few international water conflicts being decided in the International Court of Justice.

An alternative to resolving international disputes through the courts is to create effective institutional capacity that can adapt to change in order to prevent conflict. Schmeier (2012) documented over 120 river basin organizations worldwide that often are effective in managing for changes in the system and in helping resolve disputes or, better, prevent them from arising. State practice too has evolved over time, with over 500 treaties being negotiated over water resources since World War II (TFDD, 2016)

There has been slow progress on codifying principles on non-navigational water-courses in international law. Opinions have varied regarding how to treat international water. At one end of the spectrum is the 1895 Harmon Doctrine, which has been cited as supporting the unqualified right of the upstream state to utilize and dispose of the waters of an international river flowing through its territory. Other approaches have included the 1966 Helsinki Rules, which established the rule of 'equitable and reason-able utilization' as a customary international river law. Building on the establishment of the Helsinki Rules, the United Nations adopted the Convention on International Watercourses in 1997 (Salman, 2007), which entered into force in 2014. The convention includes a definition of the term 'international watercourse' which is significant for including groundwater connected to surface water systems in addition to surface water, a framework in which international water agreements can be interpreted, and the obligation of not causing significant harm to other states in a party's water use (McCaffrey, 1998).

In addition to the UN convention, nations have engaged in regional-level coopera-tion over water, the most prominent effort being through the United Nations Economic Commission for Europe (UNECE). Through efforts to develop legal and international frameworks to promote transboundary environmental cooperation, UNECE developed the Convention on the Protection and Use of Transboundary Watercourses and International Lakes, signed in 1992 and entering into force in 1996. The convention included provisions meant to protect transboundary surface water and groundwater by reasonable and equitable use of transboundary waters, ecosystem conservation/restora-tion, and controlling and reducing pollution (Bosnjakovic, 1998). While the UNECE convention has already influenced transboundary water management in Europe in a number of ways (Wouters & Vinogradov, 2003), its effectiveness has yet to be deter-mined, though the increasing number of nations that have acceded to this convention suggests that the principles espoused in each convention are being adopted in more countries.

Other informal agreements also hold significance in international watercourses, including the Johnston negotiations in the Jordan River basin (Wolf & Newton, 2008) and an unofficial memorandum of understanding (MoU) between subnational entities for the Hueco Bolson aquifer underlying Juárez and El Paso on the Mexico–United States border (Eckstein & Sindico, 2014). The UN convention is also influencing the still-developing Law of Transboundary Aquifers, which is becoming customary law even while being drafted (Eckstein & Sindico, 2014).

Third-party involvement

Treaties and other efforts towards transboundary cooperation can be more resilient to conflicts through the involvement of third parties. Third-party involvement is a concept that can describe diplomatic, economic or virtual engagement on a shared water resource. Third-party involvement indicates a level of influence from an outside party on the decisions and agreements made between riparian countries on shared waters.

Diplomatic third-party involvement

Diplomatic third-party involvement has had different iterations throughout recent history. Historical third-party involvement often came in the guise of colonialism. More recently, third-party involvement has often come in the guise of a neighbouring country's assistance, a Western country's consultation, or an international organization's mediation.

Several treaties and agreements listed in the TFDD include countries that are geographically distant from the shared basin. Many of these treaties are relics of the colonial era, when Europe played a heavy hand in resource engineering decisions in their foreign occupations. The Nile Waters Treaty, drafted before the present-day Nile Basin countries were independent nation-states, is an example of how treaties from the colonial era can be impediments in present-day negotiations due to allocations for Egypt and Sudan.

Third-party involvement comes from entities such as embassies, the United Nations, the US State Department, the North Atlantic Treaty Organization, or a neighbouring neutral country's government. The third party may be publicly recognized or kept secret. The engagement can be invited, offered, or in some cases, forced. In the former two cases, the involvement is appreciated in order to keep the conversation going, to assist with expertise through consultants, or to host talks and meetings or negotiations. Largely, third parties are sought for their ability to witness and bridge dialogue. In the last case, forced involvement may come in the form of sanctions or coercion, but is typically hard to measure directly. Examples of third-party involvement in management of a shared basin are the Lake Ohrid watershed, Indus River basin and Lake Victoria sub-basin. The World Bank served as a bridge between the countries of Albania and Macedonia to develop and sign an MoU in the Lake Ohrid watershed, which had been without any diplomatic contact for decades due to historical political circumstances (Spirkovski, Avramovski, & Kodzoman, 2001). The International Bank for Reconstruction and Development is a co-signatory to the Indus Waters Treaty between India and Pakistan, bringing an element of third-party mediation (Sarfraz, 2013). Similarly, the East African Community, through the Lake Victoria Basin Commission, helped Kenya and Tanzania sign an MoU on water management on the Mara River (Kenya's Ministry of Labour and East African Affairs, 2015). While these MoUs are brief and limited, official dialogue can start from this place in a step-by-step process towards more detailed attention to the complexity of watershed management.

Economic third-party involvement

Third-party involvement in the area of economics is common in developing-country negotiations with interested developed countries or with two countries that share an economic interest. Economic investment or trade agreements allow influence on water management decisions directly, as has been the case in some dam construction, and sometimes indirectly, by negotiations on an unrelated issue, such as compromises on water allocation in exchange for another economic good, such as electricity. This can be controversial or welcomed by basin riparians. A recent example of direct foreign investment in a shared water resource is five Thai banks' investing in Laos's controversial Xayaburi Dam. The availability of investment from the banks is what made this dam project possible, as Laos itself does not have the funds for such projects. The resulting dam, which impacts the entire Lower Mekong basin, will generate electricity, 80% of which will go to Thailand (Veilleux, 2014).

Virtual stakeholders

Virtual stakeholders are those stakeholders who do not live in the shared basin and perhaps do not directly use the shared waters, but nevertheless have great influence on how the shared waters are managed and allocated. The concept of 'virtual water' is one that assigns the embedded water in an economic product to the consumer (Allen, 1998). Similarly, a virtual third party is one that exhibits consumption patterns or high-visibility interests that are linked with the shared water resources.

One type of virtual stakeholder as a third party is that of money lenders and their associated 'experts'. Large international banks, such as the World Bank and other development banks, often weigh in on the feasibility of projects in otherwise underdeveloped areas. The World Bank, in particular, has not provided funding for projects that do not have the full support of all riparian countries in the area, such as Turkey's Ataturk Dam in the Tigris-Euphrates basin (Starr, 1991). When water infrastructure is proposed in high-risk environments to invest, such as developing nations, where most of the world's water infrastructure is currently in development or under construction, typically an associated feasibility study is conducted. Such studies have value for national governments as validation for projects, but include economic projections that are speculative and contain negative social and environmental externalities, and in this way, often do not present the total cost/benefit of given projects. By doing this, money lenders can alter and influence the management of shared water resources, even if they do not offer the loan to do so.

Future directions and techniques for addressing new problems

As demonstrated above, scholars have documented the leading causes of water conflict. The debate over how to mitigate the things which cause and exacerbate water conflict, however, continues. There are still new techniques with which transboundary water resources management can be improved. Below, we offer some suggestions for new directions.

Identifying risk

Scholars have identified and analyzed the major causes of transboundary water conflict. More recently, given the history of conflict in shared watersheds, the focus has shifted towards identifying risks for future hydro-political conflict. The shifts in hydrology due to climate change, new development of water resources from population growth, irrigation, dams and diversions, and the lack of effective institutional capacity to manage these shifts could lead to future conflict. These traditional sources of water conflict may lead to new conflicts, and there may be new forms, perhaps in the form of consequences of desalination as the technology becomes more efficient and affordable. Regardless of the source, predicting where water conflict might occur continues to be a challenge.

Scale and basin management

No two river basins, watersheds, aquifers, lakes, or any bodies of water are alike. Each has its own climate, demographic makeup, hydrology, industry, topography and cultural divisions. Despite this, it is natural to wish to find patterns from which universal patterns can be extracted. While it is true that patterns can be found to a certain extent, each basin/watershed/aquifer is unique, and any attempt towards management should be adapted to these unique dimensions.

Increasingly, with globalization, virtual water will become more important than ever. Dependence on agricultural and economic goods coming from outside a watershed means that the dependent countries or regions will not be immune to ripple effects from times of drought, scarcity or conflict. This implies that water conflicts may become more frequent between entities from both inside and outside basins.

Managing change

Complicating management, every watershed is dynamic – volumes and conditions constantly change. New uses develop; weather patterns alter runoff and discharge regimes; people migrate into and out of watersheds. Effective transboundary water management must be prepared for, and even embrace, this dynamism. This requirement for flexibility challenges current, static policy regimes (perhaps guaranteed water flows in particular).

Improving cooperative frameworks

As discussed earlier, parties are willing to engage in cooperation when opportunities exceed the risks and benefits exceed the costs. To reach this point is difficult. But if one agrees that cooperation, rather than conflict, is more beneficial for all parties, it must be accomplished. Third parties can play a role in mediating negotiations between contesting parties and alleviating the risks of cooperation.

As river basin organizations become more prevalent as a management structure on the international stage, the concept must be re-examined. With many donors requiring the building of institutional capacity (e.g. through the creation of river basin organizations) as a prerequisite for the distribution of funds to developing countries in transboundary basins (Alaerts, 1999; Mukhtarov & Gerlak, 2013), the question is raised of how effective the mere presence of institutional capacity is in rendering effective governance and preventing conflict, particularly when donor-induced projects and initiatives do not align with stakeholders' concerns. For instance, donors encouraged the Red River Basin Organization in Vietnam to identify and focus on integrated water resources management issues, despite stakeholders' voicing different concerns (Molle & Hoanh, 2011), creating confusion within institutional frameworks.

Improving baseline information and data exchange

To a certain extent, institutional capacity may be only as strong as the basic information available about the watershed in question. Lack of data in some basins on, for instance, historical hydrological patterns or water quality can make management decisions much more difficult. There are certain cases where this is due to a lack of monitoring capacity, but it can also be due to a lack of information sharing between countries or even between agencies in one country. Reliable, basic information about the watershed in question is crucial for parties to buy into cooperative frameworks. What's more, this backbone of basic information about the watershed in question is crucial in making management plans; the foundational aspects of a watershed – the hydrology, the people, the uses, the biology, the topography, etc. – must be accounted for in any effective plan.

Conclusion

Being part of a transboundary basin makes a user interconnected with the rest of its users. Every downstream user has the potential to be affected by events occurring upstream of them. To a lesser extent, upstream users are also interconnected with their downstream counterparts through downstream demands and actions. This inter-connectedness strengthens the case for the need for cooperative frameworks to manage transboundary waters. It is important to note that not all cooperative frameworks are ideal or just for all parties involved, but given a list of alternatives that includes violence, cooperation might be a preferred path.

New challenges in managing transboundary water will surely emerge, particularly with the advent of increased variability due to climate change and the growing globalized economy. To be prepared for this, parties should participate in conflict-resolution mechanisms and invest in institutional capacity with their neighbours. Systemic, holistic water management can provide the opportunity for more users to meet their basic needs and become economically resilient with respect to whatever new variables regarding management they face, thereby increasing water security.

Disclosure statement

No potential conflict of interest was reported by the author.

References

Alaerts, G. J. (1999, October 27–29). *Institutions for river basin management. The role of external support agencies (International Donors) in developing cooperative arrangements.* International workshop on river basin management – Best management practices, Delft University of Technology/River Basin Administration (RBA), The Hague.

Allen, T. (1998). Watersheds and problemsheds: Explaining the absence of armed conflict over water in the Middle East. *Middle East, 2*(1), 50.

Blomquist, W., & Ingram, H. M. (2003). Boundaries seen and unseen: Resolving transboundary groundwater problems. *Water International, 28*(2), 162–169. doi:10.1080/02508060308691681

Bosnjakovic, B. (1998). UN/ECE strategies for protecting the environment with respect to international watercourses: The Helsinki and Espoo conventions. In S. Salman & L. Boisson de Chazournes (Eds.), *International watercourses – Enhancing cooperation and managing conflict.* Technical Paper No. 414. Washington DC: World Bank.

Conca, K., Wu, F., & Mei, C. (2006). Global regime formation or complex institution building? The principled content of international river agreements. *International Studies Quarterly, 50* (2), 263–285. doi:10.1111/isqu.2006.50.issue-2

Cooley, H., & Gleick, P. H. (2011). Climate-proofing transboundary water agreements. *Hydrological Sciences Journal, 56*(4), 711–718. doi:10.1080/02626667.2011.576651

De Stefano, L., Edwards, P., De Silva, L., & Wolf, A. T. (2010). Tracking cooperation and conflict in international basins: Historic and recent trends. *Water Policy, 12*(6), 871–884. doi:10.2166/wp.2010.137

Eckstein, G., & Sindico, F. (2014). The law of transboundary aquifers: Many ways of going forward, but only one way of standing still. *RECIEL, 23*(1), 32–42. doi:10.1111/reel.12067

Eldering, M. (2010). Measuring human (In-)security. *Human Security Perspectives, 7*, 17–49.

Espey, M., & Towfique, B. (2004). International bilateral water treaty formation. *Water Resources Research, 40*, 5. doi:10.1029/2003WR002534

Fischhendler, I. (2008). Ambiguity in transboundary environmental dispute resolution: The Israeli-Jordanian water agreement. *Journal of Peace Research, 45*(1), 91–109. doi:10.1177/0022343307084925

Giordano, M. A., & Wolf, A. T. (2003). Sharing waters: Post-Rio international water management. *Natural Resources Forum, 27*(2), 163–171. doi:10.1111/narf.2003.27.issue-2

Gleditsch, N. P., Furlong, K., Hegre, H., Lacina, B., & Owen, T. (2006). Conflicts over shared rivers: Resource scarcity or fuzzy boundaries? *Political Geography, 25*(4), 361–382. doi:10.1016/j.polgeo.2006.02.004

Gleick, P. H. (1993). Water and conflict: Fresh water resources and international security. *International Security, 18*(1), 79–112. doi:10.2307/2539033

Grey, D., & Sadoff, C. W. (2007). Sink or swim? Water security for growth and development. *Water Policy, 9*(6), 545–571. doi:10.2166/wp.2007.021

Heuler, H. (2013). Tensions mount as Uganda proceeds with Nile River agreement. *Voice of America*. Retrieved from http://www.voanews.com/content/tensions-mount-as-uganda-proceeds-with-nile-river-agreement-cfa/1692986.html

Homer-Dixon, T. F. (1994). Environmental scarcities and violent conflict: Evidence from cases. *International Security, 19*(1), 5–40. doi:10.2307/2539147

International Groundwater Resources Assessment Centre. (2015). Transboundary aquifers of the World Map 2015. Retrieved from http://www.un-igrac.org/sites/default/files/resources/files/TBAmap_2015.pdf

Kenya's Ministry of Labour and East African Affairs. (2015). Tanzania vice president H.E M. G. Bilal officiates at the signing of the memorandum of understanding on the joint water resources management of trans boundary Mara River Basin during ceremony held at Butiama in Mara region of TZ (MOU) between Kenya and TZ. *Kenya's Ministry of Labour and East African Affairs*. Retrieved from http://www.meac.go.ke/index.php/2-uncategorised/101-tanzania-vice-president-h-e-mohammed-ghatib-bilal-officiates-at-the-signing-of-the-memorandum-of-understanding-on-the-joint-water-resources-management-of-the-trans-boundary-mara-river-basin-during-a-colorful-ceremony-held-at-butiama-in-mara-region-of-tanz

Kucera, J. (2011). Did Uzbekistan bomb its own railway? *Eurasianet, The Bug Pit: Military and Security in Eurasia*. Retrieved from http://www.eurasianet.org/node/64617

Lautze, J., & Giordano, M. (2005). Transboundary water law in Africa: Development, nature, and geography. *Natural Resources Journal, 45*(4), 1053–1087.

Lowi, M. R. (1995). Rivers of conflict, rivers of peace. *Journal of International Affairs, 49*(1), 123.

MacQuarrie, P. (2004). *Water security in the Middle East: Growing conflict over development in the Euphrates–Tigris Basin* (Master's thesis). Univeristy of Dublin, Dublin. Retrieved from http://www.transboundarywaters.orst.edu/publications/related_research/MacQuarrie2004_abstract.htm

McCaffrey, S. C. (1998). *The UN convention on the law of the non-navigational uses of international watercourses: Prospects and pitfalls* (World Bank Technical Paper), 17–28. Washington, DC: World Bank.

McCaffrey, S. C. (2003). The need for flexibility in freshwater treaty regimes. *Natural Resources Forum, 27*, 156–162. doi:10.1111/narf.2003.27.issue-2

Mitchell, S. M., & Zawahri, N. A. (2015). The effectiveness of treaty design in addressing water disputes. *Journal of Peace Research, 52*(2), 187–200. doi:10.1177/0022343314559623

Molle, F., & Hoanh, C. T. (2011). Implementing integrated river basin management in the Red River Basin, Vietnam: A solution looking for a problem? *Water Policy, 13*, 518–534. doi:10.2166/wp.2011.012

Mukhtarov, F., & Gerlak, A. K. (2013). River basin organizations in the global water discourse: An exploration of agency and strategy. *Global Governance, 19*(2013), 307–326.

Nickson, R. A. (1982). The Itaipú Hydro-electric project: The Paraguayan perspective. *Bulletin of Latin American Research, 2*(1), 1–20. doi:10.2307/3338386

Oestigaard, T. (2012). *Water security and food security along the Nile: Politics, population and climate change.* Uppsala: Nordiska Afrikaininstitutet.

Oregon State University Transboundary Freshwater Dispute Database (TFDD). (2016). Retrieved from http://transboundarywaters.orst.edu

Petersen-Perlman, J. D., & Fischhendler, I. (forthcoming). The weakness of the strong: Re-examining power in transboundary water dynamics. Submitted manuscript.

Putz, C. (2015). Uzbekistan still hates Rogun Dam project. *The Diplomat.* Retrieved from http://thediplomat.com/2015/08/uzbekistan-still-hates-the-rogun-dam-project/

Salman, S. M. (2007). The Helsinki Rules, the UN watercourses convention and the Berlin Rules: Perspectives on international water law. *Water Resources Development, 23*(4), 625–640. doi:10.1080/07900620701488562

Sarfraz, H. (2013). Revisiting the 1960 Indus Waters Treaty. *Water International, 38*(2), 204–216. doi:10.1080/02508060.2013.784494

Schmeier, S. (2012). *Governing international watercourses: River basin organizations and the sustainable governance of internationally shared rivers and lakes.* London and New York: Routledge.

Song, J., & Whittington, D. (2004). Why have some countries on international rivers been successful negotiating treaties? A global perspective. *Water Resources Research, 40*, 5. doi:10.1029/2003WR002536

Spirkovski, Z., Avramovski, O., & Kodzoman, A. (2001). Watershed management in the Lake Ohrid region of Albania and Macedonia. *Lakes & Reservoirs: Research & Management, 6*(3), 237–242. doi:10.1046/j.1440-1770.2001.00153.x

Starr, J. (1991). Water wars. *Foreign Policy, 82*, 17–36. doi:10.2307/1148639

Subramanian, A., Brown, B., & Wolf, A. T. (2012). *Reaching across the waters: Facing the risks of cooperation in international waters.* Washington, DC: The World Bank.

Thomas, V., & Warner, J. (2015). Hydropolitics in the Harirud/Tejen River Basin: Afghanistan as hydro-hegemon? *Water International, 40*(4), 593–613. doi:10.1080/02508060.2015.1059164

Toset, H. P. W., Gleditsch, N. P., & Hegre, H. (2000). Shared rivers and interstate conflict. *Political Geography, 19*(8), 971–996. doi:10.1016/S0962-6298(00)00038-X

United Nations-Water. (2013). *Water security and the global water agenda: A UN-Water analytical brief* (p. 1). Hamilton: United Nations University.

Veilleux, J. C. (2014). *Is dam development a mechanism for human security? Scale and perception of the Grand Ethiopian Renaissance Dam on the Blue Nile River in Ethiopia and the Xayaburi Dam on the Mekong River in Laos* (Unpublished doctoral dissertation). Oregon State University, Corvallis.

Veilleux, J. C. (2015).Water conflict case study - Ethiopia's Grand Renaissance Dam: Turning from conflict to cooperation. In S. A. Elias (Ed.), *Reference module in earth systems and environmental sciences.* Amsterdam: Elsevier.

Wegerich, K. (2008). Hydro-hegemony in the Amu Darya Basin. *Water Policy, 10*(S2), 71–88. doi:10.2166/wp.2008.208

Weinthal, E. (2002). *State making and environmental cooperation: Linking domestic and international politics in Central Asia.* Cambridge, MA and London: MIT Press.

Wolf, A. T. (1995). *Hydropolitics along the Jordan River; Scarce water and its impact on the Arab-Israeli conflict* (Vol. 99). Tokyo, New York, and Paris: United Nations University Press.

Wolf, A. T. (1999). 'Water wars' and water reality: Conflict and cooperation along international waterways. In S. Lonergan Ed., *Environmental change, adaptation, and security* (NATO ASI Series Vol. 65). Dordrecht: Kluwer Academic Press.

Wolf, A. T. (2000). Trends in transboundary water resources: Lessons for cooperative projects in the Middle East. In D. Brooks & O. Mehmet (Eds.), *Water balances in the Eastern*

Mediterranean. Ottawa: The International Development 219 Research Centre: Science for Humanity.

Wolf, A. T. (2007). Shared waters: Conflict and cooperation. *Annual Review of Environment and Resources, 32,* 3.1–3.29. doi:10.1146/annurev.energy.32.041006.101434

Wolf, A. T., Kramer, A., Carius, A., & Dabelko, G. D. (2005). Managing water conflict and cooperation. In E. Ayres (Ed.). *State of the world 2005: Redefining global security* (pp. 80–95). Washington, DC: Worldwatch Institute.

Wolf, A. T., Natharius, J., Danielson, J., Ward, B., & Pender, J. (1999). International river basins of the world. *International Journal of Water Resources Development, 15*(4), 387–427. doi:10.1080/07900629948682

Wolf, A. T., & Newton, J. (2008). *Case study of transboundary dispute resolution: The Jordan River Johnston negotiations 1953–1955; Yarmuk mediations 1980s.* Corvallis: Oregon State University.

Wolf, A. T., Stahl, K., & Macomber, M. F. (2003). Conflict and cooperation within international river basins: The importance of institutional capacity. *Water Resources Update, 125,* 1–10.

Wolf, A. T., Yoffe, S. B., & Giordano, M. (2003). International waters: Identifying basins at risk. *Water Policy, 5*(1), 29–60.

Wouters, P., & Vinogradov, S. (2003). Analysing the ECE water convention: What lessons for the regional management of transboundary water resources? *Yearbook of International Co-Operation on Environment and Development, 4,* 55–63.

Yoffe, S. B., Fiske, G., Giordano, M., Giordano, M. A., Larson, K., Stahl, K., & Wolf, A. T. (2004). Geography of international water conflict and cooperation: Data sets and applications. *Water Resources Research, 40*(5), 1–12. doi:10.1029/2003WR002530

Yoffe, S. B., Wolf, A. T., & Giordano, M. (2003). Conflict and cooperation over international freshwater resources: Indicators of basins at risk. *Journal of the American Water Resources Association, 39*(5), 1109–1126. doi:10.1111/j.1752-1688.2003.tb03696.x

Zeitoun, M., & Mirumachi, N. (2008). Transboundary water interaction I: Reconsidering conflict and cooperation. *International Environmental Agreements: Politics, Law and Economics, 8*(4), 297–316. doi:10.1007/s10784-008-9083-5

Zeitoun, M., & Warner, J. (2006). Hydro-hegemony-a framework for analysis of trans-boundary water conflicts. *Water Policy, 8*(5), 435–460. doi:10.2166/wp.2006.054

China's transboundary river policies towards Kazakhstan: issue-linkages and incentives for cooperation

Selina Ho

ABSTRACT

A significant aspect of China's power is its position as upstream riparian on many of Asia's international rivers. China participates in creating and building institutions with riparians of some of these rivers more than others. This article argues that China's relatively higher level of institutionalized cooperation with Kazakhstan on transboundary water issues is due to the interdependence between the two countries, which facilitates linkages between water issues and a cluster of political, economic, security and strategic issues. China is incentivized to cooperate and accommo-dates Kazakhstan's concerns over their shared waters because a reciprocal relationship exists between them.

Of the 24 rivers that China and Kazakhstan share along their 1700-kilometre border, the Irtysh and Ili Rivers are the largest and the most significant (Map 1.). Both are important lifelines for Kazakhstan, while at the same time playing a critical role in China's plans to boost the economy of the Xinjiang Uighur Autonomous Region. Although only 4% of the basin area of the 4248-kilometre-long Irtysh River lies in China, the 10 km^3 estimated average annual flow into Kazakhstan from China may constitute the bulk of the Irtysh's flow (Baizakova, 2015; Sievers, 2002). About one-third of the area of the 1440-kilometre-long Ili River lies in China. Within Kazakhstan, its waters support the vital rich fisheries of Lake Balkhash, and the agricultural and industrial sectors (Baizakova, 2015).

China's water diversion projects on the Irtysh and Ili have been cited as a major cause of the decline in the flow of these rivers and its consequent impact on the agricultural and aquatic ecosystems.[1] Notwithstanding the impact of Chinese activities, in managing these rivers China has adopted a relatively cooperative stance towards Kazakhstan compared to its other riparian neighbours in South and South-East Asia. Cooperation with Kazakhstan takes the form of setting up institutions, such as river commissions, and agreements to manage their shared water resources.[2] This relatively higher level of institutionalized cooperation with Kazakhstan became more visible from the early 2000s onwards. For instance, China set up a joint river commission with Kazakhstan in 2003 as part of an agreement to cooperate with Kazakhstan on the use and protection of transboundary rivers. This is significant because China is a member of only two other river commissions, with

Map 1. The Ili and Irtysh Rivers.

Source: *Indicative Overview of Irtysh and Ili Rivers* [map]. 1:126,600,000 approx. Data layers: Food and Agriculture Organization of the United Nations: AQUAmaps Regional and Global River Layers; Bernhard Lehner and Petra Döll, Center for Environmental Systems and Research: Global Lakes and Wetlands Database; David T. Sandwell, Joseph J. Becker, and Walter H. F. Smith, Regents of the University of California: SRTM15_PLUS: Data Fusion of SRTM Land Topography with Measured and Estimated Seafloor Topography; Robert J. Hijmans, Regents of the University of California: Global Level: A New File with the (2011) Global Country Boundaries [computer files]. National University of Singapore: NUS Libraries, 11 October 2016. Using: QGIS [GIS]. Version 2.4. QGIS Development Team, 2014.

Russia and Mongolia. Since the first transboundary water agreement between China and Kazakhstan, in 2001, several landmark agreements have followed, including those that touch on water quality protection, an issue China has avoided discussing with its other neighbours in South and South-East Asia. China and Kazakhstan also started preparatory work in 2010 to discuss the distribution of shared water resources between them. Even though these negotiations on water allocation have been slow, the fact that China has agreed to put this issue on the table is an achievement in itself, since China has assiduously avoided discussing water-sharing arrangements with its other riparian neighbours.

In a study of the water agreements that China has signed, including those with Kazakhstan, Chen, Rieu-Clarke, and Wouters (2013, p. 227) wrote that "China embraces the notion of cooperation" even though there remains considerable scope for further developing the agreements. In addition, although a dispute-settlement mechanism is not specified in the agreements, there are nevertheless provisions for inter-party negotiations based on consultations (Chen et al., 2013). While there have been delays in the implementation of some of these agreements, these delays cannot be solely attributed to China: Kazakhstan lacks skilled specialists, which has delayed implementation of its side of the agreements, and some projects remain forgotten in the drawers of Kazakh officials (Baizakova, 2015). In this regard, China has been helping train Kazakh specialists in water resource management in its higher educational institutions (Kazakhstan Government News, 2016).

By contrast, in the international river systems in South and South-East Asia that China is a part of, there is a lower level of institutionalization, mainly confined to non-binding memoranda of understanding (MOUs), expert-level mechanisms, exchange of hydrological information, and participation in navigation, tourism, and infrastructure projects. Against this backdrop, it is important to understand why China is willing to cooperate at a higher level with Kazakhstan. Kazakhstan is a smaller and weaker state downstream from China – how does Kazakhstan persuade China to cooperate on transboundary water issues? What are the incentives for Chinese cooperation? Kazakhstan's relative success in getting Chinese cooperation provides important lessons for weaker downstream states seeking cooperation from stronger upstream states.

This article argues that China's relatively higher level of institutionalized cooperation with Kazakhstan on water issues is the result of issue-linkages, defined as the linking of upstream–downstream issues with issues 'beyond' the river basin that facilitates quid pro quos and side payments (Daoudy, 2009). Chinese cooperation is quid pro quo for Kazakhstan's significant cooperation with China on a range of strategic, security, political and economic issues that are vital to China's domestic and foreign policy interests.

These issue-linkages are not explicit. It is not necessary for Kazakhstan to explicitly link issues during negotiations since China's desire to build up goodwill with Kazakhstan so as to gain the latter's cooperation on a host of non-water-related security, economic and strategic issues is a key consideration behind Chinese coopera-tion. Moreover, making explicit linkages can be counter-productive for Kazakhstan – China is critical to Kazakhstan's economic and strategic interests, so it would not be to Kazakhstan's advantage to strain relations with China.

However, there are indications of issue-linkages in Kazakhstan's negotiations with China even if explicit evidence is lacking in the various water agreements and official discourse between China and Kazakhstan.[3] In 2008, the joint river commission was brought under the umbrella of the China-Kazakhstan Cooperation Committee, which is co-chaired by the vice-premiers of both sides (Joint Communique, 2008) and which oversees the development and deepening of all-round cooperation between the two sides. Such a move suggests that the work of the river commission is interlinked with the entire spectrum of cooperation between China and Kazakhstan. Moreover, in joint statements from top Chinese and Kazakh leaders, transboundary water cooperation is often mentioned together with a list of cooperation programmes in the political, security, economic and energy fields.

This article is divided into the following sections. The first section sets up the puzzle by comparing China's relatively higher level of cooperation with Kazakhstan with its coopera-tion in the Mekong and Brahmaputra Rivers. The second and third sections review China's role as a hydro-hegemon and show why issue-linkage is a suitable theoretical framework for explaining China–Kazakhstan cooperation. The fourth section examines two periods in Sino–Kazakh relations, 1992–2001 and 2002–2015, to explicate in detail how growing interdependence and issue-linkages help explain the increased level of institutionalized riparian cooperation between them. The timeline of the various water agreements suggests that China has acquiesced to Kazakhstan's requests to step up transboundary water cooperation in return for Kazakhstan's support and cooperation on a range of political, strategic and economic issues.

Comparing China's institutionalized cooperation

Wolf, Yoffe, and Giordano (2003) have developed a Basins at Risk (BAR) event intensity scale to examine the spectrum of riparian conflict and cooperation (Table 1), ranging from the highest level of conflict, "formal declaration of war and extensive war acts" (−7), to the highest level of cooperation, "voluntary unification into one nation" (7). Cooperation and conflict at the river-basin level often exist on a spectrum between these two extremes. To compare China–Kazakhstan transboundary water cooperation with China's cooperation with the Mekong states and with India in the Brahmaputra River, the BAR scale is used to categorize the different forms of Chinese cooperation on water issues. Chinese cooperation ranges from non-binding MOUs and expert-level mechanisms to joint development, agreements on environmental and water quality protection, joint river commissions, and discussions on water-sharing arrangements. On the BAR scale, MOUs and expert-level mechanisms are grouped together as minor official exchanges, talks or policy expressions (1), cultural or scientific agreement or support (non-strategic) (2), or non-military economic, technological or industrial agreements (3). Joint development, agreements on environmental and water quality protection, and joint river commissions are categorized as non-military economic, technological or industrial agreements (4) or military, economic or strategic support (5). Water-sharing agreements fall under 6, as they form a critical component of international freshwater treaties.

Based on the above categorization of Chinese cooperation according to the BAR scale, China has cooperated most with Kazakhstan, followed by the Mekong states, and least with India. In 2001, China and Kazakhstan signed a landmark agreement on the use and protection of transboundary rivers. Under this agreement, the Sino-Kazakh Joint Commission on the Use and Protection of Transboundary Rivers was established

Table 1. Basins at Risk event intensity scale.

Basins at Risk rating	Event description
−7	Formal declaration of war; extensive war acts causing deaths, dislocation, or high strategic costs
−6	Extensive military acts
−5	Small-scale military acts
−4	Political-military hostile actions
−3	Diplomatic-economic hostile actions
−2	Strong verbal expressions displaying hostility in interaction
−1	Mild verbal expressions displaying discord in interaction
0	Neutral or non-significant acts for the inter-nation situation
1	Minor official exchanges, talks or policy expressions – mild verbal support
2	Official verbal support of goals, values or regime
3	Cultural or scientific agreement or support (non-strategic)
4	Non-military economic, technological or industrial agreement
5	Military, economic or strategic support
6	International freshwater treaty; major strategic alliance (regional or international)
7	Voluntary unification into one nation

Source: Wolf et al. (2003)

in 2003. This is a significant achievement, considering that China is party to only two other river commissions, with Russia and Mongolia. In 2011, China signed two agreements with Kazakhstan, the first on water quality protection and the second on environmental protection. These agreements and the establishment of the joint river commission demonstrate Chinese cooperation with Kazakhstan at 4 and 5 on the BAR scale. In 2011 as well, China finalized an agreement with Kazakhstan for a joint water-diversion and water-sharing project on a border river, the Khorgos. China and Kazakhstan have also started preparatory work since 2010 to discuss water allocation for all its transboundary rivers (Kazakhstan Ministry of Foreign Affairs, 2014), and in 2015 both sides began discussions for a draft of a water allocation agreement (PRC Website, 2015). On the BAR scale, China is moving from 5 to 6.

By comparison, there are no water-sharing discussions with India and the Mekong states. China is also not involved in any river commissions in the Mekong or the Brahmaputra. On the BAR scale, China's cooperation with Mekong states ranges from 1 to the low end of 5 (sans agreements on water quality protection and joint river commissions). China participates in joint development along the Mekong but eschews joining the Mekong River Commission as a full member. It has been a dialogue partner of the commission since 1998 but has resisted becoming a member so that it would not be subject to the commission's strict aquatic environmental standards and dam-building restrictions. China is a member of the Greater Mekong Subregion. The project is attractive to China because it focuses on infrastructure development, improving power, transportation and communication networks between Yunnan Province and the Indochina states. It does not impose strict aquatic environmental standards or dam-building restrictions on its members, so China's sovereignty and freedom of action are not compromised.

China also actively cooperates on navigation. It is a signatory of the Lancang–Upper Mekong River Commercial Navigation Agreement, which aims to expand trade and tourism among China, Laos, Myanmar and Thailand. China also signed a Greater Mekong Subregion agreement with Laos, Myanmar and Thailand for a trial programme shipping oil along the Mekong during the wet season. This is a strategic move on China's part as the shipping route might become more significant in the future as a cheaper and safer alternative to the Malacca Straits. China is also in joint patrols with Thailand, Myanmar and Laos, allowing China to establish a military presence in the lower Mekong. Cooperation in these areas is categorized as 5 on the BAR scale.

Chinese cooperation with India on the Brahmaputra is dismal, limited to 1–3 on the BAR scale. Apart from an expert-level mechanism set up in 2006 and a number of MOUs on hydrological data sharing, there has been little cooperation between China and India. Occasionally, Indian experts and media have even engaged in −1 and −2 on the BAR scale, that is, mild to strong verbal expressions of hostility. Public protests have also been held in India over Chinese dam-building activities in the upper reaches of the Brahmaputra. In 2009, China blocked India's request for a loan from the Asian Development Bank because it was earmarked for a watershed-development project in Arunachal Pradesh – this is categorized as diplomatic-economic hostile action, −3 on the BAR scale.

It is clear from this comparison using the BAR scale that China and Kazakhstan have a higher level of institutionalized cooperation compared to the other riparian states of the Mekong and the Brahmaputra. Why is China more willing to cooperate with Kazakhstan? The next two sections examine the theoretical framework for answering this question.

Hydro-hegemony

Hegemony is defined as a preponderance of military, economic, cultural and demographic power that enables a state to impose its preferences on other states through a mix of coercive and persuasive means. Mark Zeitoun and Jeroen Warner (2006, p. 435) define hydro-hegemony as "hegemony at the basin level". According to them, hydro-hegemons exercise both positive and negative forms of leadership; the former gives rise to cooperative outcomes for all, while the latter seeks to dominate by adopting coercive means, leading to conflict. The behaviour of hydro-hegemons therefore determines whether there is cooperation or conflict at the river-basin level.

China is clearly the hydro-hegemon in Asia. It is endowed with two key qualities that establish its dominance: power and geographical position. First, there is a substantial asymmetry of military, economic, political and demographic power between China and most of its neighbours. Second, this power asymmetry is reinforced by China's upstream position. It has been described as the "upstream superpower" of Asia because of its position at the headwaters of the major international rivers that flow through Asia (Nickum, 2008, p. 230). With less than 1% of its water flowing in from other countries, but outflows of more than 40 times its inflows (Nickum, 2008), China can choose to ignore the concerns of downstream neighbours.

Scholars who study upstream–downstream relations generally predict that in river basins where the upstream state is also the hegemon, cooperation is least likely, as the upstream state will have little incentive to cooperate (Lowi, 1993). This is based on realism, and is a variant of the hegemonic stability theory – summed up by Miriam Lowi (1993, p. 8) as "a hegemonic theory of cooperation". According to Lowi (1995), it is unrealistic to expect the powerful upstream riparian to undertake the creation of cooperative water distribution regimes that will constrain its manoeuvrability. The powerful upstream riparian is most likely to develop its river resources according to its domestic needs, and export the externalities to weaker downstream riparians.

Yet, while China has displayed the negative behavioural characteristics of the upstream hegemon, for instance in engaging in resource-capture strategies such as unilateral dam construction and water diversion, it has also cooperated with some of its downstream neighbours. Given the asymmetries of power between China and most of its riparian neighbours, what are the incentives for Chinese cooperation?

Patricia Wouters (2014, p. 73) argues that China imposes constraints on itself with respect to its riparian neighbours because it observes "limited territorial sovereignty", which "means that national sovereign interests and actions are prescribed and circumscribed through rules of international law, both substantive and procedural". Even though China is not a signatory of any international water treaties or conventions, "from an international legal perspective, China's approach to dealing with its riparian neighbours is based on dialogue, consultation and peaceful negotiations, and crafted around the notion of restricted territorial sovereignty" (p. 72). According to this argument, China faces an "upstream dilemma": how to meet its domestic water needs while taking into account the needs of neighbouring countries in line with its policy of good neighbourliness (p. 67; see also Wouters & Chen, 2015).

There is, however, little concrete evidence that China observes limited sovereignty in practice. In objecting to the draft text of the 1997 UN Convention on the Law of the

Non-navigational Uses of International Watercourses, China declared that "territorial sovereignty is a basic principle of international law. A watercourse state enjoys indisputable territorial sovereignty over those parts of international watercourses that flow through its territory" (qtd. in Sievers, 2002, p. 26). China's good-neighbourliness policy does not preclude its assertions of absolute sovereignty with respect to territorial and resource issues.

China uses its water resources as it deems fit. For instance, it is unilaterally building a cascade of eight dams in the upper Mekong without consulting or informing its downstream neighbours. Eric Sievers (2002, p. 2) also argues that China's diversion of the Irtysh waters "violates customary international law both in its conception and in China's dealings with its co-riparians". In non-river-related territorial and resource issues as well, for instance on the South China Sea disputes, China has shown a clear and consistent stance that is congruent with the principle of absolute sovereignty. Contrary to Wouters's claim that China is committed to the important roles of the UN and the International Court of Justice, it has chosen to ignore the UN tribunal's ruling in July 2016 rejecting Chinese claims in the South China Sea.

Issue-linkages and incentives for cooperation

I argue that China cooperates when it is in its interest to do so and when it perceives benefits from cooperation, and not because it observes limited sovereignty. Interdependence between upstream and downstream countries can produce incentives for cooperation. In the case of China and Kazakhstan, growing interdependence facilitates issue-linkages and reciprocity, and helps explain Chinese cooperation with Kazakhstan on transboundary water issues. Linkages between water and broader political, strategic and economic issues incentivize China, as the hydro-hegemon, to cooperate with Kazakhstan, a weaker downstream neighbour. China sees a strong partnership with Kazakhstan on a range of issues as critical to its interests. These issues include domestic stability issues involving the Uighur minorities in Xinjiang, bilateral Sino–Kazakh economic and energy cooperation, and at the regional level, the competition for influence and power in Central Asia among Russia, China and the United States. In addition, China wants to avoid coming across as a bully in its dealings with Kazakhstan, in line with its good-neighbourliness policy. In this instance, the foreign policy considerations of powerful countries play a restraining role on how power is displayed. The desire to create a reservoir of goodwill with the Kazakh government and people was a strong motivating factor in China's decision to embark on water negotiations with Kazakhstan.

Recent scholarship on the relationship between upstream and downstream countries has shown that, contrary to the realist prediction that cooperation is least likely when the upstream riparian is also the hegemon, there are conditions under which cooperation may ensue when the upstream riparian is the hegemon (Daoudy, 2009; Dinar, 2009). Cooperation results when riparian states perceive benefits from joint action and "strategic interactions", defined as "the manner by which party payoffs are altered so as to encourage cooperation" (Dinar, 2009, p. 330). States cooperate when there are absolute gains from cooperation, and not only when there are relative gains. Cooperation in international environmental institutions is particularly important as they address transboundary and transnational problems that cannot be effectively managed by a single country alone (Haas, Keohane, & Levy, 1993).

According to Shlomi Dinar (2009), weaker downstream states can also bring about cooperation by influencing the bargaining context to change the incentives for the stronger state to cooperate. Strategies that weaker downstream states can use to alter the payoffs and encourage cooperation from the hegemonic upstream state "include, but are not limited to, reciprocity, foreign-policy considerations and issue-linkage, and side payments" (p. 334). In a series of case studies on the Euphrates, Colorado, Rio Grande, Parana and Rhine Rivers, Dinar shows that weaker downstream countries use issue-linkage and side payments to get stronger upstream countries to cooperate.

Likewise, Marwa Daoudy (2009) has argued that weaker downstream states can invert situations of power asymmetry by acting on the dominant state's interest, primarily through the use of issue-linkages. By linking water issues with benefits 'beyond' the river basin, the weaker downstream state can narrow the power gap with the stronger state and induce the latter to cooperate. The "clustering of issues" facilitates side payments and quid pro quos (Keohane, 2005, p. 91). The benefits that accrue from cooperation on water can therefore go 'beyond' the river basin. Daoudy's study shows how downstream Syria applies linking strategies to border, security, economic and societal issues that affect upstream Turkey to alter Turkey's payoffs for cooperation in the Euphrates/Tigris negotiations.

Issue-linkage typically occurs during the negotiation process. The quid pro quos or side payments are usually set informally through trade deals or other forms of cooperation. And quid pro quos do not always take the form of material exchanges. For a stronger state, maintaining a good reputation and winning friends may be the benefits of cooperating with the weaker side. This was a key consideration for the United States, as the powerful upstream riparian, when it concluded an agreement with Mexico on managing the Colorado River in 1973:

> Since the immediate economic incentives to cooperate were not clear to the United States, and since the externality was unidirectional in any case, coordination did not seem likely. However, not only did the United States not want to be considered a belligerent bully by its southern neighbor and the rest of Latin America by rejecting cooperation, but it also considered cooperation on the water issue as a form of gaining cooperation and support on other fronts such as drug trafficking and migration. While 'no explicit linkage was made between these issues in the agreement itself, doubtless the desire to build a "reservoir of goodwill" was a significant consideration behind the executive's desire to reach agreement for the salinity issue' [Le Marquand, 1977] (cited in Dinar, 2009, p. 378)

Issue-linkages and quid pro quos in Sino–Kazakh transboundary water cooperation

China began negotiations with Kazakhstan on transboundary water management in 1999. The first agreement on water cooperation was signed in 2001 as a result. The next set of key agreements were signed from the mid-2000s onwards. These include landmark agreements on water quality protection and information exchange, and joint development of the Khorgos River. Preliminary preparations for discussions on water allocation began in 2010, and by 2015, the two sides began consultations on a draft of the Agreement on Water Distribution of Cross-Border Rivers. See Table 2 for the key

Table 2. Key events in China-Kazakhstan transboundary water management.

Date	Event
May 1999	China and Kazakhstan began discussion on transboundary rivers
12 September 2001	China-Kazakhstan Agreement Concerning Cooperation in the Use and Protection of Transboundary Rivers (also provides for the establishment of a joint river commission)
4 July 2005	Agreement for Emergency Notification of the Parties about Natural Disasters on Transboundary Rivers (inter-ministerial)
20 December 2006	Agreement on the Mutual Exchange of Hydrological and Hydro Chemical Information (Data) of Border Gauging Stations on Major Transboundary Rivers (inter-ministerial)
20 December 2006	Agreement on the Development of Scientific-Research Cooperation (inter-ministerial)
13 November 2010	Agreement on Cooperation in the Construction of Joint Waterworks 'Dostyk' (Khorgos River)
13 November 2010	Implementation Plan of Water Allocation Technical Work of Transboundary Rivers
22 February 2011	Agreement on Water Quality Protection of Transboundary Rivers
13 June 2011	Agreement on Cooperation on Environmental Protection
7 September 2013	Agreement on the Management and Operation of 'Dostyk' Joint Hydro Unit on the Khorgos River
March 2015	Consultations begin on a draft of the Agreement on Water Distribution of Cross-Border Rivers

Sources: Zhulduz Baizakova, 'The Irtysh and Ili Transboundary Rivers: The Kazakh-Chinese Path to Compromise', *Voices from Central Asia* 21, January 2015, pp. 3–4; 'Sino-Kazakhstan Joint Declaration on Further Deepening Comprehensive Strategic Partnership,' 7 September 2013, China International Water Law Programme (unofficial translation), http://internationalwaterlaw.org/documents/regionaldocs/China-Kazakhstan_declaration_7_Sept_2013-CIWL_unofficial_translation.pdf; 'The Cooperation between the Republic of Kazakhstan and People's Republic of China in the Field of Use and Protection of Transboundary Water Resources', Kazakhstan Ministry of Foreign Affairs, http://www.mfa.kz/index.php/en/foreign-policy/current-issues-of-kazakhstan-s-foreign-policy/transboundary-rivers/the-cooperation-between-the-republic-of-kazakhstan-and-the-peoples-republic-of-china-in-the-field-of-use-and-protection-of-transboundary-water-resources.

events and agreements in Sino–Kazakh transboundary water cooperation. Each subsequent agreement represents a step up in institutionalized riparian cooperation between the two countries, in tandem with the overall intensification of Sino–Kazakh ties, as China turns to Kazakhstan to deal with its separatist problems and as an alternative source of energy imports (India Today, 2015).

Scholars have pointed to the existence of issue-linkages in transboundary water cooperation between China and Kazakhstan. According to Elizabeth Economy, Kazakhstan's "apparent relative success" in getting China to the bargaining table to discuss water allocation may have resulted from Kazakhstan's value to China as a source of cooper and oil (cited in Stone, 2012). She added that other downstream countries "might consider adopting a strategy of linking access to their commodities or energy to Chinese willingness to negotiate water-allocation issues". Similarly, Deng Mingjiang (2012, p. 373), a Chinese scholar, has pointed to the presence of issue-linkages in negotiations:

> Kazakhstan was keen to speed up the cooperation process by linking the transboundary river issue with security, economic and trade, and energy issues. Particularly when the opposition in Kazakhstan widely publicized the China water threat theory, the current Kazakh government was forced to adopt a 'hard stance' towards the Chinese government to accelerate the process of finalizing agreements, and to strive for breakthroughs in the key issues of environmental protection, water pollution prevention, and water allocation.

In addition, a Kazakh scholar has privately emphasized that the Silk Road Economic Belt (SREB), of which Kazakhstan is the central node, provides an invaluable opportunity for the two sides to resolve their outstanding transboundary water issues, as substantial water resources will be needed to support the development of the SREB. He further argued that with the growing number of Chinese industries in

Kazakhstan, it is in China's interest to work towards safeguarding the quality and quantity of water in Kazakhstan (author's notes, workshop on China's international rivers, Singapore, 2016).

The following subsections discuss the increase in cooperation between China and Kazakhstan in the security, political, energy, strategic and economic spheres in the 1990s and 2000s. Increased cooperation in these areas is followed by agreements and institutions being set up to manage their shared water resources. The discussion is divided into two time periods: 1992–2001 and 2002–2015.

1992–2001

Since diplomatic relations were established in 1992, the Kazakh government has tried to nudge China to begin negotiations on the management of shared water resources. These efforts met with little success until the late 1990s. China's decision to come to the negotiating table in May 1999 suggests that China's cooperation on transboundary waters was quid pro quo for Kazakhstan's support in the mid-to-late 1990s for China's campaign against Uighur separatists as well as its facilitation of Chinese access to its energy resources. China wanted to ensure goodwill with the Kazakh government and people so as to secure continued Kazakh cooperation on these national security and energy issues. Its repeated assurances and use of inclusive rhetoric in its discussions with Kazakhstan demonstrate that a key aim of talks with Kazakhstan was to retain a reservoir of goodwill and maintain a positive image in Kazakhstan. In an interview with a Kazakh newspaper, then Chinese ambassador to Kazakhstan Yao Peisheng said, "I want to say to your readers that China will not resolve any environmental problems at the expense of its neighbours, that is, it will not sacrifice them for its own benefit" (BBC Monitoring Central Asia, 2002).

Anti-separatism

The Han-minority problem is very much intertwined with China's management of its transboundary rivers, since many of these rivers are in border areas where large populations of non-Han ethnic groups reside (Nickum, 2008). Of China's provinces and autonomous regions, Xinjiang, together with Tibet, presents some of the most serious national security problems for China. Uighurs in Central Asia have strong cross-border ties with Xinjiang's Uighurs, and are often a source of support and funding for Xinjiang's separatist movements. Among Central Asian countries, Kazakhstan is China's most important partner in managing the Uighur issue because Kazakhstan has the largest Uighur population in Central Asia, and Uighur separatists are most active in Kazakhstan. These violent separatist movements have the potential to destabilize China internally and impair its territorial cohesion. China's management of the Ili and Irtysh Rivers is therefore critical to Chinese plans to grow Xinjiang's economy as part of its campaign to blunt the separatist movements.

Since 1992, Kazakhstan has worked closely with China to contain the Uighur separatist movements in Xinjiang: "a major component of China's foreign policy regarding Kazakhstan concerns extracting guarantees from Astana that it will not tolerate any expatriate Uighur organizing on its territory" (Sievers, 2002, p. 3). In 1995, the term 'separatism' appeared for the first time in a joint declaration by China and Kazakhstan. In the joint declaration, each party agreed to oppose separatist forces on its soil that threatened the territorial integrity of the other party (Valeev & Kadyrova, 2015). Cross-border anti-separatist cooperation with Kazakhstan is vital to China's strategy to contain the separatist threat. In 1996, China applied pressure on the Kazakh government to dissolve Uighur associations and political parties, and close down the Institute of Uighur Studies within the Institute of Oriental Studies (Laurelle & Peyrouse, 2015). The Chinese intelligence services also entered Kazakhstan, more or less with the consent of the Kazakh government, to track down Uighur separatists and bring them back to China (Laurelle & Peyrouse, 2015). Under pressure from China, Kazakhstan also closed newspapers and arrested militants, particularly after serious riots in Yining, Xinjiang, which is close to the Kazakh border, in 1997 (Human Rights Watch, 2001). In 1999, upon China's request, Kazakhstan repatriated three Uighur refugees to China (Human Rights Watch, 2001). The Kazakh government has unstintingly supported China's anti-separatist campaign, stating that "we categorically do not tolerate any forms of it [separatism] and we are for the principle of non-separation and territorial integrity of China" (*Kazakhstanskaya Pravda*, 1998).

Energy cooperation

Apart from securing Kazakh cooperation in cracking down on Uighur separatists, China's motivation for cooperating with Kazakhstan on water issues is also the result of strategic energy resource considerations. It was around the mid-1990s that China and Kazakhstan started to cooperate in the energy sector. Central Asia provides an alternative solution to China's efforts to secure its energy supplies. With close to 40 billion barrels in proven reserves, Kazakhstan has 3.2% of total world oil reserves, and is likely to be the dominant player in the Caspian Basin area (Laurelle & Peyrouse, 2012). Sino–Kazakh negotiations over oil began in 1994, during the visit of then Chinese Premier Li Peng to Kazakhstan. Actual oil cooperation began in 1997, when the two countries signed their first agreement in the oil and gas sectors. The agreement allowed China National Petroleum Corporation and its subsidiaries to invest in Kazakh oil fields, and provided for the construction of an oil pipeline between western Kazakhstan and Xinjiang (Van Wie Davis, 2008). Kazakh President Nursultan Nazarbayev declared the agreement to construct the oil pipeline the "contract of the century" (*Kazakhstanskaya Pravda*, 1998).

Catalyst for talks

To ensure Kazakhstan's cooperation on containing Uighur separatists and giving China access to Kazakhstan's energy sector, it is in the Chinese interest to retain the goodwill of the Kazakh government and people. The catalyst that prompted China's decision to begin water negotiations with Kazakhstan in 1999 was the negative reaction to a string of Kazakh press reports on China's construction of the Kara Irtysh–Karamay Canal.[4] China's turnabout at this point in time following years of dragging its feet on transboundary water talks suggests that it was concerned the news reports would damage China's relations with Kazakhstan, with implications for Kazakhstan's cooperation in the security and energy arenas. The rest of this section focuses on the circumstances that led to the signing of the first water agreement between China and Kazakhstan in 2001.

China's interest in the Ili and Irtysh Rivers lies in the diversion of waters for the development of China's western provinces, specifically Xinjiang's economy. The approaches adopted to develop Xinjiang's economy have the effect of increasing Xinjiang's water consumption. The rapid expansion of Xinjiang's cotton and wheat industries, which are water-intensive crops, adds to its water demands. The development of oil reserves, as well as the mass migration of millions of Han Chinese to Xinjiang as a result of the Chinese government's policy to increase Han presence in the region, have also led to greater water consumption. These developments are a severe strain on Xinjiang's already scarce water resources. To meet the rising demand, the Chinese government decided to increase water extraction from the Ili and Irtysh Rivers (Peyrouse, 2007b).

Kazakhstan is worried about the economic and ecological consequences of China's water extraction and diversion projects. Kazakhstan requires a large amount of water to support its own economic development, so the falling level of the Ili and the Irtysh Rivers due to Chinese water withdrawals is a deep area of concern. Both rivers provide water for the agricultural, industrial and fishery sectors of central and eastern Kazakhstan. Moreover, in the Kazakh government's estimation, the Irtysh is the main source of water for around 4 million people, out of the country's total population of 16 million (Laurelle & Peyrouse, 2012). The Irtysh supplies Astana, Kazakhstan's capital. Important north-eastern towns like Karaganda, Semei and Pavlodar draw some of their freshwater directly from the Irtysh. The fall in the level of the Irtysh also impairs industrial potential. Three hydroelectric stations in the north-east, Bukhtarma, Ust-Kamenogorsk and Shulbinsk, which are important for industrial growth, depend on Soviet-era reservoirs that in turn rely on water flows from the Irtysh. The Irtysh is also important for navigation between April and October each year, and for commercial exchange between Kazakhstan and Omsk, a Russian town. Ecologically, Kazakhstan is concerned that a reduction in the flow of the Irytsh could significantly damage Lake Zaysan. As for the Ili River, it provides water for the metal and energy industries in Kazakhstan. It supplies the Kapchagay hydroelectric station, which delivers electricity to the south of Kazakhstan, an area suffering from energy deficit. Furthermore, between the Chinese border and Lake Balkhash, the Ili is the chief source of irrigation for the fields lining the Grand Almaty Canal and is crucial to rice growing in the Akdalinsk region.

The Kazakh government has raised the problem of transboundary waters with China since 1992; for instance, the Kazakh side sent the Chinese a draft 'Proposals for Consultation on the Joint Use of Transboundary Rivers' (Valeev & Kadyrova, 2015). However, these initial attempts by Kazakhstan to move water negotiations with China forward were not successful. In the early 1990s, the Chinese government had begun construction of the 300-kilometre-long and 22-metre-wide Kara Irtysh–Karamay Canal, expected to divert 10–40% of the Irtysh River to Ulungur Lake (Central Asia News, 2012). The canal was completed in 1999. There were two objectives: to irrigate 140,000 hectares of new agricultural land, and to transport water to the Karamay oil fields, which has confirmed oil reserves of 1.7 billion tons (Laurelle & Peyrouse, 2012). The canal diverts about 500 million cubic metres of water per year, although the figure is expected to exceed 1 billion cubic metres when it reaches full operational capacity in 2020 (Laurelle & Peyrouse, 2012).

When China first started constructing the canal, there was little media or public attention on China's water diversion plans. However, as a result of the negative press reports in 1998, the water diversion project drew national attention in Kazakhstan, and the opposition, the press, non-governmental organizations and the Kazakh public began to pressure the Kazakh government. For instance, Murat Auezov, one of Nazarbayev's political opponents and the former ambassador to China, portrayed the Chinese government as "stonewalling prevaricators and the Kazakhstan government as incompetent collaborators" (Sievers, 2002, p. 5).

Its increasingly negative image in Kazakhstan and fear of souring diplomatic ties with Kazakhstan eventually incentivized the Chinese to begin negotiations with Kazakhstan. During the first round of talks, for instance, in May 1999, China repeatedly emphasized that Kazakhstan's interests had been fully taken into account in its water diversion projects, and the facilities that were being built would not inflict any damage on China, Kazakhstan, or the environment (BBC Monitoring Central Asia, 1999). It also reassured the Kazakh side that the transboundary water issues would be resolved to the "full satisfaction of both sides" (BBC Monitoring Central Asia, 1999). These statements suggest that the desire to contain any diplomatic fallout and repair the damage to its image in Kazakhstan was a key motivation for beginning water talks with Kazakhstan.

Five rounds of discussions were held in 1999, 2000 and 2001, and eventually the Agreement on Cooperation on the Use and Protection of Transboundary Rivers, which provided for the setting up of a joint river commission, was signed in 2001. Particularly during the fourth round of negotiations, Kazakhstan began to adopt a tougher attitude to press China to come to an agreement. The state newspaper *Kazakhstanskaya Pravda* said that if China proceeded with plans to divert 40% of the Irtysh's flow, a "global environmental catastrophe" would ensue, words which government officials and the mainstream media had avoided using till then (qtd. in Sievers, 2002, p. 11). Kazakhstan also accused China of plans to divert the Ili, which could turn Lake Balkhash into another Aral Sea (Sievers, 2002).

2002–2015

Following the signing of the 2001 agreement, the Sino-Kazakh Joint Commission on the Use and Protection of Transboundary Rivers was established in 2003. Its goals include the implementation of the 2001 agreement; the coordination, monitoring and measuring of the content and quality of water; and joint research, among others (Laurelle & Peyrouse, 2012). The institutionalization of Sino–Kazakh transboundary water cooperation in the form of the joint commission provides a focal point for coordination, facilitates information exchange, and helps reduce transaction costs in negotiations. From 2003 to 2016, 14 meetings were held (Interfax-Kazakhstan General Newswire, 2016), which resulted in many of the agreements listed in Table 2. These agreements touch on issues that China considers sensitive, including water quality protection and discussions on water sharing.

The joint commission's ability to achieve significant breakthroughs in institutionalizing cooperation in the 2000s is concomitant with the overall expansion of ties. In 2002, China and Kazakhstan fully settled their border issues by signing a protocol on border demarcation, and in 2005, Kazakhstan was the first among Central Asian states to establish a strategic partnership with China. In exchange for Kazakhstan's cooperation on a variety of issues critical to China's interest, China cooperates with Kazakhstan on water quality, joint development, and water-allocation negotiations. These issues include the intensified campaign against Uighur separatists following the 11 September 2001 terrorist attacks on the United States, increased Sino–Kazakh collaboration in the energy sector as well as the development of Xinjiang's economy, and Kazakhstan's importance as a strategic partner in the Shanghai Cooperation Organisation (SCO).

Counter-terrorism

The September 11 terrorist attacks provided justification for China to step up its campaign against the Uighur separatists. The Chinese government rallied behind the US government and offered strong support for actions to combat terrorism. Prior to the September 11 attacks, the Chinese government had long claimed that the Uighur separatists are "religious extremist forces" and "violent terrorists" (Human Rights Watch, 2001). These claims gained some traction after September 11, and in 2002, the Chinese government succeeded in getting the East Turkestan Islamic Movement, a separatist group in Xinjiang and Central Asia, labelled a terrorist organization by the United States.

Kazakhstan's role as a key partner in China's 'counter-terrorism' campaign grew during this period. In the immediate aftermath of the September 11 attacks, senior leaders in the Chinese and Kazakh governments met to pledge unity in counter-terrorism measures (Ramani, 2015). In December 2002, China and Kazakhstan signed a bilateral agreement to combat terrorism, separatism and extremism. The terms of the agreement reflect the terms in the Shanghai Convention on Combating Terrorism, Separatism, and Extremism, signed in 2001. The intelligence cooperation and information exchanges that began in the 1990s were formalized in this agreement. China also began to donate communications equipment and other technology to the Kazakh military, and started military exercises with Central Asian countries and Russia to combat terrorism in 2002 (Ramani, 2015).

The Shanghai Cooperation Organisation

Also, in the early 2000s, China launched several initiatives in which Kazakhstan's cooperation and support was invaluable. Originally known as the Shanghai Five, the SCO was formed when Uzbekistan joined the grouping in June 2001. Security engagement is at the core of SCO activities. Initially set up to settle border issues, it has since evolved into conducting joint military and anti-terrorism exercises. In addition to military exercises, a Regional Anti-Terrorist Structure was set up in Tashkent in 2004. Anti-terror exercises are conducted under its umbrella. Among its achievements are the extradition treaty among member states and the creation of a 'blacklist' of about a thousand people and 40 organizations (Laurelle & Peyrouse, 2012). Securing Kazakhstan's support within the ambit of the SCO was critical to Chinese efforts to coordinate counter-terrorism activities in Central Asia and Xinjiang.

In addition, Kazakhstan's support in the SCO was important for Chinese strategic interest in Central Asia *vis-à-vis* the United States and Russia. The United States had stepped up its presence in the region as a result of its counter-terrorism campaign. By maintaining close ties with Kazakhstan, China had hoped to dilute and limit US influence in the region. Russia, which had traditionally dominated Central Asia but whose influence had waned as a result of its domestic troubles in the 1990s, also began to reassert its presence in the region in the 2000s. Kazakhstan's partnership with China is an important counter-balance against Russia in the SCO.

Energy cooperation

Sino-Kazakh cooperation in the energy sector reached a new high point in the 2000s. This was signified by the opening of the Atasu–Alashankou oil pipeline, running from central Kazakhstan to western China, in 2006 (Dwivedi, 2006). The pipeline has a capacity of 10 million tons per year, which is set to gradually increase over the years (Valeev & Kadyrova, 2015). Chinese oil companies have also made further inroads into Kazakhstan – Sinopec and China National Offshore Oil Corporation are also now active in Kazakhstan (Valeev & Kadyrova, 2015). They operate oil fields in Kazakhstan and exercise substantial control over important Kazakh oil companies; in 2005, the China National Petroleum Corporation bought Petrokazakhstan, a leading energy firm (Omelicheva, 2011). China is now the second-largest importer of Kazakh oil. It is projected that by 2020, the export of Kazakh oil to China could reach 13 million tons (Valeev & Kadyrova, 2015). China hopes to boost its stake in the Kazakh oil and gas industry from 22% to 24% (Guschin, 2015).

Economic cooperation

Kazakhstan's economic cooperation with Xinjiang is critical to the Chinese government's efforts to develop the western provinces – the Go West (*xibu dakaifa*) campaign was officially launched in 2000. Xinjiang is a key prong of the Go West project, and there are close economic ties between Xinjiang and Kazakhstan. The increased cooperation between China and Kazakhstan on water issues is in tandem with the overall increase in economic cooperation between the two in the 2000s. Trade between Kazakhstan and Xinjiang jumped substantially in the early 2000s – the volume of trade between Kazakhstan and Xinjiang grew from a little more than USD 1 billion between 1990 and 1993 to more than USD 10 billion between 2000 and 2004 (Laurelle & Peyrouse, 2012). Sino–Kazakh bilateral trade is particularly important in the northern part of Xinjiang, which accounts for 70% of regional production due to its considerable resources (Laurelle & Peyrouse, 2012). In March 2015, China and Kazakhstan signed 33 deals worth USD 23.6 billion (Tiezzi, 2015). Chinese companies based in Xinjiang are also investing in infrastructure and industries in Kazakhstan (Tiezzi, 2015). China accounted for 9% (Russia accounted for 5.4%) of Kazakhstan's gross inflow of foreign direct investment in 2013 (Guschin, 2015). China is currently Kazakhstan's second-largest trading partner and largest export market. China and Kazakhstan also jointly established the Khorgos Special Economic Zone, the Khorgos International Centre for Boundary Cooperation and the Khorgos Dry Port to facilitate trade between them. These projects have taken on a new significance with China's launching of the SREB initiative when Chinese President Xi Jinping visited Kazakhstan in 2013.

Increased transboundary water cooperation

To reciprocate Kazakhstan's support and cooperation in these political, economic and strategic initiatives of the early 2000s, China further acceded to Kazakhstan's requests to step up cooperation in transboundary water management. In 2006, the Agreement on the Mutual Exchange of Hydrological and Hydro Chemical Information of Border Gauging Stations on Major Transboundary Rivers was signed. Cooperation on water quality took a step further when the Agreement on Water Quality Protection of Transboundary Rivers was signed in 2011.

China and Kazakhstan are also cooperating in the joint development of water resources, particularly in the development of hydropower. A major achievement is the 2010 agreement to build the Dostyk (Friendship) hydro-engineering complex on the Khorgos River for joint use and management. Joint development of the Khorgos River is linked to the cross-border development of the Khorgos area, including the setting up of a special economic zone, free trade zone and dry port. The 2010 agreement provides for the fair and equitable division of the Khorgos's waters, which is a significant step forward in Sino–Kazakh transboundary water relations, although it must be noted that negotiations in this case are less controversial because the Khorgos River is a border river and not a cross-border river. Construction of the complex began in 2011 and was completed in 2013. In 2013, another agreement was signed to manage and operate the Dostyk project. A joint service responsible for operating the hydraulic power system was set up to manage the distribution of water (*Astana Times*, 2014).

Besides the Dostyk project, China and Kazakhstan are involved in two other hydro-electric projects. The Moinak project, estimated at USD 310 million, is partially financed by a USD 200 million credit from Beijing (Peyrouse, 2007a). It is situated on the Charyn River and is the first 'turnkey' construction project for a new station since Kazakhstan's independence. Its capacity of 300 MW partly makes up for the electricity shortfall in the south of Kazakhstan. And in 2014, China and Russia agreed to provide USD 400 million in loans to construct a third electric power generation block at the Ekibastuz station (Tengri News, 2014).

Even more significantly, steps were taken in the 2000s to begin negotiations on a water-allocation arrangement between China and Kazakhstan. Kazakhstan has repeatedly raised the issue of a proper allocation of water resources with China. Several working groups have been established to deal with water-sharing arrangements. These include the Joint Control Commission of the Division of Water of the Khorgos River (2002) and the Permanent Water Commission in the Field of Use and Water Division of the Sumbe River and Kayshybulak River (2008).

However, China's response to Kazakhstan's requests for a comprehensive allocation agreement that covers all shared water resources was slow, and the pace picked up only in the late 2000s, suggesting that Chinese cooperation is aimed at reciprocating increased Kazakh cooperation on security, energy and economic issues. In December 2009, Kazakh officials submitted to their Chinese counterparts a draft Concept for Water Distribution along the Irtysh and Ili Rivers (Economic Commission for Europe Convention, 2009). In 2010, a Working Group of Experts was set up under the Joint River Commission to look into the Implementation Plan of Water Allocation Technical Work of Transboundary Rivers (Kazakhstan Ministry of Foreign Affairs). Although progress on water-allocation negotiations has been slow and halting (Council of Foreign Relations Website, 2012), the two sides managed to start consultations on a draft of the Agreement on Water Distribution of Cross-Border Rivers in 2015. If the working group succeeds and a comprehensive water-sharing agreement is signed with Kazakhstan, it will represent an unprecedented step by China to share water resources with a riparian neighbour.

Conclusion

China's size, population and economic and military prowess put it in an extremely favourable position when it comes to navigating riparian relations. However, weaker downstream states are not without leverage. As demonstrated above, China has tended towards cooperation in setting up institutions to manage transboundary waters with Kazakhstan, a smaller and weaker downstream neighbour, because the growing inter-dependence between them since the mid-1990s and 2000s has allowed both sides to derive benefits 'beyond' the river basin. The timeline of the various water agreements suggests that transboundary water cooperation is quid pro quo for Kazakhstan's cooperation on a range of political, strategic, security and economic issues that are vital to Chinese interests.

This does not mean, however, that problems do not exist between China and Kazakhstan in managing their shared water resources. China has confined its engagement of Kazakhstan on transboundary river issues to bilateral platforms, and has refused to expand these bilateral arrangements to include Russia despite Kazakhstan's promptings. Discussions on water issues at the SCO remain at a superficial level. At the SCO meetings in 2013 and 2014, Kazakhstan proposed the setting up of a water commission to address water security issues, but to little avail (South Asia Monitor, 2014).

A key issue that China appears to be dragging its feet on is the "rational and equitable use" of water as laid out in Article 4 of the 2001 agreement. A major stumbling block is the lack of provisions for the requisite institutional mechanism to implement this article. The Sino–Kazakh Joint Commission is supposed to discuss this issue, but little progress has been made. There is also a lack of monitoring and compliance mechanisms, and of dispute-settlement mechanisms. Moreover, volume intakes by both sides are not specified; the agreement does not stipulate any rules for the specific treatment of the Ili and the Irtysh, going no further than calling for "measured" utilization of common waters (Peyrouse, 2016).

Although negotiations on a water-allocation plan have begun, the negotiations between the two countries have been slow, and Chinese officials have reportedly asked for more time to come to terms on a water-sharing agreement for the Ili and Irtysh, shifting the deadline from the end of 2014 to the end of 2016 (Stone, 2012). China is said to have put forth the principle that water allocation should be according to the number of people who actually live along the river banks (Clarke, 2014; Radio Free Asia, 2013). If this principle is applied, most of the water allocation will go to the Chinese side, an arrangement that is clearly unacceptable to Kazakhstan.

Nevertheless, the fact that both sides are now discussing a draft of a water-sharing agreement is an encouraging sign that a water-allocation arrangement between China and Kazakhstan could be finalized in the near future. Recent developments in China's economic outreach programme suggest that Kazakhstan may be able to gain further leverage in issue-linkage; Kazakhstan is the central node for China's plan to develop the SREB, as it sits on most of the routes linking China and Europe. This explains why Xi Jinping chose the occasion of his visit to Kazakhstan in 2013 to announce the initiative for the first time. While no explicit statements were made in public to link the SREB with cooperation over transboundary waters, the SREB provides an additional opportunity for Kazakhstan to urge China to speed up negotiations on the water-sharing agreement.

While it is not within the scope of this article to examine the limitations of issue-linkage strategies, it has nevertheless alluded to some of them. They are listed here as areas which require further research. First, issue-linkage is generally more effective in a bilateral context. Issues become more complex and quid pro quos more difficult in a multilateral setting in which multiple interests intertwine and could contradict each other. It is therefore harder to employ issue-linkage strategies in a river basin like the Mekong, where there are six riparian states. Second, there are limits to how much a weaker state can exercise issue-linkage strategies against a stronger state. Overt and hard-line issue-linkage by a weaker state may be perceived by the stronger state as a threat, which may cause the stronger state to dig in its heels. Moreover, the weaker state may be dependent on the stronger state for trade and investments – Kazakhstan's economic reliance on China limits how far it will attempt to push China on water issues. Third, for a weaker state to succeed in issue-linkage, the stronger side must perceive substantial gains from cooperation. There has to exist a common set of interests between the two sides. These interests must extend not only to economic interests but also to broader security and strategic interests, as is the case between China and Kazakhstan. These security and strategic interests may be less clear in the context of the river basins that China shares with South and South-East Asia. Finally, issue-linkage is a double-edged sword. As much as it can be an effective tool for weaker states to employ, the stronger state may use issue-linkage as a strategy to extract concessions from the weaker side. The weaker side may have to make sacrifices in certain areas to gain cooperation from the stronger side on other areas. For instance, to win Chinese goodwill and cooperation, Kazakhstan has opened up its strategic energy sector to Chinese companies. Furthermore, its own national security concerns aside, Kazakhstan's policy towards its Uighur population is very much dictated by Chinese concerns – the closing down of Uighur schools and associations under Chinese pressure is an example. Likewise, allowing Chinese intelligence to operate within its border is a substantial concession by Kazakhstan.

Notes

1. For a detailed account of the environmental impact of Chinese activities, see Stone (2012). However, the environmental problems in the Ili and Irtysh are not the results of China's actions alone. Kazakhstan has been criticized for its irrational usage of water (Baizakova, 2015).
2. Douglass North (1999, p. 3) defines institutions as "the humanly devised constraints that shape human interaction". They can be either formal or informal. Formal institutions

include constitutions, laws and property rights, while informal ones are customs, tradi-tions, taboos and norms. This article is focused on formal institutions at the river-basin level, which include water treaties and agreements, joint river commissions, MOUs, institutionalized exchanges of information, working groups and expert-level mechanisms.

3. Chinese treaties and agreements are long on broad principles but short on detailed imple-mentation plans. Some of these broad principles include 'fair and reasonable use' (*gongping heli*), 'non-interference in internal affairs' (*hubuganshe neizhen*), and 'consideration of mutual interest' (*kaoludao shuangfang de liyi*).

4. See Sievers (2002, pp. 7–13) for an account of the pressures the press reports exerted on the Kazakh government, which in turn increased the pressure on China to come to the negotiating table.

Acknowledgments

The author would like to thank the reviewers and the editor-in-chief of *Water International* for their helpful comments. She is indebted to Terence Lee for looking through the drafts and offering suggestions for improvements. She also thanks, Jing Bo-jiun for his assistance in researching this article, Khasan Redjaboev, for his translation of Russian-language Kazakh newspapers and Feng Yikang for putting together the map for this article.

References

Astana Times. (2014). Kazakhstan, China to joint operate Dostyk hydraulic power system. Retrieved from http://www.astanatimes.com/2014/11/kazakhstan-china-jointly-operate-dostyk-hydraulic-power-system/

Baizakova, Z. (2015). The Irtysh and Ili Transboundary Rivers: The Kazakh-Chinese path to compromise. *Voices from Central Asia*, 21.

BBC Monitoring Central Asia. (1999). *First round of Kazakh-Chinese talks on use of transboundary rivers ends.*

BBC Monitoring Central Asia. (2002). *Chinese envoy upbeat on prospects for Kazakh-Chinese cooperation.*

Central Asia News. (2012). *Central Asian water issues will not be a part of SCO agenda but China is prepared to discuss it bilaterally with Kazakhstan.*

Chen, H., Rieu-Clarke, A., & Wouters, P. (2013). Exploring China's transboundary water treaty practice through the prism of the UN watercourses convention. *Water International, 38*(2). doi:10.1080/02508060.2013.782134

Clarke, M. (2014). Kazakh responses to the rise of China: Between Elite Bandwagoning and societal ambivalence? In N. Horesh & K. Emilian (Ed.), *Asian thought on China's changing international relations*. Basingstoke, UK: Palgrave Macmillan.

Council on Foreign Relations Website. Testimony by Elizabeth Economy before the US-China Economic and Security Review Commission. (2012, January 26). *China's global quest for resources and implications for the United States*. United States Congress, Second Session, 112th Congress. Retrieved from http://www.cfr.org/china/chinas-global-quest-resources-implications-united-states/p27203

Daoudy, M. (2009). Asymmetric Power: Negotiating Water in the Euphrates and Tigris. *International Negotiations, 14*, 361–391.

Deng, M. (2012). hasakesitan kuajie heliu guoji hezuo wenti [International Cooperation Problems on Transboundary Rivers in Kazakhstan]. *ganhanqu dili* [Arid Land Geography], *35*(3), 365–376.

Dinar, S. (2009). Power asymmetry and negotiations in international river basins. *International Negotiations, 14*, 329–360.

Dwivedi, R. (2006). China's Central Asia policy in recent times. *The China and Eurasia Forum Quarterly, 4*, 4.

Economic Commission for Europe Convention on the Protection and Use of Transboundary Watercourses and International Lakes. (2009). *River basin commissions and other institutions for transboundary water cooperation: Capacity for water cooperation in Eastern Europe, Caucasus and Central Asia.* New York, NY: United Nations.

Guschin, A. (2015). China, Russia and the tussle for influence in Kazakhstan. *The Diplomat.* Retrieved from http://thediplomat.com/2015/03/china-russia-and-the-tussle-for-influence-in-kazakhstan/

Haas, P., Keohane, R., & Levy, M. (Eds.). (1993). *Institutions for the earth: Sources of effective international environmental protection.* Cambridge, MA: The MIT Press.

Human Rights Watch. (2001). China: Human rights concerns in Xinjiang: A human rights watch backgrounder. Retrieved from https://www.hrw.org/legacy/backgrounder/asia/china-bck1017.pdf

India Today. (2015). *The big story: Bend it like Beijing.* New Delhi.

Interfax-Kazakhstan General Newswire. (2016). *Kazakhstan and China to discuss use of transboundary rivers in November.* Astana.

Joint Communique. (2008, October 31). *zhonghua renmin gongheguo he hasakesitan gongheguo lianhe gongbao* [Joint Communqiue of the People's Republic of China and the Republic of Kazakhstan].

Kazakhstan Government News. (2016). *Cooperation between Kazakhstan and China flourishes – MFA.*

Kazakhstan Ministry of Foreign Affairs. (2014). The Cooperation between the Republic of Kazakhstan and People's Republic of China in the field of use and protection of transboundary water resources. Retrieved from http://www.mfa.kz/index.php/en/foreign-policy/current-issues-of-kazakhstan-s-foreign-policy/transboundary-rivers/the-cooperation-between-the-republic-of-kazakhstan-and-the-peoples-republic-of-china-in-the-field-of-use-and-protection-of-transboundary-water-resources

Kazakhstanskaya Pravda. (1998). Kazakhstan-Kitay: iest neobhodimost 'sverit chasy' [Kazakhstan-China: There is a need to "set the clocks"]. Astana.

Keohane, R. (2005). *After hegemony: Cooperation and discord in the world political economy*, 2nd edn. New Jersey: Princeton University Press.

Laurelle, M., & Peyrouse, S. (2012). *The Chinese question in Central Asia: Domestic order, social change and the Chinese factor.* London: Hurst & Company.

Laurelle, M., & Peyrouse, S. (2015). *Globalizing Central Asia: Geopolitics and the challenges of economic development.* London: Routledge.

Le Marquand, D. (1977). *International rivers: The politics of cooperation.* Vancouver: Westwater Research Centre.

Lowi, M. (1993). *Water and power: The politics of a scarce resource in the Jordan River Basin.* Cambridge: Cambridge University Press.

Lowi, M. (1995). Rivers of conflict, rivers of peace. *Journal of International Affairs, 49*(1), 123–144.

Nickum, J. (2008). The upstream superpower: China's international rivers. In O. Varis, C. Tortajada, & A. K. Biswas (Eds.), *Management of transboundary rivers and lakes.* Berlin: Springer-Verlag.

North, D. (1999). *Institutions, institutional changes and economic performance.* Cambridge: Cambridge University Press.

Omelicheva, M. (2011). *Counterterrorism policies in Central Asia.* London: Routledge.

Peyrouse, S. (2007a). The hydroelectric sector in Central Asia and the growing role of China. *China and Eurasia Forum Quarterly, 5*(2), , 131–148.

Peyrouse, S. (2007b). Flowing downstream: The Sino-Kazakh water dispute. *China Brief, 7*, 10.

Peyrouse, S. (2016). Discussing China: Sinophilia and Sinophobia in Central Asia. *Journal of Eurasian Studies, 7*(1).

PRC Ministry of Foreign Affairs Website. 2015. Joint declaration on new stage of comprehensive strategic partnership between the people's Republic of China and the Republic of Kazakhstan.

Retrieved August 31, 2015, from http://www.fmprc.gov.cn/mfa_eng/wjdt_665385/2649_665393/t1293114.shtml

Radio Free Asia. (2013). Rivers threatened as China, Kazakhstan water pact remains elusive. Retrieved from http://www.tibet.hu/tibetpressen/veszelyben-a-folyok-mig-kina-es-kazahsztan-nem-jut-egyezsegre

Ramani, S. (2015). The emerging China-Kazakhstan defense relationship. *The Diplomat.* Retrieved from http://thediplomat.com/2015/12/the-emerging-china-kazakhstan-defense-relationship/

Sievers, E. (2002). Transboundary jurisdiction and watercourse law: China, Kazakhstan, and the Irtysh. *Texas International Law Journal, 37*(1), 1–42.

South Asia Monitor. (2014). *Water security at SCO.* Retrieved from http://southasiamonitor.org/detail.php?type=sl∋

Stone, R. (2012). For China and Kazakhstan: No meeting of the minds on water. *Science, 337,* 27. doi:10.1126/science.337.6093.405

Tengri News. (2014). China and Russia to spend $400 million on Kazakhstan Power Plant. Retrieved from https://en.tengrinews.kz/finance/China-and-Russia-to-spend-400-million-on-Kazakhstan-power-23474/

Tiezzi, S. (2015). China, Kazakhstan sign $23 billion in deals. *The Diplomat.* Retrieved from http://thediplomat.com/2015/03/china-kazakhstan-sign-23-billion-in-deals/

Valeev, R., & Kadyrova, L. (2015). Formation and development of bilateral relations of the Republic of Kazakhstan and the People's Republic of China in 1990–2000s. *Journal of Sustainable Development, 8,* 4. doi:10.5539/jsd.v8n4p277

Van Wie Davis, E. (2008). Uyghur muslim ethnic separatism in Xinjiang, China. *Asian Affairs: An American Review, 35,* 1.

Wolf, A. T., Yoffe, S. B., & Giordano, M. (2003). *International waters: Indicators for identifying basins at risk.* Paris, France: UNESCO.

Wouters, P. (2014). The Yin and Yang of International Water Law: China's transboundary water practice and the changing contours of state sovereignty. *Review of European Community and International Environmental Law, 23,* 1.

Wouters, P., & Chen, H. (2015). Editors' introduction. 'The China water papers' – Transboundary water cooperation. *Water International, 40,* 1. doi:10.1080/02508060.2014.990144

Zeitoun, M., & Warner, J. (2006). Hydro-hegemony – a framework for analysis of transboundary water conflicts. *Water Policy, 8,* 435–460. doi:10.2166/wp.2006.054

China's "old and new" Mekong River politics: the Lancang-Mekong Cooperation from a comparative benefit-sharing perspective

Sebastian Biba

ABSTRACT
This article analyzes China's Mekong River politics before and after the establishment of the Lancang-Mekong Cooperation (LMC) from a comparative benefit-sharing perspective. China's pre-LMC approach focused too much on the creation of economic benefits from and beyond the river while neglecting ecological benefits to the river. Moreover, despite the problems this 'old' approach caused for China and its downstream neighbours, China's current LMC strategy seems to essentially replicate its former approach. While sustainable water resources management is identified as a priority area, actual cooperation and benefit sharing in this field remain insufficient.

Introduction

China plays the role of an unrivalled 'multidirectional, transboundary water provider' (Chellaney, 2011, p. 2) around its entire periphery. However, China's hydro-political record in its shared river basins has to date remained highly ambiguous. One the one hand, China has not been overly cooperative in terms of joint river governance. China was one of only three countries that voted against the 1997 United Nations Convention on the Law of Non-navigational Uses of International Watercourses. China has not inked any comprehensive river treaty covering the issue of water distribution among riparian states. China has engaged unilaterally in dam building and river diversion activities along several of its international rivers, even though all of this bears the potential for negative ecological, economic and political impacts on its neighbours. And China has been circumspect about joining multilateral river management institutions as a full member and has been cautious about empowering multilateral organizations of which it actually is a member to play a bigger role in questions of joint water governance (Biba, 2014). On the other hand, China has not been completely uncooperative regarding its shared rivers. It has concluded water-specific bilateral treaties, which include water quality issues, with three riparian neighbours (Kazakhstan, Mongolia and Russia), and it has agreed to memoranda of understanding according to which it provides hydrological data, especially during flood seasons, to several downstream neighbours (Biba, 2014; Su, 2014).

Against this backdrop, it is interesting and noteworthy that inside China there is increasing understanding of the potential negative effects of its river politics on its foreign relations. He (2015, p. 312), for example, has argued that the upstream–downstream situation prevalent in many of China's shared rivers 'has resulted in diplomatic tensions between China and its riparian neighbours'. In a similar vein, Feng et al. (2015, p. 329) have maintained that 'China's transboundary water resources management continues to evolve as one of the most important emerging geopolitical issues for the nation and the region'. As a result, there appears to be a growing consensus amongst Chinese experts that the country's international hydro-politics should be modified (personal communications, July–August 2016). He (2015) summarized what appears to be the majority position when she proposed that China needed to shift from 'responsive diplomacy' to 'preventive diplomacy'. What she was pithily referring to was that China's past river politics were very reactive, usually a 'passive response after a request or dispute' (p. 316; also see Biba, 2014, 2016a). This strategy would have to be transformed into a 'more active and open approach', 'with a view to anticipating and preventing disputes' (He, 2015, p. 321).

The one river basin in which China has already begun trying to implement this 'new' river politics is the Mekong, mainly through a recently launched multilateral mechanism, the Lancang-Mekong Cooperation (LMC). This article sets out to evaluate the LMC in light of a potential Chinese turn towards a more active and more preventive hydro-political strategy regarding the Mekong by comparing it to China's pre-LMC approach. Conceptually, the article uses the framework of benefit sharing introduced by Sadoff and Grey (2002); methodologically, it draws on a number of personal interviews with Chinese experts. The main argument the article seeks to advance is that while the LMC certainly indicates a more active Chinese approach, it is questionable whether the LMC in its current form also represents an expression of promising preventive diplomacy. This is mainly because like China's 'old' and failed Mekong River politics, the LMC currently appears to be preoccupied with creating economic benefits *from* and *beyond* the river and is as of now unlikely to provide sufficient and stable ecological benefits *to* the river. In an interesting side note that is, however, not discussed further in this article, the LMC's preoccupation with economic benefits is also indicative of the fact that among the so-called 'nine dragons' – that is, the various institutions until recently governing water resources management in China (Feng et al., 2015) – those favouring economic development at whatever cost (such as the National Development and Reform Commission) tended to have the upper hand over those focusing on ecological preservation (such as the Ministry of Environmental Protection) or those championing smooth diplomatic relations (such as the Ministry of Foreign Affairs). It remains to be seen if the March 2018 ministry reform will affect this constellation.

The remainder of this article proceeds along the following lines. The next section outlines the key aspects of Sadoff and Grey's benefit-sharing framework. In view of this framework, the article then delineates and assesses China's approach to Mekong River politics in the era before the LMC. It then discusses the LMC as the embodiment of China's 'new' Mekong approach and compares its benefit-sharing rationale with that of China's 'old' approach. A brief conclusion summarizes the major findings and alludes to some scenarios that could lie ahead.

The benefit-sharing framework

As Sadoff and Grey (2002, p. 389) have aptly pointed out, 'International rivers can elicit cooperation or conflict. The choice between the two will in large part be determined by perceptions of their relative benefits.' On the one hand, river management is often complicated by the fact that rivers flow across political boundaries and their resources have to satisfy multiple demands for various users. On the other hand, benefit sharing is geared towards reducing those management problems and aims to promote increased cooperation. It does so by focusing on allocating the outputs from water use – such as electricity, food, and environmental services – and disregarding the allocation of water itself (Alam, Dione, & Jeffrey, 2009). According to Lee (2015, p. 140), benefit sharing can therefore be regarded as a 'practical tool that can be utilized to promote cooperation between countries with transboundary rivers' because this mechanism emphasizes a state of 'hydro-interdependency' between the different stakeholders. It is, however, not only a 'practical tool'. In fact, it can also be said to include a normative, even an idealistic, angle. This is because riparian countries sharing a basin do not always freely acknowledge that they are in a state of hydro-interdependency. As Alam et al. (2009, p. 94) rightly pointed out,

> Implementing the benefit-sharing principle in an international setting is likely to carry significant political risk for the incumbent governments. Therefore, we would expect most governments to maximize their autonomy by focusing on water allocation and unilateral control, rather than implementing a sovereignty bargain that may lead to greater payoffs in the long term but reduces states' autonomy in the short term.

In this sense, it could be argued that other theoretical approaches such as hydro-hegemony, with its focus on the (various forms of) power riparian countries bring to the table (Zeitoun & Warner, 2006), and even hydro-politics, defined as the systematic study of the nature and conduct of cooperation *and* conflict between states over transboundary water resources (Elhance, 2000), represent more realistic frameworks for studying international river politics. This might be even more the case when it comes to China, which has been labelled an 'upstream superpower' (Nickum, 2008). However, the benefit-sharing approach should certainly not be discredited simply because of its focus on the positive aspects of riparian interaction. Rather, a theoretically informed look specifically at how riparian cooperation could be improved can be quite helpful. What is more, several Chinese scholars doing research on the country's river politics have explicitly mentioned benefit sharing as a potential way to move things between China and its riparian neighbours forward (personal communications, July–August 2016).

Following Sadoff and Grey's (2002) use of the concept, benefit sharing in international river basins can be defined as 'the development of water uses in their "optimal" locations, and the distribution of these benefits, rather than the water, to users across the basin' (Alam et al., 2009, p. 93). For Sadoff and Grey, who introduced the framework to the field of water resources management, there are four different types, or broad categories, of benefits to be derived from international cooperation in a shared river: Type 1 refers to increasing benefits *to* the river. The river is regarded as a complex ecosystem, and international cooperation is meant to improve the management of this ecosystem. Type 1 therefore comprises environmental benefits, such as improved water quality, enhanced biodiversity or overall sustainability. Type 2 refers to increasing benefits *from* the river. The river is treated as an economic system which international cooperation can manage more efficiently. Type 2 thus comprises direct economic benefits – for example, in the form of enhanced hydropower or food production or increased navigability. Type 3 describes the reduction of costs *because of* the river. The river is attributed political signifi-cance in that international cooperation regarding the river has a positive influence on overall relations between the riparian states. Consequently, Type 3 comprises political benefits – for instance, through the alleviation of international tensions. Type 4, finally, refers to increasing benefits *beyond* the river. International coopera-tion in relation to the river is understood as a catalyst for more comprehensive cooperation in other areas. Type 4 particularly comprises indirect economic benefits, such as the integration of regional infrastructure and trade.

Furthermore, Sadoff and Grey's four types of benefits are in a dynamic relationship with each other. That is to say, there are linkages between them, and cooperation in relation to one type of benefit can have an impact on cooperation in others. Sadoff and Grey reveal these dynamic relationships very clearly (see also Figure 1): Type 1 environ-mental cooperation is considered to underpin sustainable economic development (Type 2 and Type 4 benefits) and is viewed as a means to build trust conducive to reducing political tensions (Type 3 benefits). Type 1 cooperation is explicitly treated as the basis for all other types of benefits to be derived. Type 2 economic cooperation is regarded as motivating joint stewardship of the river's resources (Type 1 benefits), easing political tensions (Type 3 benefits) and increasing opportunities for more comprehensive economic cooperation (Type 4 benefits). Type 3 political cooperation is not only facilitated by Type 1 and Type 2 cooperative actions, but is also seen as 'returning the favour' by further enabling enhanced environmental and economic cooperation (Type 1, Type 2 and Type 4 benefits). Last but not least, opportunities for Type 4 cooperation are believed to arise from benefits previously derived from the other three types, without any (explicit)

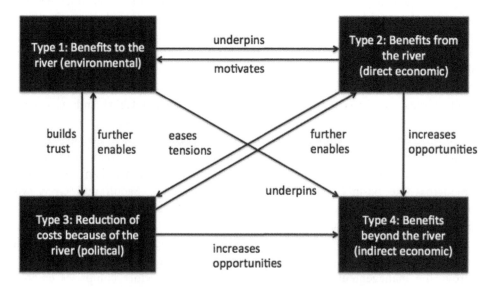

Figure 1. The dynamics between Sadoff and Grey's four types of benefit sharing.
Source: Author, based on information from Sadoff and Grey (2002).

recognition of positive feedback loops from Type 4 cooperation to the other types. At the same time, if certain types of cooperative actions do not take place or if benefits are distributed very unequally, then the positive dynamics mentioned can fail to materialize and may even be reversed into negative relationships.

China's 'old' Mekong River politics

China is the most upstream and most powerful riparian country on the Mekong, which originates on the Tibetan Plateau and then flows for approximately 4900 km through five more countries – Myanmar, Laos, Thailand, Cambodia and Vietnam – before emptying into the South China Sea. Regarding China's 'old', that is, pre-LMC, Mekong River politics, three aspects are particularly relevant: China's dam-building activities along the Mekong; China's relationship with the Mekong River Commission (MRC); and China's engagement in the Greater Mekong Subregion (GMS).

China's dam-building activities and their potential impacts

As of this writing, China is still the only Mekong riparian country to have completed dams on the river's mainstream. In 1986, China launched construction of a dam cascade on the middle reaches of the Lancang (as the Mekong is called in China). The first of these dams became operational in 1993. At present, six dams have been finalized and one is still in the planning stage, while the most downstream dam has been cancelled (Biba, 2016a). There have been rumours for several years about a second cascade further upstream. According to Fan et al. (2015), five dams of this upper cascade are now under construction. All these dams are (primarily) meant for hydropower generation.

The problem is that large-scale dam building has often been a highly contentious undertaking worldwide (Biswas, 2012). On the Mekong, as in other cases, proponents of dam building have argued that the dams may reduce carbon and sulphur dioxide emissions, help with flood control in the wet season, increase downstream water supply for irrigation and navigation during the dry season, and enhance regional economic development by providing a new source of energy (Biba, 2012; Li, He, & Feng, 2011).

At the same time, however, there is a long list of potentially negative implications for countries downstream of China's Mekong dams. It starts with several ecological challenges. First, at least two of China's dams (the Xiaowan and Nuozhadu) have large reservoirs that generally take years to fill. During this period, water may be held back, leading to substantial falls in water levels during the dry season. Second, the capacity of China's dams to effectively control floods and droughts has been questioned. It is deemed possible that China could withhold water in the dry season to maintain its hydroelectricity output and release water to protect the dams during floods. Third, flow regulation entails fewer seasonal floods downstream during normal years. Fewer seasonal floods, however, are likely to result in a decline in soil fertility over wide areas of rice cultivation in the Lower Mekong Basin. Fewer seasonal floods would also mean less natural capacity to constrain saltwater intrusion from the sea into the Mekong Delta. Fourth, aquatic life long adapted to the Mekong ecosystems could be seriously jeopardized by the changing flow regime. Fish migration, for example, could be blocked by dam building. If the Mekong's biodiversity dropped, this would likely be accompanied by falling productivity in the wild fisheries. The economic component of soil fertility and wild fisheries is obvious, as these factors directly affect both the food supply and the economic viability of the Mekong's downstream countries. Finally, a political aspect should not be neglected. Technically speaking, China's dams on the upper Mekong put Beijing in a position to control the quantity of water flowing downstream. The dams thus represent a potentially powerful tool to exert influence on the downstream countries and pose a possible diplomatic threat (Biba, 2012; Goh, 2004).

While it is both necessary and correct to emphasize that more in-depth and long-term research is required to better understand the complexity of the Mekong ecosystems and fully judge the exact impacts of the Chinese dams (Lee, 2015), the possible negative corollaries of China's dams mentioned above have already played a role in Mekong crises (see below). Also, regardless of the actual positive and negative effects of the Chinese dams, it is important to note that China has built its dams unilaterally. That is, China has not been willing to discuss or consult on its dam building with its downstream neighbours. In fact, China has not even informed its neighbours about its plans and activities *ex ante*, but has presented them with *faits accomplis*.

China's relationship with the MRC

The one organization that has (at least until recently) been crucial to Mekong River governance and therefore could have played a role in negotiating China's dam building is the MRC. The MRC was established in 1995; its members include Laos, Thailand, Cambodia and Vietnam. According to its founding document, the Mekong Agreement, the MRC's objective is

> to cooperate in all fields of sustainable development, utilization, management and con-servation of the water and related resources of the Mekong River Basin ... in a manner to optimize the multiple-use and mutual benefits of all riparians and to minimize the harmful effects that might result from natural occurrences and man-made activities. (Mekong River Commission [MRC], 1995)

However, even though the MRC has on various occasions been inclined to admit China as a full member, Beijing has declined. Instead, China (like Myanmar) has been a dialogue partner of the MRC since 1996. Moreover, in 2002 China signed the Agreement on the Provision of Hydrological Information on the Lancang/Mekong River, under which China has provided the MRC with water-level data from its territory during each year's flood season (MRC, 2017).

The MRC is far from being a perfect river basin organization and especially over the last few years has struggled to maintain its relevance as Laos has also moved ahead unilaterally with several controversial mainstream dam projects. But why exactly has China steadfastly declined to become a full member? The reasons are manifold. First, China shares many international rivers with many neighbouring countries. It has therefore been cautious not to let the Mekong become an unwanted precedent. If China joined the MRC and made concessions to its members, other riparian neighbours in other shared river basins might well make similar demands that would then be harder to reject. What is more, China's powerful upstream position makes it difficult for it to identify incentives to allow a multilateral cooperation agreement to constrain its freedom of action. Second, the MRC is often considered in China to have a rather narrow agenda, with a focus exclusively on water resources management. China has favoured broader regional cooperation frameworks that put a premium on promoting socio-economic development. Third, if China were to become a full MRC member, it would have to enter into an arrangement with predefined rules. Article 39 of the 1995 Mekong Agreement is very clear about additional parties having to abide by the rights and obligations under the agreement. For China, this is unacceptable because in its eyes the agreement is imbalanced, particularly serving the interests of downstream countries while not recognizing the services upstream countries provide for the river – for instance, in terms of sediment accumulation. In other words, China has been anxious that it could pay unduly for MRC membership without being properly compensated. Fourth, the Mekong is located in a remote corner of China. Moreover, river politics in general have only recently gained importance for China's foreign relations. Hence, the issue of MRC membership did not formerly attract much attention from the central policy and decision makers in Beijing. Finally, MRC interest in letting China join might in fact have been lukewarm, too. More precisely, it has been suggested that some Western donor agencies that are (or at least used to be) critical for MRC funding, as well as some MRC member countries themselves, have been apprehensive that as a full member China could try to seize the reins and transform the MRC according to its own preferences (personal communications, August 2012 and July–August 2016; see also Backer, 2007).

China's engagement in the GMS

While China has not been a member of the MRC, it has been actively engaged in the GMS. The GMS is a 1992 initiative of the Asian Development Bank (ADB). Unlike the MRC, its members include all six Mekong riparian countries, including China. The GMS has the 'ultimate objective of promoting the development of GMS markets and the movement of goods and people across the common borders' (Dosch, 2011, p. 19). To realize this objective, the initiative has above all sought to enhance infrastructure linkages; enable cross-border trade, investment and tourism; and increase private-sector participation and competitiveness (Asian Development Bank, 2011).

According to Lee (2015), China has been particularly interested in pushing forward cooperation in two areas: transport and energy. In terms of transport, China has especially focused on developing the North-South Corridor, which includes its provinces of Yunnan and Guangxi and stretches southward in two different sectors to Bangkok and Hanoi, respectively. The country has been heavily involved in building roads to improve the transport infrastructure for regional trade (Lim, 2008). Furthermore, China, together with Myanmar, Laos and Thailand, in 2000 inked the Agreement on Commercial Navigation on the Mekong-Lancang River, which paved the way for blasting rocks and rapids in the Mekong (Schmeier, 2009). Once again, China's rationale for this contentious undertaking – which was soon stopped after protests from Thai fishermen (Perlez, 2005) as well as from the Cambodian and Vietnamese governments (Keskinen, Mehtonen, & Varis, 2008), but has more recently re-emerged on the

Sino-Thai agenda (RFA, 2017) – was to clear the way for its trading vessels to reach new markets in the downstream Mekong countries, thereby also promoting the development of its own hinterland.

In the energy sector, where the initiative launched its GMS Road Map for Expanded Energy Cooperation in 2009, China's dams have been an important component of GMS plans for regional power trading intended to meet growing electricity demand in its member countries (Kuenzer et al., 2013). From the outset of China's hydropower development along the Mekong, Thailand has shown great interest in electricity imports from China's upstream dams. A memorandum of understanding indicates Thailand's purchase annually of up to 3000 MW generated by Chinese dams by 2017 (Menniken, 2007). Vietnam has also imported electricity from China (Eyler, 2014). But Middleton and Allouche (2016) have pointed to the possibility that China's anticipated role within the regional power trade might be shifting from an intended exporter of electricity to downstream Mekong countries to a seeker of power imports from these countries due to increased domestic demand.

Apart from its focus on transport and energy cooperation, other notable Chinese actions have included the creation of a US\$ 20 million poverty reduction fund within the ADB after China's hosting of the GMS summit in 2005 and the gradual removal of tariffs for more than 200 items from Cambodia, Laos and Myanmar since 2006 (Lim, 2008). China has also used the GMS as an important venue for promoting Beijing's idea that all Mekong riparian countries should 'keep to the path of pursuing common development and prosperity', as then-premier Wen Jiabao put it during the 2008 GMS summit (Wen, 2008). In outlining how to follow this path, Wen listed priority items that tellingly ranked infrastructure development, transport and trade facilitation, rural development, and health cooperation ahead of ecological and environmental protection, while water resources management was completely missing.

Assessing China's 'old' Mekong River politics from a benefit-sharing perspective

When judged from a benefit-sharing point of view, how can China's river politics regarding the Mekong before the establishment of the LMC – characterized chiefly by unilateral dam building, reluctance to join the MRC and active involvement in the GMS – be assessed? What were the strengths and weaknesses of China's 'old' approach?

Focusing on benefits beyond the river

In the pre-LMC era, China primarily sought to create Type 4 benefits – that is, benefits beyond the river. For one thing, China helped promote the development of a regional market in the Mekong Basin through its active engagement in the GMS (also see Lee, 2015). As a matter of fact, China 'emerg[ed] as the top trader, investor and donor' along the Mekong (Chheang, 2010, p. 362). More precisely, Biba (2012) has stressed that China's bilateral trade with the downstream Mekong countries (excluding Myanmar) grew more than 30-fold, from US\$ 1.6 billion to US\$ 53 billion, in 1990–2007 and that China's foreign direct investment into the same countries – particularly their energy, transport, agribusiness and tourism sectors – increased more than six-fold, from nearly US\$ 93 million to over

US$ 580 million between 2003 and 2009. In terms of aid, Reilly (2012) has stated that China provided Cambodia US$ 204 million in grants and US$ 500 million in concessional loans between 2003 and 2009, while Laos had received US$ 46.5 million in Chinese development aid by 2009. Myanmar was given US$ 24.2 million in grants, US$ 482.7 million in subsidized loans and US$ 1.2 million in debt relief between 1997 and 2006.

As a result of its much enhanced economic footprint, China has certainly also contributed to the socio-economic progress that has been discernible in many places along the Mekong for years now. The huge increases in average per capita incomes throughout the region are one example of this progress. For instance, the per capita GDP of Myanmar, the poorest GMS country, rose from US$ 63 in 1992 to US$ 832 in 2011. The per capita GDP of Thailand, the richest GMS country, grew from US$ 1864 to US$ 5394 over the same period (Asian Development Bank, 2012). Similarly, in all GMS countries, poverty levels were reduced by a large margin and infant mortality rates dropped steeply between 1992 and 2011, whereas life expectancy, school enrolment and literacy rates have increased markedly (pp. 1–2).

Nevertheless, China's involvement in the GMS has not exclusively been meant to generate benefits beyond the river. In fact, China has also tried to use the GMS to facilitate activities favourable for Type 2 and Type 3 benefits. Regarding Type 2 benefits (benefits from the river), China has built its Mekong dams unilaterally, but dam building, as previously mentioned, has generally fit within GMS energy-security strategies and has moreover put downstream Mekong countries in a position to potentially benefit from the hydropower generated by the Chinese dams. Not surprisingly, then, Mehtonen (2008) has pointed out that central governments in the downstream Mekong countries have not usually been opposed to China's dam-building activities. Rather, 'all of the Mekong countries are involved in the plans for regional power trade' (p. 161).

In terms of Type 3 benefits, it should not be overlooked that Southeast Asia, including the Mekong subregion, is vital for China to achieve some of its major foreign policy goals. According to Medeiros (2009), China's approach to this region has been driven not only by the economic desire to expand China's access to markets, investment and resources, but also by the objective to reassure regional countries that China is not a threat whose influence should be constrained. A good way to convince the smaller countries of the Mekong about China's benign intentions is 'multilateral buffering', where a broader and more institutionalized framing of the relationship can help neutralize contentious issues (Womack, 2010, p. 31). Speeches like Wen Jiabao's at the 2008 GMS summit, which, as cited above, promised to focus on *common* development and prosperity, are indicative of China's aim to also promote a political relationship with the downstream Mekong countries via the GMS, with the goal of reducing costs because of the river.

In view of this constellation of benefits, and particularly because of the achievements of the GMS, an initiative in which China has played a leading role, Lee (2015, p. 149) has assessed China's pre-LMC benefit-sharing policy in a rather positive light, arriving at the conclusion that

> the usefulness of the benefit-sharing approach [between China and the downstream Mekong countries] lies in unlocking a list of common agendas in the form of nonwater issues, which have had the effect of providing fresh approaches to solving conflicting issues related to volumes of water and hydropower development within the water sector.

While not using the benefit-sharing framework explicitly, and while also not seeing China's strategy quite as positively as Lee, Biba (2012) has argued in a similar vein that China 'has been able to link its dam-building to broader issues of common develop-ment, including increased levels of trade, investment and aid as well as an institutio-nalized diplomacy' (p. 622) and had thus successfully 'nip[ped] in the bud any criticism of its hydropower projects' (p. 617).

Neglecting benefits to the river

Such assessments have been useful, but they likewise need to be amended because they have not been nuanced enough. For a more complete picture, the two following aspects and their consequences also need to be taken into consideration. First, from an empirical angle, it is striking that China's pre-LMC benefit-sharing approach regarding the Mekong was neither successful at, nor in fact particularly interested in, creating Type 1 benefits – that is, benefits *to* the river. Most illustrative of this claim is China's ongoing unwillingness to become a full member of the MRC, which – regardless of its actual accomplishments – has been the primary intergovernmental institution in the Mekong Basin to emphasize ecological matters such as 'sustainable development, utilization, management and conservation of the water and related resources' (MRC, 1995; also see above). At the same time, these environmental issues have largely been ignored by the GMS. Even Lee (2015, p. 148) has recognized that the costs from the development of the GMS, such as 'threats to livelihoods and degradation of the environment', have not been adequately addressed by the ADB initiative.

Second, China's empirical neglect of Type 1 benefits, with its concomitant focus on Type 4 benefits, has consequences on the conceptual level. That is to say, China sought to turn the benefit-sharing framework upside down. As already outlined above, Sadoff and Grey (2002) have suggested that Type 4 benefits (benefits beyond the river) usually represent the final stage of the framework, not its starting point. This is largely because it is deemed relatively more difficult to create positive feedback loops from Type 4 benefits to the other three types of benefits (see again Figure 1). In contrast, Type 1 benefits are treated as the 'cornerstone of river basin management' (p. 393), 'under-pinning all other benefits that can be derived' (p. 392).

Most significant, now, is that China's low appreciation of benefits to the river has had a negative effect on its Mekong River politics more generally. While it is true that (a) China has helped create certain benefits to be shared by all Mekong riparian countries – as argued, for example, by Lee (2015) – and that this has led to a situation where (b) downstream Mekong *governments* do not usually criticize China, at least not publicly, on its dam building, as pointed out by Biba (2012), the Chinese have failed to take into account the reactions of those local communities in the downstream Mekong countries whose livelihoods are increasingly threatened by the negative corollaries of the Mekong dams, as well as of those activist groups that advocate for local communities.

A primary example of the partial failure of China's pre-LMC Mekong River politics was evident during the spring 2010 Mekong crisis (Biba, 2016a, 2018; Yeophantong, 2014). From January to April 2010, water levels in the lower Mekong sank dramatically. The impacts comprised smaller fish catches; less water for the irrigation of agriculture, for livestock and for drinking; and a stop to river transportation, which paralyzed

Mekong trade and tourism (Middleton, 2011). Many in the downstream Mekong countries who saw their livelihoods endangered started to blame the Chinese dams on the Mekong's upper reaches for storing water and thus causing, or at least contributing to, the extremely low water levels. More concretely, local communities directly affected by the low water levels, regional and international activist and NGO groups committed to sustainable development, and the media inside and outside the region propelled a week-long series of statements, reports and articles that were harshly critical of China's dam building – even though these dams' precise and actual role in the crisis has remained ambiguous until today (Biba, 2016a, 2018).

In response to the crisis, China had to briefly switch to 'damage control mode' (Storey, 2010, p. 8). That is, it took a few unprecedented measures particularly geared to rejecting the negative role of the Chinese dams, increasing transparency and thus assuaging concerns. Above all, China shared dry-season hydrological data from the time of the crisis with the MRC; invited downstream countries' representatives to visit one of its Mekong dams; and stressed that ecological concerns were actually on its list, as exemplified by the cancellation of one its dams, supposedly to avoid hindering fish migration (Biba, 2016a, 2018). China's image abroad had already suffered, though. What is more, similar accusations against China's dams had been heard before 2010 and were made again after the crisis.

China's 'new' Mekong River politics: the LMC

The LMC is a relatively recent and certainly the newest cooperation mechanism for the Mekong, bringing together all six Mekong riparian countries: China, Myanmar, Laos, Thailand, Cambodia and Vietnam. The LMC was officially launched during an inaugural leaders' meeting held in the city of Sanya in China's Hainan Province in March 2016, after the foreign ministers of the six countries had already signed a related agreement in November 2015 in Yunnan Province and China's premier Li Keqiang had pitched the idea of the new mechanism back at the 2014 summit of the Association of Southeast Asian Nations (ASEAN) in Myanmar. The LMC follows what it calls a '3 + 5 cooperation framework', referring to three pillars and five key priority areas on which it seeks to focus its cooperation. The three pillars are political and security issues, economic and sustainable development, and social, cultural and people-to-people exchanges, and are thus modelled after the distinctive ASEAN approach. The five key priority areas, which are meant to enable more practical cooperation, are connectivity, production capacity, cross-border economic cooperation, water resources, and agriculture and poverty reduction (Biba, 2016b).

China's rationale for establishing the LMC

Despite the equality of all member states as well as the consensual decision making, both of which are emphasized in the few official LMC documents (e.g., Lancang-Mekong Cooperation, 2016a), China is clearly in the driver's seat of this new mechanism. The founding history of the LMC outlined above is proof enough, in addition to the superior material capabilities China brings to the table. Against this backdrop, the question of why China has shown such a strong interest in creating the LMC is an intriguing one.

First, the fact that China's Mekong dam-building activities were being criticized by various foreign actors in an increasingly strong manner and that the accusations China faced were receiving broad international media coverage – resulting in challenges to a positive image of China abroad – made China realize that it had to do something about its Mekong River politics (personal communication, 12 July 2016). In particular, the spring 2010 Mekong crisis and the events surrounding it represented a 'watershed moment' at which China began thinking about shifting away from its previously reactive approach (personal communication, 4 August 2016).

Second, and closely related, China has sought to gain more control and exercise greater leadership over the Mekong subregion. Existing mechanisms such as the MRC and GMS have for various reasons been regarded as ultimately 'ineffective' for encouraging regional cooperation (personal communication, 4 July 2016) and impossible for China to dominate (personal communication, 11 July 2016; also see below). The LMC, therefore, is China's attempt to increase its own institutional influence over the Mekong River (personal communication, 1 July 2016), in the hope of setting the rules for regional cooperation going forward (personal communication, 11 August 2016; also see Biba, 2016b).

Third, China's growing need to enhance its institutional 'capabilities' regarding the Mekong has been triggered not only by previous crises, which saw the buck passed to China, but also by a general trend of steadily increasing great-power presence in the Mekong Basin in recent years. Back in 2009, the United States set up the Lower Mekong Initiative and Japan established the Japan-Mekong Summit – two comprehensive subregional cooperation mechanisms with all the Mekong riparian countries except China as members. China has been wary of the intentions and influences of these initiatives led by 'external' actors with which China has difficult relations, and has used their forays into the region as a justification to create its own mechanism (personal communications, 12 July and 4 August 2016).

Fourth and finally, the establishment of the LMC has fit very well into China's much bigger and still unfolding major foreign policy vehicles – that is, the Belt and Road Initiative and the Asian Infrastructure Investment Bank. In other words, the LMC is meant to promote synergy with the Belt and Road Initiative and make use of Asian Infrastructure Investment Bank funding, thereby pushing forward China's neighbourhood diplomacy through joint development and enhancing regional integration centred on China (personal communications, 4 August and 11 August 2016; also see Middleton & Allouche, 2016).

Key features of the LMC

Why China has driven the foundation of the LMC is one matter of interest; how it has sought to design it is another – though they are related. It may be helpful to point out a few noteworthy features of this new mechanism.

The first is membership. As mentioned, the LMC membership consists of all six Mekong riparian countries. More important, however, is the fact that there are no external – that is, non-riparian – countries involved. This is very much in contrast to both the MRC, which has (at least until recently) been funded largely by a group of mostly Western donors, and the GMS, which is an initiative of the ADB, where Japan and the United States have the greatest voting power. Consequently, it was a central objective of China's to create the LMC as a mechanism free from external influence and for Mekong riparian countries only. China's foreign minister, Wang Yi, has tellingly referred to the LMC as the 'baby of the six countries' (cited in Biba, 2016b).

The second feature is the organization's institutional design. For one thing, as the Sanya Declaration from the inaugural LMC leaders' summit stated, the mechanism will be based on a 'project-oriented model' (LMC, 2016a). Significantly, this suggests that the LMC has no intention of being an international river basin organization such as the MRC, which has codified rules and procedures of cooperation derived from the UN Watercourse Convention (Middleton & Allouche, 2016). For another thing, while the foreign ministers have called for a 'strengthening' of the LMC's institutional design, they only intend to set up a 'coordination unit/national secretariat' in each member state (LMC, 2016c). This implies that there will be no autonomous LMC secretariat, and the LMC will therefore not be given any 'actorness' of its own. While all LMC members are officially said to be equal, the loose institutional framework certainly creates options for China, as the most powerful member state, to be more equal than the others.

The third and last feature of significance at this point is the LMC agenda. The LMC is the first multilateral initiative driven by China and officially includes sustainable water resources management on its agenda. China has refused to join the MRC partly because of its (sole) focus on water resources management (see above), and it has also reportedly prevented water issues from playing a role within the GMS (Hensengerth, 2009). Meanwhile, water resources management is only one of five LMC priority areas. China has clearly sought to establish a comprehensive cooperation mechanism that combines the agendas of the MRC and the GMS and also takes up new issues of concern to China, particularly cooperation on non-traditional security problems (e.g., smuggling, drug trafficking and illegal migration; also see below).

Assessing China's 'new' Mekong River politics from a benefit-sharing perspective

Due to the recent founding of the LMC and a shortage of official information, any assessment of the LMC must be seen as preliminary for the time being. Nevertheless, it is already possible to discern certain tendencies and preferences that China appears to be tying to the LMC. The most interesting question is what has changed, from a benefit-sharing perspective, in China's 'new' Mekong approach relative to the 'old' one and what has stayed the same.

Again: a focus on benefits other than those to the river

First of all, it seems rather obvious so far that the LMC is meant to duplicate, and in fact reinforce, part of China's 'old' Mekong approach, in that a strong focus on Type 4 benefits cannot be overlooked. Three of the five LMC priority areas – connectivity, production capacity and cross-border economic cooperation – can be directly related to creating economic benefits, mostly beyond the river. As highway infrastructure has already made a lot of headway through the GMS, China is now seeking to use the LMC to move ahead, particularly with upgrading the railway network and expanding the power grid and the cable connection between China and mainland Southeast Asia (Lu, 2016). The fact that plans for blasting rocks and rapids in the Mekong have recently re-emerged (see above) also fits into the picture of developing the regional infrastructure.

Moreover, China is interested in building cross-border economic cooperation zones and cross-border industry zones to expedite the region's industrialization (p. 10). It is not surprising, then, that apart from the Sanya Declaration, the inaugural LMC leaders' summit produced only one other document – dealing with production-capacity cooperation and highlighting the need to improve regional trade, investment cooperation, and the industrial-development abilities and manufacturing capacities of all members (LMC, 2016b).

China also appears to be anxious to enhance Type 2 and Type 3 benefits. Regarding benefits from the river (Type 2), it is crucial to understand that dam building in the Mekong has entered a new phase. Debates are no longer so much about whether dams should be built, but rather about how to operate dams and how to coordinate their operation schemes. As dam building on the Chinese stretches of the Mekong is nearly complete (at least with regard to the lower cascade), the focus of attention is increasingly being directed downstream, where Laos in particular has in recent years started building its own mainstream dams. China has played a noteworthy role in this regard. First, China's dam building upstream has set an important precedent for downstream countries. Second, Chinese companies have been among the major investors in downstream dam projects. According to some estimates, Chinese state-owned enterprises will implement up to 40% of the proposed mainstream and tributary hydropower schemes in the downstream Mekong countries in the coming years. These schemes include four mainstream dams: three in Laos (Pak Beng, Pak Lay, Sanakham) and one in Cambodia (Sambor) (Hirsch, 2011; International Rivers, 2014). Middleton and Allouche (2016, p. 112) have therefore argued that 'with the inclusion of China, the LMC offers the possibility for cascade coordination serving a transnational network of state and non-state actors whose interests … converge around the optimisation of hydropower dam operation'.

In terms of reducing costs because of the river (Type 3), the multilateral buffering strategy China has previously followed through the GMS is applicable to the LMC as well. Also, the LMC's comprehensive agenda, which includes politico-security matters, aims to create more Type 3 – that is, political – benefits. After an incident on 5 October 2011 in which two Chinese cargo ships were attacked on the Mekong and all 13 Chinese crew members were killed by suspected Myanmar-based drug traffickers, non-traditional security issues along the Mekong gained in importance for China. To guarantee security for river trade, a joint law enforcement and security cooperation mechanism between China, Myanmar, Laos and Thailand was established in December 2011. China has also sent advisors to Myanmar and Laos, whose police forces it helps train and equip, and has set up a joint police command office close to the Mekong in its Yunnan Province. Between December 2011 and July 2016, 48 joint river patrols were conducted by the four countries (China Daily USA, 2016). The Sanya Declaration announced plans to further deepen cooperation in this and related areas to prevent them from straining bilateral relations in the future (LMC, 2016a).

Benefits to the river: insufficient progress so far

While China is therefore using the LMC to further strengthen the creation of those types of benefits it already put a premium on in its pre-LMC Mekong River politics, the striking question is whether China is also ready for an additional sea change in terms of

Type 1 benefits – that is, benefits to the river. After all, it was the neglect of Type 1 benefits that generated considerable problems for China's pre-LMC strategy.

To begin with, and as previously highlighted, the LMC has indeed put sustainable water resources management on its agenda. More specifically, the March 2016 Sanya Declaration stated that it is the LMC's goal to

> enhance cooperation among LMC countries in sustainable water resources management and utilization through activities such as the establishment of a center in China for Lancang-Mekong water resources cooperation to serve as a platform for LMC countries to strengthen comprehensive cooperation in technical exchanges, capacity building, drought and flood management, data and information sharing, conducting joint research and analysis related to Lancang-Mekong river resources. (LMC, 2016a)

However, it was then almost two years, until mid-December 2017, before the setup of the centre was confirmed (Ministry of Foreign Affairs, 2017). What is more, it still remains largely unclear what *concrete* measures conducive to creating benefits for the river the enumerated cooperation areas may actually include and how far China is really willing to increase the scope of cooperation beyond what already existed before the establishment of the LMC, speaking of the China–MRC agreement on flood-season data provision, for example. Zhang Jiuhuan, vice chairman of the China Public Diplomacy Association, has vaguely remarked that downstream countries 'concerned about the implications of the dam construction upstream … could raise the issue for consultation with countries upstream'. But China's foreign minister, Wang Yi, has pointedly stated that the LMC should 'discuss easy issues first' (cited in Biba, 2016b).

Joint water resources management is not generally treated as an 'easy' cooperation field in China, though. In fact, water issues continue to be frequently tied to national security considerations, and national law continues to restrict related data sharing (personal communication, 5 July 2016). Consequently, while Chinese experts increasingly understand the need for joint water resources management in China's shared rivers and therefore echo LMC official documents' statements that sustainable water resources management should be one of its priority areas, they likewise tend to believe that the LMC will be a mechanism largely for economic development through connectivity and production capacity enhancement and that water resources management will effectively be a secondary factor at best (personal communications, 1 July, 4 July, 3 August, 4 August, and 11 August 2016).

What is more, the reality on the ground is that there is mostly only talk, if anything, about joint sustainable water resources management, with much more concrete action in economic development (see above). The gap between water cooperation rhetoric and water cooperation practice surrounding China's 'new' Mekong River politics is due to the fact that deals on agriculture, energy and transportation, for example, are not only usually made first, but also tend to be decoupled from environmental concerns that could create Type 1 benefits. In the end, sustainable water resources management therefore becomes an isolated and hollowed-out issue area of low importance (personal communication, 11 July 2016).

Part of the reason for this situation is that from China's perspective it is more difficult to see the possibility of benefit sharing in the realm of sustainable water resources management than in other LMC priority areas. For instance, as the most upstream Mekong riparian country, China does not need hydrological data from downstream; nor is China particularly dependent on its downstream neighbours in terms of Mekong drought and flood management (personal communication, 1 and 4 July 2016). In other words, whereas all LMC members have the same dream when it comes to achieving economic modernization (Type 2 and Type 4 benefits), China often sleeps in a different bed than the others as far as sustainable water resources management is concerned (Type 1 benefits).

From China's vantage point, the solution could be some form of 'benefit compensation'. This would mean that if China creates benefits to the river for the downstream countries, the latter should compensate China through other types of benefits. In China's view, those benefits could include increased trade and investment opportunities for Chinese companies, favourable food exports to China, and in the case of Vietnam even a pro-China agreement on the South China Sea, where Beijing and Hanoi have overlapping maritime territorial claims (personal communication, 5 July 2016). However, while such benefit-compensation models may seem quite advantageous, and even natural, from China's point of view, they are not only hard to practically measure against each other, but might also be viewed by downstream countries as having a threatening undercurrent.

In the meantime, and despite the negative experiences of previous crises, China remains relatively reluctant to substantially and permanently increase the transparency and bindingness of joint sustainable water resources management for the Mekong. This could also be witnessed in March 2016, a few months after the first LMC foreign ministers' meeting and shortly before the inaugural LMC leaders' summit. China offered to release water from its Mekong dams to alleviate the severe drought in the lower basin, following a request from Vietnam. In Thailand, however, China's water releases were so unpredictable that they had a damaging impact on some riverside communities. At the same time, it was regarded as unlikely that the releases would actually reach Vietnam and improve the situation all the way downstream. Therefore, while China's act of hydro-diplomacy was hailed by its domestic state press as 'benefit[ting] 60 million people' in the Lower Mekong Basin (Xinhua, 2016), other observers had a less triumphant take. Pongsudhirak (2016), for example, emphasized that China's water release was primarily a 'harbinger for the downstream countries' dependence on China's goodwill and generosity'. Similarly, for Middleton and Allouche (2016, p. 112), it was indicative of a 'new arrangement for managing the Lancang-Mekong, but rather than institutionalised within a rules-bound organisation such as the MRC, it instead is apparently dependent upon the good will of China'.

Very recently, it was reported that together with the water cooperation centre, the LMC also had inaugurated the Lancang-Mekong Environmental Protection Centre in Kunming in March 2018 and that part of the centre's future work could be water pollution control (Xinhua, 2018). This might indeed become an area of cooperation beneficial to downstream Mekong countries as far as sustainable water management is concerned. However, it is far too early to tell whether and how this will actually materialize.

Conclusion

This article has analyzed China's Mekong River politics before and after the establishment of the LMC from a comparative benefit-sharing perspective. China's pre-LMC approach focused too much on the creation of economic benefits while neglecting ecological benefits by not putting joint sustainable water resources management on the agenda. However, despite the problems this situation caused for China and its downstream neighbours, China's current LMC strategy appears to essentially replicate the country's former approach. On the one hand, the creation of economic benefits from and beyond the river (Type 2 and Type 4) is being further strengthened through the LMC, and the political benefits (Type 3) are also being expanded. On the other hand, however, progress on ecological benefits to the river (Type 1) still remains slow and insufficient to date.

Even though sustainable water resources management has now been identified as a priority area for cooperation, the actual emphasis of the LMC on this area to date remains secondary at best. In following this approach, China continues to disregard the order of the different types of benefits to be derived from cooperation on international rivers, with Type 1 benefits to the river logically underpinning all other types of benefits (Sadoff & Grey, 2002). More significantly, it is unclear how China intends to win over threatened riverside communities and produce trust among Mekong countries more generally if sustainable water resources management continues to largely fall through the cracks or, where related action does occur, continues to be non-transparent and non-binding and therefore unpredictable and unstable. In sum, therefore, China's 'new' Mekong River politics still lacks a comprehensive understanding that water is the gossamer that links all socio-economic activities with the environment and that socio-economic development cannot occur, especially in the long run, without sustainable water resources management.

That said, the LMC certainly has the potential to actually implement sustainable water resources management for the Mekong. To do so, however, China has to first identify adequate solutions to the following issues. First, what will the LMC's future relationship with the MRC look like – supplement or replacement? Chinese experts are divided over this question (personal communications, July–August 2016), but it clearly carries a lot of weight – because the LMC and the MRC appear to represent very different sets of norms and rules. Second, to what extent will the LMC be ready to involve non-state actors? China's foreign policy has not generally been famous for its inclination to involve non-state actors. Yet, the exclusion of these actors, particularly all kinds of river activist networks, has spelled trouble for both the MRC and China's Mekong hydro-politics in the past (Yeophantong, 2014). Third, and above all, what kind of hegemon does China intend to be? While Lee (2015) has stressed that (successful) benefit sharing creates a state of hydro-interdependency, the LMC – despite all the rhetoric to the contrary – looks much more like a hierarchical construct with China on top. As Zeitoun and Warner (2006, p. 452) have posited, 'The hydro-hegemon determines the nature of the interaction, and to what extent the benefits derived from the flows will also extend to the weaker co-riparians.' Critically, however, such interaction may range from 'a positive/leadership hydro-hegemonic configuration' to 'a dominative form of hydro-hegemony and a lingering … conflict' (p. 452). China's decisions on these issues will indicate whether China's (Mekong) river politics is only becoming more active or whether it is also becoming better at preventing crises.

Acknowledgments

The author is grateful to all the Chinese experts who kindly shared their views on the topics dealt with in the article as well as to the anonymous reviewers for valuable comments.

Disclosure statement

No potential conflict of interest was reported by the authors.

References

Alam, U., Dione, O., & Jeffrey, P. (2009). The benefit-sharing principle: Implementing sovereignty bargains on water. *Political Geography*, *28*, 90–100.

Asian Development Bank. (2011). *The greater Mekong subregion economic cooperation program strategic framework 2012–2022*. Manila: ADB.

Asian Development Bank. (2012). *The greater Mekong subregion at 20: Progress and prospects*. Manila: ADB.

Backer, E. (2007). The Mekong River Commission: Does it work, and how does the Mekong basin's geography influence its effectiveness. *SüDostasien Aktuell*, *4*, 31–55.

Biba, S. (2012). China's continuous dam-building on the Mekong. *Journal of Contemporary Asia*, *42*(4), 603–628.

Biba, S. (2014). Desecuritization in China's behavior towards its transboundary rivers: The Mekong river, the Brahmaputra river, and the Irtysh and Ili rivers. *Journal of Contemporary China*, *23*(85), 21–43.

Biba, S. (2016a). From securitization moves to positive outcomes: The case of the spring 2010 Mekong crisis. *Security Dialogue*, *47*(5), 420–439.

Biba, S. (2016b, February 1). China drives water cooperation with Mekong countries. *China Dialogue*. Retrieved from https://www.chinadialogue.net/article/show/single/en/8577-China-drives-water-cooperation-with-Mekong-countries

Biba, S. (2018). *China's hydro-politics in the Mekong: Conflict and cooperation in light of securitization theory*. London and New York: Routledge.

Biswas, A. (2012). Impacts of large dams: Issues, opportunities and constraints. In C. Tortajada, D. Altinbilek, & A. Biswas (Eds.), *Impacts of large dams: A global assessment* (pp. 1–18). Berlin and Heidelberg: Springer.

Chellaney, B. (2011). *Water: Asia's new battleground*. Washington, DC: Georgetown University Press.

Chheang, V. (2010). Environmental and economic cooperation in the Mekong region. *Asia Europe Journal*, *8*(3), 359–368.

China Daily USA. (2016, July 24). China, Laos, Myanmar, Thailand complete joint patrol on Mekong river. Retrieved from http://usa.chinadaily.com.cn/world/2016-07/24/content_26201702.htm

Dosch, J. (2011). Reconciling trade and environmental protection in ASEAN-China relations: More than political window dressing? *Journal of Current Southeast Asian Affairs*, *30*(2), 7–29.

Elhance, A. (2000). Hydropolitics: Grounds for despair, reasons for hope. *International Negotiation*, *5*(2), 201–222.

Eyler, B. (2014). The coming downturn of China-Vietnam trade relations. *East by Southeast*. Retrieved from http://www.eastbysoutheast.com/fear-change-future-china-vietnam-trade-relations/

Fan, H., He, D., & Wang, H. (2015). Environmental consequences of damming the mainstream Lancang-Mekong river: A review. *Earth Science Reviews*, *146*, 77–91.

Feng, Y., He, D., & Wang, W. (2015). Identifying China's transboundary water risks and vulnerabilities – A multidisciplinary analysis using hydrological data and legal/institutional settings. *Water International*, *40*(2), 328–341.

Goh, E. (2004). *China in the Mekong river basin: The regional security implications of resource development on the Lancang Jiang* (Working Paper No. 69). Singapore: Institute of Defence and Strategic Studies.

He, Y. (2015). China's practice on the non-navigational uses of transboundary waters: Transforming diplomacy through rules of international law. *Water International, 40*(2), 312–327.

Hensengerth, O. (2009). Transboundary river cooperation and the regional public good: The case of the Mekong river. *Contemporary Southeast Asia, 31*(2), 326–349.

Hirsch, P. (2011). China and the cascading geopolitics of lower Mekong dams. *The Asia-Pacific Journal, 9*(20). Retrieved from http://www.japanfocus.org/-Philip-Hirsch/3529

International Rivers. (2014). China's overseas dam list. Retrieved from https://www.internatio nalrivers.org/resources/china-overseas-dams-list-3611

Keskinen, M., Mehtonen, K., & Varis, O. (2008). Transboundary cooperation vs. internal ambitions: The role of China and Cambodia in the Mekong region. In N. Pachova, M. Nakayama, & L. Jansky (Eds.), *International water security: Domestic threats and opportunities* (pp. 79–109). Tokyo: UNU Press.

Kuenzer, C., Campbell, I., Roch, M., Leinenkugel, P., Tuan, V. Q., & Dech, S. (2013). Understanding the impact of hydropower developments in the context of upstream-down-stream relations in the Mekong river basin. *Sustainability Science, 8*(4), 565–584.

Lancang-Mekong Cooperation. (2016a, March 23). Sanya declaration of the first Lancang-Mekong Cooperation (LMC) leaders' meeting – For a community of shared future of peace and prosperity among Lancang-Mekong countries. Retrieved from http://www.fmprc.gov.cn/mfa_eng/zxxx_662805/t1350039.shtml

Lancang-Mekong Cooperation. (2016b, March 23). Joint statement on production capacity cooperation among Lancang-Mekong countries. Retrieved from http://news.xinhuanet.com/english/2016-03/23/c_135216863.htm

Lancang-Mekong Cooperation. (2016c, December 3). Joint press communiqué of the second Lancang-Mekong Cooperation (LMC) foreign ministers' meeting. Retrieved from http://www.fmprc.gov.cn/mfa_eng/zxxx_662805/t1426601.shtml

Lee, S. (2015). Benefit sharing in the Mekong river basin. *Water International, 40*(1), 139–152.

Li, Z., He, D., & Feng, Y. (2011). Regional hydropolitics of the transboundary impacts of the Lancang cascade dams. *Water International, 36*(3), 328–339.

Lim, T. (2008). *China's active role in the Greater Mekong Sub-Region: A 'win-win' outcome?* (Background Brief No. 397). Singapore: East Asia Institute.

Lu, G. (2016). *China seeks to improve Mekong sub-regional cooperation: Causes and policies* (Policy Report February 2016). Singapore: S. Rajaratnam School of International Studies.

Medeiros, E. (2009). *China's international behavior*. Santa Monica: RAND Corporation.

Mehtonen, K. (2008). Do the downstream countries oppose the upstream dams? In M. Kummu, M. Keskinen, & O. Varis (Eds.), *Modern myths of the Mekong: A critical review of water and development concepts, principles and policies* (pp. 161–172). Helsinki: Helsinki University of Technology.

Mekong River Commission. (1995). Agreement on the cooperation for the sustainable develop-ment the Mekong river basin. Retrieved from http://www.mrcmekong.org/assets/Publications/agreements/agreement-Apr95.pdf

Mekong River Commission. (2017). About the MRC: Upstream partners. Retrieved from http://www.mrcmekong.org/about-the-mrc/upstream-partners/

Menniken, T. (2007). China's performance in international resource politics: Lessons from the Mekong. *Contemporary Southeast Asia, 29*(1), 97–120.

Middleton, C. (2011, November 25–27). Conflict, cooperation and the trans-border commons: The controversy of mainstream dams on the Mekong river. *Conference Paper for the 3rd International Winter Symposium of the Global COE Program, Slavic Research Center.* Sapporo, Japan: Hokkaido University.

Middleton, C., & Allouche, J. (2016). Watershed or powershed? Critical hydropolitics, China, and the 'Lancang-Mekong Cooperation framework'. *International Spectator, 51*(3), 100–117.

Ministry of Foreign Affairs. (2017). Joint press communiqué of the third Lancang-Mekong Cooperation (LMC) foreign ministers' meeting. Retrieved from http://www.fmprc.gov.cn/mfa_eng/zxxx_662805/t1520022.shtml

Nickum, J. (2008). The upstream superpower: China's international rivers. In O. Varis, C. Tortajada, & A. Biswas (eds.), *Management of transboundary rivers and lakes* (pp. 227–244). Berlin: Springer-Verlag.

Perlez, J. (2005, March 19). In life on the Mekong, China's dams dominate. *New York Times*. Retrieved from http://www.nytimes.com/2005/03/19/international/asia/19mekong.html

Pongsudhirak, T. (2016, March 25). China's 'water grab' and its consequences. *Bangkok Post*. Retrieved from http://www.bangkokpost.com/print/910004/

Radio Free Asia. (2017, January 27). The Mekong part 4: Blasting the rapids in Thailand. Retrieved from http://www.rfa.org/english/thailand-mekong-01272017104723.html

Reilly, J. (2012). A norm-taker or a norm-maker? Chinese aid in southeast Asia. *Journal of Contemporary China, 21*(73), 71–91.

Sadoff, C., & Grey, D. (2002). Beyond the river: The benefits of cooperation on international rivers. *Water Policy, 4*(5), 389–403.

Schmeier, S. (2009). Regional cooperation efforts in the Mekong river basin: Mitigating river-related security threats and promoting regional development. *Austrian Journal of Southeast Asian Studies, 2*(2), 28–52.

Storey, I. (2010). China's 'charm offensive' loses momentum in southeast Asia [part II]. *China Brief, 10*(10), 7–10.

Su, Y. (2014). Contemporary legal analysis of China's transboundary water regimes: International law in practice. *Water International, 39*(5), 705–724.

Wen, J. (2008, March 31). Build a bond of cooperation and a common homeland. *Speech at the 3rd GMS summit*, Vientiane, Laos. Retrieved from http://www.china-embassy.org/eng/zt/768675/t476515.htm

Womack, B. (2010). *China among unequals: Asymmetric foreign relationships in Asia*. Singapore: World Scientific.

Xinhua. (2016, April 18). China water release benefits 60 million people along lower Mekong river: Cambodian official. Retrieved from http://news.xinhuanet.com/english/2016-04/18/c_135289171.htm

Xinhua. (2018, March 22). China deepens Lancang-Mekong environmental protection cooperation. Retrieved from http://en.people.cn/n3/2018/0322/c90000-9440739.html

Yeophantong, P. (2014). China's Lancang dam cascade and transnational activism in the Mekong region: Who's got the power? *Asian Survey, 54*(4), 700–724.

Zeitoun, M., & Warner, J. (2006). Hydro-hegemony – A framework for analysis of transboundary water conflicts. *Water Policy, 8*(5), 435–460.

Assessing the Indus Waters Treaty from a comparative perspective

Neda Zawahri and David Michel

ABSTRACT

The 1960 Indus Waters Treaty dividing the rivers of the Indus system between India and Pakistan has continued to function through two wars and numerous political tensions. Nevertheless, given mounting pressures on the Indus' waters due to population growth, climate change and mismanagement, many call for abandonment or renegotiation of the treaty. This article situates these criticisms within the quantitative literature analyzing river treaties to demonstrate that the same critiques are applicable to many treaties. Comparative analysis also reveals that while some of the treaty's weaknesses can be addressed, important structural obstacles render certain of its deficiencies difficult to correct.

Introduction

In the field of hydropolitics, which examines the prospects for cooperation and conflict over transboundary rivers, the 1960 Indus Waters Treaty (IWT) between India and Pakistan is considered an important example of cooperation between riparians with a history of animosity and conflict (Ali, 2008; Biswas, 1992; Salman & Uprety, 2001; Zawahri, 2009). For nearly 60 years, the IWT has provided its signatory states with mechanisms to peacefully share the Indus basin and manage issues that arose. This cooperation has survived two wars and numerous border clashes. More recently, however, despite this history of cooperation, the IWT has been facing mounting criticism, along with calls to revoke, revise, or abandon the accord.

Critics charge that the IWT both fails to address certain specific issues and furnishes an inadequate framework for the riparians to meet emerging challenges. In the first category, for instance, myriad analysts and policy makers have noted that the treaty makes no mention of the groundwater aquifers both countries increasingly exploit (Indus Basin Working Group, 2013; Shah & Panchali, 2017). Numerous experts likewise lament that the IWT largely neglects questions of water quality, pollution, environmental flows and ecological protections (Raman, 2017; Sarfraz, 2013). Beyond ignoring particular issues, the treaty also omits particular parties. Even as the IWT was being negotiated, some analysts foresaw that excluding Afghanistan and China would prove problematic should these riparians' claims on the river grow, a problem now realized (Chellaney, 2013; Hirsch, 1956). More broadly, many observers judge that the treaty, focused on physically dividing and regulating the Indus' waters, is ill-adapted to resolving the increasingly complex socio-economic, climate and environmental pressures confronting the riparians (Alam et al., 2011; Burgess, Owen, & Sinha, 2016; Ranjan, 2016). Indeed, as the supply of fresh water in the basin decreases and the demand increases, some observers warn of future violent conflict, even 'water wars', between India and Pakistan (Chandio, 2008; Dawn, 2016).

Although many of these assessments correctly identify the treaty's shortcomings, they are weakened by their frequent lack of comparative analysis. That is, most of these critiques are framed as an in-depth single case study, obscuring the degree to which these features and flaws of the IWT reflect and respond to not only the particularities of the Indus but also structural dynamics often shared with other transboundary basins throughout the world. This article seeks to situate the treaty's shortcomings and evaluate criticisms of the IWT in the literature examining the factors influencing treaty formation, design and effectiveness through quantitative analysis of several databases that catalogue the content and circum- stances of transboundary river treaties and record interactions between states over shared basins.[1] This contextualization will demonstrate that these criticisms are not unique to the IWT. Rather, they can be applied to many of the world's river treaties governing basins confronting difficult political and environmental realities (Ho, 2017; Petersen-Perlman, Veilleux, & Wolf, 2017). Comparative analysis suggests that some of the criticisms directed at the IWT can be addressed in the treaty or through alternative approaches. But others of the IWT's deficiencies reflect substantial structural obstacles that will be harder to over- come. Appreciating these obstacles is essential if the riparians are to forge politically acceptable means to redress the treaty's shortfalls, and in the process preserve a sustainable mechanism for managing the basin on which they both depend.

The article proceeds in four sections. The next section provides an overview of the Indus River dispute, along with a brief discussion of the IWT. The third sets the critiques of the IWT in the empirical literature analyzing transboundary river agree- ments. The fourth proposes that the negotiation of memoranda of understanding could offer the riparians a potential means to address increasing tension over the Indus and maintain transboundary cooperation. The fifth section concludes.

The Indus River dispute and the IWT

Consisting of the main Indus, two western tributaries, and three eastern tributaries, the Indus River system covers over 1.12 million km^2, making it one of the largest in the world. Of the basin's total area, Afghanistan encompasses 6%, China 8%, India 39%, and Pakistan 47% (FAO AQUASTAT, 2011). The Indus stem and Sutlej tributary rise in China, while the Jhelum, Chenab, Ravi and Beas originate in India before flowing into Pakistan, where they support the world's largest contiguous irrigation network. For Pakistan, the Indus constitutes virtually its only source of surface water. In India, the Indus system waters the arid and semi-arid north-western territory that is the country's breadbasket. All told, Pakistan withdraws 63% of the water used across the basin, while India draws 36% (Indus Basin Working Group, 2013). The region's rugged terrain has thus far complicated China's development of its stretch of the Indus, although China is building dams on the river. In Afghanistan, the Kabul tributary accounts for 30% of Afghanistan's internal renewable surface water resources before flowing into Pakistan, but years of civil conflict have considerably dilapidated the nation's water infrastructure, and withdrawals remain small (FAO, 2013; Shroder, 2014).

The Indus River dispute between India and Pakistan originates in the decolonization process. The international border established by the 1947 Partition of British India bisected the river system and the region's canals. The new boundary placed India upstream along the Indus system and Pakistan downstream. India thereby also gained control of the canal waters that sustain Pakistan's agricultural sector. After years of conflict over the Indus and eight years of intense negotiations mediated by the World Bank, New Delhi and Islamabad signed the IWT in 1960. The treaty divided the Indus tributaries that India and Pakistan shared. The three eastern tributaries – Ravi, Sutlej and Beas – were allocated to India. Pakistan received the three western tributaries – the main Indus, the Jhelum and the Chenab. The resulting quantitative allocation granted 170.27 km^3 of the western tributaries' waters to Pakistan and 62.21 km^3 to India. On the eastern tributaries, all 11.1 km^3 went to India (FAO AQUASTAT, 2011). Crucially, before reaching Pakistan, the western tributaries flow through India-controlled – and Pakistan-contested – Jammu-Kashmir. Thus, the Indus River issue is intertwined with the riparians' territorial dispute over the status of Jammu-Kashmir.

The IWT stipulates rights and responsibilities of the signatory states. On the western tributaries allotted to Pakistan, India retains the right to construct and operate irrigation works and hydroelectric infrastructure, with certain restrictions. India can develop the hydropower potential of these tributaries, and use their water to meet domestic household, municipal, industrial and irrigation needs. India must supply Pakistan with relevant information regarding these works, along with flood warnings and hydrological data. On the eastern tributaries assigned to India, Pakistan retained specific rights. Before these tributaries make their final departure from India they are fed by minor tributaries originating in Pakistan. The treaty protects the rights of Pakistani farmers residing along these streams to develop the water to meet their needs (Articles II and III, Annexure B). India also relies on Pakistan for the drainage of floodwaters and agricultural runoff from its western states (Article IV, 4 and 5; Article IV, 8,). Finally, to compensate Pakistan for the loss of the eastern tributaries, the treaty provided for constructing link canals, barrages and dams to transfer at least 20 million acre-feet of water to regions of the country cut off from previous irrigation supplies.

The IWT also crafted an effective institutional structure, creating mechanisms for the parties to manage their shared river and address their water disputes. The IWT establishes the Permanent Indus Commission (PIC) to implement the treaty, facilitate cooperation in the river's development, and manage conflicts of interests. It requires the PIC meet at least once every year to exchange data and information on member states' projects. Unusually among transboundary river treaties, the IWT institutes elaborate monitoring mechanisms designed to overcome the parties' concerns of cheating, and sophisticated conflict-resolution procedures. The PIC is thus responsible for negotiating any 'questions' arising between the parties. If the commission cannot resolve the issue, it sends the matter to their respective foreign secretaries for negotiation. If the foreign secretaries fail to settle the issue, and if the nature of the issue falls within one of 23 areas listed in the treaty, it becomes a 'difference' that can be sent to a neutral expert for resolution. Should the neutral expert so opine, or should the contested issue fall outside the 23 listed areas, the matter may be considered a 'dispute' and can be sent to a court of arbitration. The IWT specifies mechanisms for implementing these conflict-resolution mechanisms, and the process is overseen by the World Bank, which was signatory to certain portions of the treaty (IWT, 1960; Salman & Uprety, 2001).

The parties have employed the IWT's conflict-resolution mechanisms three times – in 2005, 2010 and 2016. In the first instance, Pakistan challenged the design of the Baglihar Dam, which India was constructing over the Chenab. Pakistan feared the dam would increase India's capacity to control the tributary's flow (Wirsing & Jasparro, 2007). Noting technological changes since the treaty signing and new worries about climate change, the neutral expert agreed with India's design for the dam. But to accommodate Pakistan's misgivings, the expert asked India to undertake specific structural alterations to the plan. In 2010, Pakistan invoked the court of arbitration mechanism over India's construction of the Kishanganga Dam, which would redirect water from the Kishanganga/Neelum River into the Jhelum. Pakistan contended this diversion of water from one tributary to another would reduce the flow downstream to its own planned Neelum-Jhelum Hydroelectric Project. Pakistan also opposed India's depleting the reservoir water level below the dead storage capacity for drawdown flushing to control sediment at the Kishanganga project. The court ruled that India could undertake the diversion as long as it maintains 9 m^3/s downstream flow in the tributary, while ruling against depleting the dead storage capacity to flush silt (Court of Arbitration, 2014). In 2016, Pakistan requested the creation of a court of arbitration concerning Kishanganga and another project, while India has asked a neutral expert be appointed. This dispute is ongoing.

Contextualizing critiques of the IWT

In the nearly six decades since the treaty's signature, criticisms of the IWT have become more prominent and widespread. These criticisms regard how the treaty addresses disputes arising between the parties and how it fails to address new challenges arising in the basin. In the first category, Pakistan worries over India's increasing interest in constructing hydrological infrastructure on the western tributaries. India has 33 hydro-power projects in the construction or planning phase along the western tributaries (Adeel & Wirsing, 2017). The IWT's information-sharing and conflict-resolution mechanism treats each project as an individual case, but some Pakistanis fear that, though individual Indian projects and proposals may abide by the technical letter of the IWT's restrictions, these multiple installations on the western tributaries may engender significant cumulatively detrimental impacts downstream by controlling the timing and flows of the rivers (Briscoe, 2010). In India, however, some view the persistent Pakistani challenges to planned infrastructure works on the western rivers as unjustifiably impeding India's legitimate development efforts (Chellaney, 2013). Thus, there are fundamental asymmetries in perceived risks and benefits under the IWT. Pakistan claims reasonable protection from undue harms; India claims rightful development of its allotted waters. Beyond the ongoing clash of contending interests, many observers fear that new pressures emerging since the IWT's creation will further stress riparian relations.

Increasing demand, mismanaged supply

Population growth, socio-economic development, mismanagement of existing supplies, and climate change are straining Pakistan's and India's capacity to meet ever-increasing domestic water needs. India's population is expected to increase from 1.3 billion in 2015 to 1.7 billion in 2050. Pakistan's population will soar from 189 million inhabitants today to 310 million at mid-century (UN, 2015). Projected water demands will grow accordingly. By 2025, India's domestic and industrial water withdrawals from the Indus are expected to double over 2000 levels (Sharma, Amarasinghe, & Sikka, 2008), while irrigation water use will climb 12% by 2025 compared to 1995 (Ringler, Cline, & Rosegrant, 2009). Similarly, experts anticipate that municipal and industrial water withdrawals in Pakistan will nearly triple between 2010 and 2025, even as the country will need 250 km^3 of water for agriculture (Qureshi, 2011).

Long-term renewable water supplies in the Indus Basin average 287 km^3/y, while estimates of total annual water demand across the basin are 257–299 km^3, with India withdrawing about 98 km^3 yearly and Pakistan 180–184 km^3 (Indus Basin Working Group, 2013). The Indus is increasingly considered a 'closed' basin, meaning that, under current management practices, nearly all of the basin's annually available renewable water is already allotted to various human uses and ecosystem services, leaving little buffer to accommodate new users or rising needs (Molle, Wester, & Hirsch, 2010). Assuming current policy regimes continue and existing levels of efficiency and productivity persist, models project a 52% gap between available renewable water supplies and annual demands on the Indian banks of the Indus in 2030. Pakistan will suffer nearly a 50% shortfall (2030 Water Resources Group, 2009).

Ineffective management of existing supplies in India and Pakistan drives worsening water stress. As in many other states, domestic water allocation between sectors and regions is highly political. Agriculture accounts for 93% of water withdrawals from the Indus and employs 40% of the labour force in Pakistan and 55% in India (Indus Basin Working Group, 2013). Farmers thus represent potent political forces in both nations. Both countries subsidize agricultural inputs – such as diesel or electricity for powering tube wells – skewing incentives to over-consume water resources. Likewise, the more politically powerful provinces in both countries exploit the Indus waters at the expense of the weaker provinces. These disfavoured provinces often blame the Indus treaty for their predicament. Pakistan's Sindh Province accuses upstream Punjab of usurping flows from its downstream neighbours to compensate for waters ceded to India under the IWT. In 2002, Jammu-Kashmir's legislative assembly passed a near-unanimous resolution calling for annulling the IWT on the grounds that its restrictions on the western rivers illegitimately shackle Kashmiri development. The Indian states of Haryana and Punjab also disagree on the Indus allocations (Swain, 2017).

Reallocating water from more powerful to weaker constituencies is a complicated and highly political process that regimes prefer to avoid. While addressing the unequal distribution of domestic water supplies and improving water use efficiency are essential to sustainably developing the basin in India and Pakistan, this effort belongs to the realm of domestic politics. Large-N studies of transboundary basins highlight the particular difficulties that encumber managing 'federal' rivers such as the Indus, where decision-making authority is divided between national and sub-national governments. Upstream authorities typically prefer flexible allocation approaches to risk sharing and conflict resolution in the face of challenges such as climate change, while downstream authorities prefer fixed volume distributions to ensure their water security (Garrick et al., 2013). Thus, it is difficult to hold the IWT or any other treaty governing transboundary basins accountable for not addressing domestic water disputes. Moreover, it is uncertain how renegotiation of the treaty can address the impending water shortages in India and Pakistan, which are generally issues of domestic politics. When negotiating river treaties, riparians states prefer to protect their sovereignty in managing domestic water disputes (Magsig, 2017).

Mining groundwater

As with other basins worldwide, policy makers in the Indus have preferred to focus their efforts on increasing water supply, instead of undertaking policy changes to improve the efficiency with which water is used. To meet increasing demand, India and Pakistan are depleting the basin's shared groundwater. Many users rely on groundwater to supplement or replace river sources when surface supplies are insufficient or unavailable. Across the Indus, groundwater provides 48% of total water withdrawals (Indus Basin Working Group, 2013). Such intensive exploitation of the basin's aquifers is unsustainable. In Pakistan, water tables in some areas are falling by two to three metres annually, and dropping to economically inaccessible depths in many wells. In India, aquifer overdrafts exceed natural replenishment rates in 61% of the units in Haryana State, 80% of Punjab, and 71% in Rajasthan (Central Ground Water Board, 2015). Satellite data indicate Indus Basin aquifers losing groundwater at a yearly rate of 10 km^3 in 2002–2008 (Tiwari, Wahr, & Swenson, 2009). The IWT contains no provisions for governing, sharing or protecting the basin's groundwater. A product of its time, the treaty was concluded before the Green Revolution of the 1960s spurred the dramatic expansion of irrigation demands and groundwater extraction across South Asia (Shah, 2007).

Globally, there has been a general neglect to include provisions for governing groundwater resources in river treaties or to negotiate separate agreements to regulate the world's 592 identified transboundary aquifers (Eckstein & Sindico, 2014; IGRAC, 2015). An analysis of 688 treaties governing shared water resources found that only 14% contained provisions on transboundary groundwater (Giordano et al., 2014). Indeed, there are only seven legal mechanisms focusing specifically on regulating the use of transboundary aquifers (Conti & Gupta, 2016). Thus, the IWT is not unique in ignoring groundwater supplies. Nevertheless, it is important to encourage the Indus riparians to regulate their use of this shared resource. As surface water supplies decrease, India and Pakistan will increasingly turn to their transboundary aquifers to meet growing demand. In the process, questions about the rights and obligations of states sharing common aquifers will increasingly arise, but international groundwater law is not well developed to guide the negotiation process.

Water quality

The Indus also confronts increasing challenges to water quality. Surface water quality is generally high in the less populated upper Indus Basin, but progressively deteriorates downstream as agriculture, industry and municipalities dump fertilizers, pesticides, toxic metals and human waste into rivers, canals and drains. Towns, farms and factories pour 55 km³ of wastewater into the Indus every year (Babel & Wahid, 2008). Little of this pollution is treated. A sanitation survey evaluating 423 Indian cities nationwide judged not a single city 'healthy' and classed only four as 'recovering', none of them in the Indus. All the others were assessed as either needing considerable improvements or on the brink of emergency (Ministry of Urban Development, 2010a, 2010b). Available data for Pakistan suggest that 92% of municipal wastewater and 99% of industrial effluents are discharged untreated (Azizullah, Khattak, Richter, & Häder, 2011). Consequently, nitrogen and phosphorous pollution are more than twice the Indus's assimilation capacity (Liu, Kroeze, Hoekstra, & Gerbens-Leenes, 2012). The IWT contains no specific or binding provisions regarding water quality, proffering only a hortatory passage pledging the parties' 'intention' to prevent undue pollution 'as far as practicable' (Art. 4[10]).

Analysis of issue areas covered in river treaties suggests that historically states have prioritized agreement over hydropower and water quantity issues, and largely ignored environmental issues (Hamner & Wolf, 1998). More recently, however, states have begun to include water quality and environmental concerns in their negotiations (Conca, Wu, & Mei, 2006; Giordano et al., 2014). In fact, 'almost every agreement signed in the last decade at least mentions the environment or water quality' (Giordano et al., 2014, p. 257). As more agreements cover water quality issues, empirical analysis of the factors influencing their formation becomes possible. Tir and Ackerman (2009) found that water quality agreements are more likely to form when the riparians are democratic, economically interdependent and economically developed, are confronting a water shortage and exhibit power asymmetry. India and Pakistan have several of these attributes, including power asymmetry and water shortages, but they are not both democratic states that are economically interdependent and developed, which may complicate (but not prohibit) their ability to negotiate over water quality. However, as demand for water in the basin increases, interest in addressing water quality may increase. A third-party mediator could help the riparians overcome these structural obstacles and find politically acceptable means to tackle the deteriorating water quality (Zawahri, 2009).

Climate change

Global climate change will significantly impact the Indus basin's freshwater resources, compounding the many water stresses facing the region. Global warming threatens to upset the prevailing regional precipitation patterns, shuffling the seasonal timing and geographical distribution of the rain and snow that sustain the basin's water supplies. Fully half of the precipitation nourishing the region falls during the summer monsoon (Rajbhandari et al., 2015). Climate change will disrupt key monsoon drivers such as the moisture content of the atmosphere and temperature contrasts between the ocean and the neighbouring land surface, stirring fears that global warming could scramble the monsoon regime. Similarly, more than any other major river system, the Indus depends on the mountain glaciers surrounding its headwaters. Seasonal snow and ice meltwater contributes 35–50% of the Indus' total flow (Savoskul & Smakhtin, 2013). As climate change lifts temperatures and skews precipitation patterns, Himalayan glaciers are receding. Recent analyses estimate that Indus Basin glaciers annually shed 7 billion metric tonnes of ice in 2003–2008 (Kääb, Treichler, Nuth, & Berthier, 2015). Initially, increased melting could boost river flows, exacerbating flood risks. As deglaciation continues, meltwater flows will wane, diminishing the downstream supplies available for drinking, sanitation, agriculture, hydropower, industry and ecosystems (National Research Council, 2012).

Like most river treaties, the IWT makes no mention of climate change. At the time the accord was negotiated, the phenomenon was unknown outside a tiny scientific circle. Using a newly compiled database, experts examined the factors that can help states build capacity to adapt to increased variability of water resources and future expected shortages. A treaty, along with an effectively designed river basin organization, can help riparians peacefully manage disputes arising from climate change (Dinar, Katz, De Stefano, & Blankespoor, 2015; Tir & Stinnett, 2012). Brochmann and Hensel (2009) show that when water scarcity afflicts a transboundary basin, a treaty diminishes the prospects for conflict between the riparians. The Indus has a treaty, and although the IWT does not mention climate change, it does give the riparians mechanisms, such as the PIC, to address challenges stemming from climatic variability (Zawahri, 2009).

Fragmented governance

The IWT has been critiqued for institutionalizing the fragmented governance of a multilateral basin, because it excludes China and Afghanistan (Chellaney, 2013). Sub-basin accords that result in fragmented, or divided, governance contradict the tenets of Integrated Water Resources Management (IWRM), which is advanced by hydrologists, environmentalists and engineers who argue that to provide riparians a collective good, basin states must recognize the watershed as an ecological whole and respect the interdependence of different users (Global Water Partnership, 2000). It is unquestionable that the IWT does not develop the basin in an integrated manner. Several structural factors impede adaptation of an IWRM, including the history of conflict and lack of trust between riparians, unwillingness to share sufficient data to support integrated development, and a strong sense of national sovereignty over domestic territory (Waterbury, 1997).

Another major risk of fragmented governance concerns the threat of future tension with excluded riparians. Notably, China and Afghanistan will continue to develop the tributaries of the Indus basin in their territory, increasing their ability to control the river's flow (Chellaney, 2013). Research suggests that the construction of dams or other hydrological infrastructure without any institutional mechanism that facilitates cooperation can contribute to conflict between riparians (De Stefano, Petersen-Perlman, Sproles, Eynard, & Wolf, 2017).

Despite the importance of negotiating basin-wide treaties, globally, fragmented governance of multilateral basins prevails. Examination of 688 transboundary river treaties reveals that bilateral treaties (between just two states) are more prevalent than multilateral agreements (three or more states). Basin-wide accords are extremely rare, representing only 13% of the existing treaties (Giordano et al., 2014). Empirical investigation of the factors contributing to the negotiation of bilateral treaties over multilateral basins finds that they are more likely to form where riparians are highly dependent on the river, there is an asymmetry of power in the basin, and there is a need to avoid the high costs associated with negotiating and cooperating in a multilateral setting (Zawahri & Mitchell, 2011). To achieve a basin-wide agreement, riparians require an interest in cooperation driven by dependence on the river system, as well as balance of power; hydro-hegemons tend to prefer bilateral accords. In the Indus, India and Pakistan depend heavily on the basin, China and Afghanistan much less so. Moreover, the basin has two hydro-hegemons, India and China, which reduces the incentive for multilateral or basin-wide treaties (Zeitoun & Warner, 2006; Zawahri & Mitchell, 2011).

Whether on the Indus, Ganges, or Brahmaputra, India has always favoured negotiating bilateral accords over its multilateral basins (Crow & Singh, 2000). These bilateral agreements have permitted India to secure its own interests and prevent the formation of coalitions between Nepal and Bangladesh, which could upset the existing distribution of power in the basins (Elhance, 1999). Similarly, China prefers to conclude bilateral agreements on its transboundary rivers and adheres to a concept of 'restricted territorial sovereignty' over natural resources, an approach which tempers cooperation with the qualification that collaboration on shared waters must not sacrifice China's self-perceived national interests (Wouters & Chen, 2013). Thus, the prospects for a multilateral or basin-wide accord along the Indus basin confront structural obstacles.

Emerging challenges and cooperation spillovers

Reviewing the IWT's many shortcomings, numerous analysts believe that the geographical allocation of the tributaries between India and Pakistan essentially severed the hydrological relationship between them, making cooperation minimal and passive. The treaty establishes a 'riparian iron curtain', more like a 'divorce settlement' than a collaborative agreement (Alam et al., 2011, p. 6). These critics see the treaty as a static technical instrument, ill-suited to dealing with complex evolving challenges like climate change and sustainable development, and call for reinterpreting, revising, or renegotiating the IWT to better meet these tests (Adeel & Wirsing, 2017; Burgess et al., 2016; Raman, 2017; Ranjan, 2016; Sarfraz, 2013; Verghese, 2005). In fact, the treaty does explicitly provide that if both states agree, they can cooperate in joint engineering projects along the river (Article VII, 1c). Appropriately designed and operated, such joint works could offer one approach to mitigating water variability stemming from climate change. That India and Pakistan have never enacted Article VII cooperation reflects mutual lack of trust and political will more than lack of a treaty mechanism.

Many observers had hoped that the IWT could foster greater comity between the parties. On its signature in 1960, the IWT was hailed by World Bank president Eugene Black (1960) as laying to rest a bitter dispute that had threatened war between India and Pakistan. Writing in their respective countries' major papers, Pakistan's president Ayub Khan and Indian prime minister Jawaharlal Nehru expressed their hopes that the treaty would bring better future relations (Ahmad, 2011). Nearly six decades on, many critics point out the IWT's general failure either to enhance water cooperation or to improve the overall adversarial relationship between India and Pakistan (Ali, 2008; Sarfraz, 2013; Sridharan, 2005). Indeed, assessing the rising socio-economic and environmental pressures, worsening water stress, hydro-political frictions and strained governance institutions in the basin, the United Nations Environment Programme (UNEP, 2016) identifies the Indus as a 'hotspot' for increasing transboundary tensions. Rather than smoothing relations between India and Pakistan, their shared river is a potential source and possible tool of broader conflict. In 2009, then Pakistani president Asif Ali Zardari asserted that failure to resolve this resource challenge fuels the discontent that can fire extremism and terrorism (Zardari, 2009). Some Indian analysts and policy makers consider that Delhi should condition continued compliance with the IWT – and leverage the latent ability to control the Indus's flow conferred by its upstream position – to compel Pakistan to crack down on its domestic extremists (Chellaney, 2011; IDSA, 2010; Times of India, 2016).

Such criticisms of the IWT are accurate, but they are not without historical irony. While the treaty negotiations explicitly sought to defuse grave strategic tensions between Pakistan and India, they did so by deliberately detaching the contentious issues from politics and framing them as technical questions to be resolved by engineers (Biswas, 1992). The same critique, moreover, could also be applied to many, if not most, of the world's river treaties. Few agreements designed to govern transboundary rivers between adversarial states, such as those between Israel and Jordan, or among Egypt, Sudan, and the downstream nations of the Nile, have succeeded in creating wider lasting harmony. It is questionable whether river treaties should be evaluated by their ability to improve overall riparian relations. Although the IWT has perhaps not generated spillover comity between India and Pakistan, it is equally valid (and valuable) to argue that the treaty has very likely contributed to preventing water disputes from degenerating and 'spilling over' into larger conflicts between them (Kraska, 2009).

Adapting the IWT to a changing world

When compared to the rest of the world's river treaties, the IWT is a structurally sound agreement (Salman & Uprety, 2001). Negotiated by engineers, it is highly detailed and technically precise in its 12 articles and eight annexures. Yet the IWT architecture has weaknesses regarding its adaptability. Unlike such instruments as the 1964 Columbia River treaty between the US and Canada, which contains an option for renegotiation after a 50-year period by either of its signatories, the IWT is a permanent treaty, without any end date. The IWT likewise lacks a formal means for easily amending the treaty or adding new issues. The IWT's Final Provisions, in Article XII, allow that the 'Treaty may from time to time be modified', but set a high hurdle: 'a duly ratified treaty concluded for that purpose between the two Governments' (IWT, 1960).

Previous amendments

Even so, over the years the IWT has seen minor adjustments when the parties were in agreement or whenever it was necessary to implement portions of the treaty. In 1964, the World Bank negotiated a supplemental agreement with the donors that funded the infrastructure for transferring Pakistan's dependence from the eastern to the western tributaries (Salman, 2008). The amount of agricultural land that India can irrigate along the western tributaries was modified in 1982, when the PIC agreed that India can irrigate 260,000 hectares (The Hindu, 1982). In 1989, the PIC updated the method by which India delivers flood warnings to Pakistan (Gupta, 2002).

It could be argued that the IWT has undergone further amendment via the World Bank's unique participation in overseeing implementation of the conflict-resolution mechanisms. In 2005, when Pakistan requested appointment of a neutral expert, the bank realized that the treaty was vague about the procedures for selection, appointment, and funding. The bank searched for mechanisms used by other organizations, ultimately copying those of the International Centre for Settlement of Investment Disputes, thereby redefining the IWT processes for the future (Salman, 2008). By the same token, it could be argued that the neutral expert's Baglihar decision, affirming India's right to draw on the latest technology in engineering hydrological infrastructure on the western rivers, also effectively amended the implementation of certain treaty provisions. Moreover, the expert noted that the impact of climate change should be considered when riparians are designing infrastructure (Lafitte, 2007). The award of the Kishanganga Court of Arbitration could be said to have similarly reshaped the interpretation of specific treaty terms. Yet the bank's mediation represents a limited approach on which to base treaty revisions. More importantly, the treaty's conflict-resolution mechanisms are designed to facilitate management of disputes between the riparians. They are neither designed for nor applicable to consensually integrating new issues such as climate change and ecosystem protection into the IWT.

In principle, more significant changes, such as the joint construction of drainage or 'engineering works', could be made by mutual agreement under the Indus Treaty's Article VII, on Future Co-operation. Other modifications might be attempted via the IWT's Final Provisions, in Article XII. Yet, revisiting the IWT this way to address challenges such as groundwater and pollution risks reopening issues that have already been settled. Officially negotiating new agreements under the IWT umbrella could all too easily invite parties dissatisfied with the present accord to engage in quid pro quo tactics, conditioning acceptance of the new treaty on revisions to the old one. Pakistan could demand more waters from the eastern rivers, and India could require greater freedom to develop the western tributaries, potentially undermining the whole edifice. Frustrations with the existing treaty have led some to call for rescinding the treaty altogether (Financial Express, 2016). Such a cure would be worse than the disease. By the terms of Article XII, the IWT may only be terminated by an express treaty agreement between the parties, a highly unlikely prospect. Unilateral annulment poses further problems. For Pakistan, revoking the IWT would send a dramatic signal of discontent. But as the downstream riparian, withdrawing from the IWT would free Islamabad from few obligations while freeing India from all restrictions. For India, abandoning the treaty would lift restraints on developing the western tributaries, but would bring slight immediate advantage and severe international condemnation (Kugelman, 2016; Parvaiz, 2016). On the one hand, it would require years for India to construct the infrastructure necessary to capitalize on its newfound liberty to control the western rivers. On the other, India's other neighbours on major transboundary basins, Bangladesh, Bhutan, China and Nepal, would be forced to reconsider their relations in light of Delhi's unilateral reversal. China in particular could take India's action as offering licence to divert Indus waters for its own purposes (Telegraph, 2016). Pakistan has warned that it would view Indian withdrawal as an act of war (Dawn, 2016).

Memoranda of understanding

We propose that instead of revising or revisiting the IWT using its formal renegotiation procedures, the states should pursue memoranda of understanding (MOUs) and other cooperative avenues outside the official accord that address issues as they arise, while using the treaty as a structure to organize their development of the basin. Employing MOUs and other alternative approaches outside the IWT to address issues like climate change avoids reopening questions of rights and allocations. Explicitly adopting strategies that do not invoke the international treaty or its modification mechanisms could allow decision makers to highlight the powerful *national* rationales for the parties to tackle shared problems such as water quality and groundwater aquifers, as it is in fact domestic consumers who are most directly affected by rising water pollution and falling water tables. Using MOUs to organize cooperation in areas of mutual concern offers the further attraction that, under international law, barring express stipulation otherwise, MOUs are typically considered not legally binding on the parties (Aust, 2010). Thus, MOUs could foster coordination between otherwise reluctant riparians by removing the spectre of legal liability.

Other riparians have used MOUs to tackle transboundary water issues. India and China, for instance, maintain an MOU for flood-season data sharing on the Sutlej River (Ministry of Water Resources, 2015). Other countries have used MOUs to address transboundary groundwater aquifers. In 2009, Mali, Niger and Nigeria signed an MOU, Consultative Mechanism for the Management of the Iullemeden Aquifer (Eckstein, 2011). Crucially, India and Pakistan could achieve considerable water cooperation in many areas, such as data sharing, natural disaster resilience, capacity building, agricultural water use efficiency and conservation practices, through the use of MOUs, while maintaining the integrity of the IWT. Indeed, national policies in both countries – and co-riparians Afghanistan and China – explicitly call for greater international cooperation to address climate change impacts on water resources (Michel, 2017). Similarly, much international water cooperation can be realized between actors and stakeholders outside the confines of state-to-state relations, such as joint scientific activities among universities, joint capacity building by civil society organizations, and sharing of best practices and policy experiences between provinces, cities and other localities (Alam et al., 2011; Indus Basin Working Group, 2013; Shah & Panchali, 2017).

Conclusion

The Indus basin is undergoing substantial change through the construction of hydrological infrastructure, population growth, socio-economic development, impacts of climate change and contamination of water supplies. As a result, the riparians face increasing challenges in managing their shared water resources, raising growing concern of future conflict or even 'water wars' in the basin. For six decades, the IWT has regulated water relations between India and Pakistan, but many experts believe that the IWT ill-equips the riparians to address new problems and new realities that differ significantly from those confronting the basin in 1960, when the treaty was signed.

To evaluate these criticisms, this article has placed them in a comparative context to appreciate how the IWT's weaknesses compare with other river treaties and consider means by which some problems can be addressed. We find that while the critiques are legitimate, many of them can be applied to most transboundary river treaties around the world. Historically, few river treaties include provisions protecting water quality, regulating groundwater resources or responding to climate change. Yet, as climate change continues to impact the basin, water quality deteriorates and groundwater is over-exploited, the riparians will need to respond to these pressures. The article proposes that the parties could address these issues through MOUs and other alternative means of cooperation, without needing to revise or renegotiate the IWT.

Correcting other IWT deficiencies – failing to improve Indo-Pakistani relations, neglecting IWRM, or excluding other riparians – confronts substantial structural obstacles. Remedying these problems will require a level of cooperation, along with structural shifts in basin dynamics, that may be beyond the ability of riparians, with their history of animosity and conflict. Moreover, the basin has two hydro-hegemons (India and China), presenting significant hurdles to the formation of basin-wide treaties.

Examination of the IWT from a comparative perspective suggests important lessons, along with directions for future research for other basins. Research reveals that treaties and institutions are important for averting conflict in basins that are undergoing rapid environmental or development changes (De Stefano et al., 2017). River treaties can also help states stabilize their relationship and interaction by creating rules, regulations, decision-making procedures and conflict-resolution mechanisms (Ho, 2017; McCaffrey, 2003). This has largely been the Indus experience. Yet the IWT shares the flaws of most river treaties worldwide, neglecting to manage groundwater resources, protect water quality or improve the basin's capacity to adapt to climate change (Giordano et al., 2014). Future research needs to consider how river treaties in general can adapt to these challenges while preserving mechanisms and institutions that have operated effectively. Another avenue of research is needed to illuminate how third parties, such as the World Bank, can help riparians negotiate often highly political issues. Finally, research is needed on finding politically acceptable means to overcome the significant structural obstacles confronting riparians' ability to incorporate IWRM into river treaties and to build basin-wide accords.

Note

1. Examples of these databases include the Transboundary Freshwater Dispute Database, which contains 688 treaties from 1820 to 2007. Advances in data technology have enabled construction of event-based databases to track news events concerning river disputes and how they are resolved. Examples of these databases include the Issue Correlates of War and the International Water Events Database.

References

2030 Water Resources Group. (2009). *Charting our water future: Economic frameworks to inform decision-making.* McKinsey and Company.

Adeel, Z., & Wirsing, R. G. (2017). Introduction. In Z. Adeel & R. G. Wirsing (Eds.), *Imagining Industan: Overcoming water insecurity in the Indus Basin* (pp. 3–20). Switzerland: Springer International.

Ahmad, A. (2011). Indus Waters Treaty: A dispassionate analysis. *Policy Perspectives, 8*(2), 73–83.

Alam, R., Shah, M. A. A., Gardezi, M., Sohail, S. B., D'Souza, R., Singh, I, Jain, A. K., ... Main, T. S. (2011). *Re-imagining the Indus*. New Delhi: Observer Research Foundation/Lahore University School of Management.

Ali, S. (2008). Water politics in South Asia: Technocratic cooperation and lasting security in the Indus Basin and beyond. *Journal of International Affairs, 61*(2), 167–182.

Aust, A. (2010). *Handbook of international law* (2nd ed.). Cambridge: Cambridge University Press.

Azizullah, A., Khattak, M. N. K., Richter, P., & Häder, D.-P. (2011). Water pollution in Pakistan and its impact on public health–A review. *Environment International, 37*(2), 479–497.

Babel, M., & Wahid, S. (2008). *Freshwater under threat: South Asia*. Nairobi: UNEP.

Biswas, A. (1992). Indus Water Treaty: The negotiating process. *Water International, 17*(4), 201–209.

Black, E. (1960). The Indus: A moral for nations. *New York Times*, December 11.

Briscoe, J. (2010). Troubled waters: Can a bridge be built over the Indus. *Economic and Political Weekly, 45*(50), 28–32.

Brochmann, M., & Hensel, P. (2009). Peaceful management of international river claims. *International Negotiation, 14*(2), 393–418.

Burgess, J. P., Owen, T., & Sinha, U. K. (2016). Human securitization of water? A case study of the Indus Waters Basin. *Cambridge Review of International Affairs, 29*(2), 382–407.

Central Ground Water Board. (2015). *Ground water year book India: 2014-2015*. Faridabad: Government of India.

Chandio, M. (2008). Renegotiate the Indus Treaty. *Dawn*, November11.

Chellaney, B. (2011). *Water: Asia's new battleground*. Washington, DC: Georgetown University Press.

Chellaney, B. (2013). *Water, peace, and war: Confronting the global water crisis*. Lanham, MD: Rowman & Littlefield.

Conca, K., Wu, F., & Mei, C. (2006). Global regime formation or complex institution building? The principled content of international river agreements. *International Studies Quarterly, 50* (2), 263–285.

Conti, K., & Gupta, J. (2016). Global governance principles for the sustainable development of groundwater resources. *International Environmental Agreements, 16*(6), 849–871.

Court of Arbitration. (2014). Indus waters Kishenganga arbitration partial award. In E. Lauterpacht, C. Greenwood, & K. Lee (Eds.), *International Law Reports* (Vol. 154). Cambridge: Cambridge University Press.

Crow, B., & Singh, N. (2000). Impediments and innovation in international rivers: The waters of South Asia. *World Development, 28*(11), 1907–1925.

Dawn. (2016). Revocation of Indus Waters Treaty can be taken as an act of war: Sartaj Aziz. September 27.

De Stefano, L., Petersen-Perlman, J. D., Sproles, E. A., Eynard, J., & Wolf, A. T. (2017). Assessment of transboundary river basins for potential hydro-political tensions. *Global Environmental Change, 45*(45), 35–46.

Dinar, S., Katz, D., De Stefano, L., & Blankespoor, B. (2015). Climate change, conflict, and cooperation: Global analysis of the effectiveness of international river treaties in addressing water variability. *Political Geography, 45*(45), 55–66.

Eckstein, G. (2011). Managing buried treasure across frontiers: The international law of Transboundary Aquifers. *Water International, 36*(5), 573–583.

Eckstein, G., & Sindico, F. (2014). The law of transboundary aquifers: Many ways of going forward, but only one way of standing still. *Review of European, Comparative and International Environmental Law, 23*(1), 32–42.

Elhance, A. (1999). *Hydropolitics in the Third World: Conflict and cooperation in international river basins*. Washington, DC: US Institute of Peace Press.

FAO. (2013). *Afghanistan* (Water Report 39). Rome: FAO.

FAO AQUASTAT (2011). *Indus Basin* (Water Report no.37). Rome: FAO. Retrieved from http://www.fao.org/nr/water/aquastat/basins/indus/index.stm

Financial Express. (2016). India should end Indus Waters Treaty with Pakistan: Yashwant Sinha. September 27.

Garrick, D., De Stefano, L., Fung, F., Pittock, J., Schlager, E., New, M., & Connell, D. (2013). Managing hydroclimatic risks in federal rivers: A diagnostic assessment. *Philosophical Transactions of the Royal Society A, 371*(2002), 20120415.

Giordano, M., Drieschova, A., Duncan, J. A., Sayama, Y., De Stefano, L., & Wolf, A. T. (2014). A review of the evolution and state of transboundary freshwater treaties. *International Environmental Agreements, 14*(3), 245–264.

Global Water Partnership. (2000). *Integrated water resources management* (TAC Background Papers no.4). Stockholm: GWP.

Gupta, A. C. (2002). India's commissioner to the permanent Indus commission, Zawahri Interview, New Delhi, India, May 13. doi:10.1044/1059-0889(2002/er01)

Hamner, J., & Wolf, A. (1998). Patterns in international water resource treaties: The transboundary freshwater dispute database. *Colorado Journal of International Environmental Law and Policy, 8,* 157–177.

Hirsch, A. (1956). From the Indus to the Jordan: Characteristics of Middle East International River Disputes. *Political Science Quarterly, 71*(2), 203–222.

Ho, S. (2017). Introduction to 'Transboundary river cooperation: Actors, strategies and impact.' *Water International, 42*(2), 97–104.

IDSA. (2010). *Water security for India: The external dynamics* (DSA Task Force Report). New Delhi: Institute for Defence Studies and Analysis.

Indus Basin Working Group. (2013). *Connecting the drops: An indus basin roadmap for cross-border water research, data sharing, and policy coordination.* Washington, DC: Observer Research Foundation/Stimson Center/Sustainable Development Policy Institute.

Indus Waters Treaty. (1960). Signed in Karachi, Pakistan. September 19. 419 U.N.T.S. 126.

International Groundwater Resources Assessment Centre/UNESCO International Hydrological Programme. (2015). *Transboundary Aquifers of the World.* Delft, Netherlands: IGRAC.

Kääb, A., Treichler, D., Nuth, C., & Berthier, E. (2015). Brief Communication: Contending estimates of 2003–2008 glacier mass balance over the Pamir–Karakoram–Himalaya. *The Cryosphere, 9*(2), 557–564.

Kraska, J. (2009). Sharing water, preventing war – Hydrodiplomacy in South Asia. *Statecraft & Diplomacy, 20*(3), 515–530.

Kugelman, M. (2016). Why the India-Pakistan war over water is so dangerous. *Foreign Policy,* September 30.

Lafitte, R. (2007). *The Indus Waters Treaty 1960 – Baglihar hydroelectric plant: Expert determination executive summary.* Islamabad/New Delhi: Government of Pakistan/Government of India.

Liu, C., Kroeze, C., Hoekstra, A. Y., & Gerbens-Leenes, W. (2012). Past and future trends in grey water footprints of anthropogenic nitrogen and phosphorous inputs to major world rivers. *Ecological Indicators, 18*(18), 42–49.

Magsig, B. O. (2017). The Indus Waters Treaty: Modernizing the normative pillars to build a more resilient future. In Z. Adeel & R. G. Wirsing (Eds.), *Imagining Industan: Overcoming water insecurity in the Indus Basin* (pp. 69–90). Switzerland: Springer International.

McCaffrey, S. (2003). The need for flexibility in freshwater treaty regimes. *Natural Resources Forum, 27*(2), 156–162.

Michel, D. (2017). Managing the Indus in a warming world: The potential for transboundary cooperation in coping with climate change. In Z. Adeel & R. G. Wirsing (Eds.), *Imagining Industan: Overcoming water insecurity in the Indus Basin* (pp. 91–120). Switzerland: Springer International.

Ministry of Urban Development. (2010a). *National rating and award scheme of sanitation for Indian cities.* New Delhi: Government of India.

Ministry of Urban Development. (2010b). *Rank of cities on sanitation 2009-2010: National urban sanitation policy.* New Delhi: Government of India.

Ministry of Water Resources. (2015). *Status of Memorandum of Understanding (MoU) signed with foreign countries in the water sector.* New Delhi: Government of India.

Molle, F., Wester, P., & Hirsch, P. (2010). River basin closure: Processes, implications and responses. *Agricultural Water Management, 97*(4), 569–577.

National Research Council. (2012). *Himalayan glaciers: Climate change, water resources, and water security.* Washington, DC: National Academies Press.

Parvaiz, A. (2016). India simply cannot afford to scrap the Indus Water Treaty with Pakistan. *Quartz,* September 25.

Petersen-Perlman, J., Veilleux, J., & Wolf, A. (2017). International water conflict and cooperation: Challenges and opportunities. *Water International, 42*(2), 105–120.

Qureshi, A. (2011). Water management in the Indus Basin in Pakistan: Challenges and opportunities. *Mountain Research and Development, 31*(3), 152–260.

Rajbhandari, R., Shrestha, A. B., Kulkarni, A., Patwardhan, S. K., & Bajracharya, S. R. (2015). Projected changes in climate over the Indus River Basin using a high resolution regional climate model (PRECIS). *Climate Dynamics, 44*(1), 339–357.

Raman, D. (2017). Damming and infrastructural development on the Indus River Basin: Strengthening the provisions of the Indus Waters Treaty. *Asian Journal of International Law,* 1–31. doi:10.1017/S2044251317000029

Ranjan, A. (2016). Disputed waters: India, Pakistan and the transboundary rivers. *Studies in Indian Politics, 4*(2), 191–205.

Ringler, C., Cline, S. A., & Rosegrant, M. W. (2009). Water supply and food security: Alternative scenarios for the Indian Indo-Gangetic River Basin. *International Journal of River Basin Management, 7*(2), 167–173.

Salman, M. A. S. (2008). The Baglihar difference and its resolution process – A Triumph for the Indus Waters Treaty? *Water Policy, 10*(2), 105–117.

Salman, M. A. S., & Uprety, K. (2001). *Cooperation and conflict on South Asia's international rivers: A legal perspective.* Washington, DC: World Bank.

Sarfraz, H. (2013). Revisiting the 1960 Indus Waters treaty. *Water International, 38*(2), 204–216.

Savoskul, O., & Smakhtin, V. (2013). *Glacier systems and seasonal snow cover in six major asian river basins: Hydrological role under changing climate* (IWMI Research Report 150). Colombo, Sri Lanka: International Water Management Institute.

Shah, M. A. A., & Panchali, S. (2017). Stakeholder perspectives on transboundary water cooperation in the Indus River Basin. In D. Suhardiman, A. Nicol, & E. Mapedza (Eds.), *Water governance and collective action.* London: Routledge.

Shah, T. (2007). The groundwater economy in South Asia. In M. Giordano & K. G. Villholthi (Eds.), *The agricultural groundwater revolution: Opportunities and threats to development.* Wallingford: CABI Publishing.

Sharma, B., Amarasinghe, U. A., & Sikka. A. (2008). Indo-Gangetic river basins: Summary situation analysis, international water management institute. *New Delhi Office,* July.

Shroder, J. (2014). *Natural resources in Afghanistan: Geographic and geologic perspectives on centuries of conflict.* San Diego: Elsevier.

Sridharan, E. (2005). Improving Indo-Pakistan relations: International relations theory, nuclear deterrence and possibilities for economic cooperation. *Contemporary South Asia, 14*(3), 321–339.

Swain, A. (2017). Water insecurity in the Indus Basin: The costs of noncooperation. In Z. Adeel & R. G. Wirsing (Eds.), *Imagining Industan: Overcoming water insecurity in the Indus Basin.* Switzerland: Springer International.

The Hindu. (1982). Jammu-Kashmir main beneficiary under amended Indus Waters Treaty. June 23.

The Telegraph. (2016). Water weapon cuts both ways. September 24.

Times of India. (2016). PM Modi reviews Indus Water Treaty, says 'blood and water can't flow together.' September 26.

Tir, J., & Ackerman, J. (2009). Politics of formalized river cooperation. *Journal of Peace Research, 46*(5), 623–640.

Tir, J., & Stinnett, D. (2012). Weathering climate change: Can institutions mitigate international water conflict? *Journal of Peace Research, 49*(1), 211–225.

Tiwari, V. M, Wahr, J, & Swenson, S. (2009). Dwindling groundwater resources in northern india, from satellite gravity observations. *Geophysical Research Letters, 36.* doi:10.1029/2009GL039401

UN. (2015). *World population prospects: The 2015 revisions – Vol.1: Comprehensive tables.* New York: United Nations.

UNEP-DHI/UNEP. (2016). *Transboundary river basins: Status and trends.* Nairobi: UNEP.

Verghese, B. G. (2005). It's time for Indus II: A new treaty will help the peace process too. *The Tribune,* May 26.

Waterbury, J. (1997). Between unilateralism and comprehensive accords: Modest steps toward cooperation in international river basins. *International Journal of Water Resources Development, 13*(3), 279–290.

Wirsing, R., & Jasparro, C. (2007). River rivalry: Water disputes, resource insecurity and diplomatic deadlock in South Asia. *Water Policy, 9*(3), 231–251.

Wouters, P., & Chen, H. (2013). China's 'soft-path' to transboundary water cooperation examined in the light of two UN global water conventions – Exploring the 'Chinese Way'. *Journal of Water Law, 22*(2–3), 229–247.

Zardari, A. (2009). Partnering with Pakistan. *The Washington Post,* January 28.

Zawahri, N. (2009). Mediating international river disputes: Lessons from the Indus River. *International Negotiation, 14*(2), 281–310.

Zawahri, N., & Mitchell, S. (2011). Fragmented governance of international rivers: Negotiating bilateral versus multilateral treaties. *International Studies Quarterly, 55*(3), 835–858.

Zeitoun, M., & Warner, J. (2006). Hydro-hegemony–A framework for analysis of trans-boundary water conflicts. *Water Policy, 8*(5), 435–460.

The Heilongjiang (Amur) River in Sino-Russian relations: from conflict towards cooperation

Wan Wang and Xing Li

ABSTRACT
The Heilongjiang River has played the role of wind vane and barometer in the long history of Sino–Russian relations. From the seventeenth century, as a focus of disputes and conflicts, it has evolved to the present-day mutually beneficial cooperation between China and Russia. The current status of Sino–Russian interactions over the Heilongjiang River is basically cooperation due to the continuous improvement of bilateral relations. In the context of the Belt and Road Initiative, cooperation over the Heilongjiang River will provide an impetus to further deepen Sino–Russian relations.

Introduction

In the context of increasingly prominent global water scarcity and cross-border water resource conflicts, transboundary rivers play an increasingly important role in international relations. China is facing challenges from cross-border water disputes in the north-west and south-west, and the international community has blamed China for some of the development activities in the international rivers in its vicinity. The problem of cross-border rivers has become one of the important issues in the relationships between China and its neighbouring countries. There are many studies and research results on Sino–Russian relations, but these studies are often conducted from the perspective of national interests, ideology and nationalism; very few research results have been obtained from the perspective of cross-border rivers. The 2005 Songhua River water pollution incident began to attract the attention of the international community to the Heilongjiang River.[1] The attention has its inevitability; the Heilongjiang River's cross-border problem can be seen as an embodiment of China's emerging cross-border river problem in the north.

The literature offers many studies on transboundary water resources using specific cases, statistics or game theory. Regardless of their approach, most of the studies are inseparable from the discussion of conflict or cooperation, which is explained as the nature of water politics (Dinar, Dinar, Mccaffrey, & Mckinney, 2007). There is a wealth of analysis of factors influential in water-related relations among or between riparian countries. Since resources and environment have played increasingly crucial roles in international security, scarcity has become a main factor driving both water-related conflict and cooperation. In places where water is scarce, competition for limited supplies makes water an issue of national security, increasing water-related conflict among riparians (Gleick, 1993; Hensel, Mitchell, & Ii, 2006). The more significant a shared river basin is to a riparian country, the more likely this country is to sign a treaty regarding basin management (Espey & Towfique, 2004). Riparians tend to cooperate more when the shared river is 'contiguous' than when it is 'successive' (LeMarquand, 1977). The geographical asymmetry of 'successive' rivers provide less incentive for riparians to cooperate, except when the downstream country is a hegemon (Lowi, 1995).

Research on the Heilongjiang River's transboundary water resources is also centred on the theme of water-related conflict and cooperation. Many valuable studies on the Heilongjiang River have been conducted by Russian scholars, such as S. F. Nosova, (2007), V. M. Bolgov (Bolgov, Demin, & Shatalova, 2016), and N. V. Prohorova (2011); Chinese scholars, such as Wang Zhijian (Wang & Zhai, 2007), Li Chuanxun (Li, 2006), Teng Ren (Teng, 2007), Jia Shengyuan (Jia, 2002), Jia Dexiang (Jia, Bai, & Liang, 2010), and Zhou Haiwei (Zhou, Ying, & Qian, 2013); and other research institutions, such as PatriciaWouters (Wouters & Chen, 2013). Russian scholars pay more attention to conflicts over the Heilongjiang River, regarding China as a threat that might trigger future conflicts, and believing that China bears more blame for the problems that have emerged or may emerge in the future regarding transboundary water resources and the ecological environment. For example, I. V. Gotvanskii (2007) argued that the growing population and economic pressure on the Chinese side has led to the deterioration of the ecological environment and human living conditions on the river. And Natalia Pervushina (2012) contended that the river could become a factor in cross-border conflicts through the reduction of water volume on the Chinese side. V. L. Gorbatenko (2012) claimed that the negative impact on the ecological environment in Russia near the river was caused by China's construction of a hydropower station on the river's main stream. V. P. Karakin alleged that China is the main 'contributor' of water pollution in the Heilongjiang River (Darman et al., 2011). Karakin also asserted that the vast majority of the pollutants, such as waste-water contamination, was from the Chinese side and that the pressure of human activities in north-eastern China on the river's water was 10 times the pressure of human activities in Russia (Darman et al., 2011). Nevertheless, Sino–Russian cooperation over the transboundary river water resources of the Heilongjiang River has also been recognized by Russian scholars. Bolgov et al. stated that although there have been differences and disputes over Heilongjiang River issues, the two countries have shown mutual respect and trust and have made many positive achievements in their cooperation over the river.

Chinese scholars tend to address the improvement of cooperation with Russia over the Heilongjiang River with respect to non-traditional security or eco-environmental management (Jia et al., 2010; Jia, 2002; Li, 2006; Teng, 2007; Wang & Zhai, 2007; Zhou et al., 2013). Studies of the river mainly focus on aspects such as the river's natural profile, water pollution, floods, hydropower development, ecological environment, and laws. Studies from the perspective of international relations have been rare.

This article is focused on international relations. It considers the role of the Heilongjiang River in Sino–Russian relations and examines the historic and current trends of conflict and cooperation in the Heilongjiang River basin.

The Heilongjiang River in china

China and its neighbouring countries share more than 40 transboundary rivers, including cross-border rivers,[2] border rivers[3] and mixed rivers. Among these, the cross-border rivers are mainly distributed in the south-west and north-west regions of China, such as the Lancang-Mekong, the Yarlung Zangbo, the Yuanjiang, and the Eerqisi; and the border rivers are mainly distributed in North-East China, such as the Yalu, the Tumen, and the Wusuli. In most of the cross-border rivers, China sits at the upper reaches and enjoys an advantageous geographical location. In contrast, the symmetry of border rivers means that the sharing of water resources by China and its neighbour involved two-way influences, and the two sides have equal rights to and obligations over the river. Also, because state borders delineate national territory, construct and maintain national identity, and involve national sovereignty and territorial issues, border security is extremely important to countries on both sides of the boundary river. Furthermore, rivers are fluid, and soil erosion can lead to river diversion, and drifting of the channel centreline, and such changes can cause border disputes.

The Heilongjiang River travels through North-East China. It has two sources, in the north and in the south. The southern source is the Argun River, which originates in the Greater Khingan Range of China. The northern source is the Shilka River, which originates at the foot of Mount Kent, in the north of Mongolia. The Argun and the Shilka are called the Heilongjiang after their confluence in Luoguhe Village, Mohe County, China. Therefore, generally, the Heilongjiang refers to the section from the confluence of the Argun and the Shilka to the estuary in the Sea of Okhotsk, a distance of 2,840 km. The upper and middle reaches of the Heilongjiang are shared by China and Russia, and the downstream section is in Russia. According to Chinese data, the river basin has an area of 1,843,000 km^2 and is shared by four countries: China, Russia, Mongolia and North Korea.[4] According to the United Nations Environment Programme, the total area of the Heilongjiang River basin is 2,093,000 km^2, with 889,000 km^2 (42.5%) in China and 1,008,000 km^2 (48.2%) in Russia.[5]

The Heilongjiang is a mixed-type river, because part of it forms the Sino–Russian border and another part of it makes China and Russia upstream and downstream countries. Cooperation over mixed-type rivers is more complex than for pure boundary or transboundary rivers. Issues include hydropower development, water quality protection, boundary issues, and water allocation, among others.

The Heilongjiang River in Sino–Russian relations: a historical perspective

Among a dozen major cross-border rivers in China, the Heilongjiang River pervades almost the entire process of the development of Sino–Russian relations from their beginning to the present; thus, history has given the Heilongjiang River a unique position in Sino–Russian relations. In the course of the historical development of Sino–Russian relations, the Heilongjiang River has been a site of either conflict or cooperation between the two sides and served as a wind vane of the relations between the two countries. It was originally an inland river of China; three treaties – the Treaty of Nerchinsk, the Treaty of Aigun and the Treaty of Peking – made it into a border river. Its present position reflects not mere historical accident but the ongoing power struggle between China and Russia and their international relations. Of course it was also influenced by many other factors, such as the international situation at the time, domestic politics and national character. The treaties helped maintain a relative peace in the area of the river, but there were still many conflicts in the border area. For instance, in the 1900s, Russia violated China's inland navigation rights on the Songhua River while building the Middle East Railway in China. In July 1900, Russia slaughtered the Chinese residents in Hailanpao and in the 64 villages of Jiangdong. In 1929, China and the Soviet Union were involved in a large-scale armed conflict over ownership of the Middle East Railway, and the Soviet Union attacked and occupied Heixiazi Island. In the early days of the founding of the People's Republic of China, the Sino–Soviet alliance was friendly, and the border was relatively peaceful. Deterioration in relations between the two countries brought the Treasure Island incident, and border negotiations dragged on for a long time. Obviously, during this period the Heilongjiang River was a locus of dispute and conflict between the two sides. But with the evolution of the international pattern, the balance of power between China and Russia and their relative international status has changed in favour of China, and the river has changed from a disputed boundary to a platform for cooperation. Especially when China and Russia were facing the emerging international system and international order after the disintegration of the Soviet Union, their identical historical mission and similar international situation changed people's conception. They came to realize that conflict was not the only way to obtain national interests. Mutual benefit, win-win cooperation and win-win compromise became the consensus between China and Russia, embodied in the highest level of leadership on both sides. The strategic partnership of cooperation between China and Russia has been developing steadily in a positive direction. The two countries gradually started to cooperate over navigation, fisheries, and the development and utilization of the ecological environment along the Heilongjiang River; they also established some cooperation mechanisms. Cooperation in the areas of navigation and fisheries on the Heilongjiang River had an early start, and steady progress has continued to the present.[6] In terms of ecological and environmental protection, China and Russia commenced their cooperation in 1994 and have been improving the status of environmental protection cooperation in their high-level meetings since 1997. Especially since the 2005 Songhua River water pollution incident, the two countries have made substantial progress in environmental cooperation. This progress was apparently embodied in the long-acting Sino–Russian environmental cooperation mechanism, which was rapidly established in 2006 under the framework of the Committee of the Regular Meeting between Chinese Premier and Russian Prime Minister. Step by step, a joint monitoring mechanism for transboundary water quality and a reporting and information exchange mechanism for emergent transboundary environmental incidents were established. In 2008, China and Russia signed the Agreement on Rational Utilization and Protection of Transboundary Waters, which was named a historic agreement by the Russian Ministry of Natural Resources.

On the other hand, the two countries' cooperation on the development and utilization of the Heilongjiang River started early but has only made slow progress. The joint Sino–Soviet exercise started in 1956 to examine the economic potential of the Heilongjiang and Argun Rivers was interrupted by a Sino–Soviet split (Vinogradov, 2013). Thanks to Sino–Soviet detente, the exercise resumed in 1986 but never produced a concrete result.[7] Since the beginning of the twenty-first century, the historically undecided territorial issues that so easily led to conflict have been completely resolved. The flooding in the Heilongjiang River in 2013 also contributed to cooperation between the two sides on flood control.

The Heilongjiang River is the wind vane of Sino–Russian relations. When the relationship between the two countries is unstable, the river region is the frontier of the dispute; when this relationship is positive and stable, the river provides a stage for cooperation. At present, Sino–Russian relations are the best in their history and have good momentum for further development. Since 1996, when their relations rose to 'strategic cooperation partners', the two sides have reached a new height of political mutual trust and deepening military cooperation, demonstrating considerable economic and trade cooperation potential and enhancing public contacts. This provides a harmonious environment for cooperation over the Heilongjiang River; even when contradictions emerge between the two sides, they can be resolved.

Is the Heilongjiang River basin at risk for future water conflict?

Among a dozen major transboundary rivers in China, the current status of China and Russia regarding the Heilongjiang River is basically cooperation due to the continuous improvement of bilateral relations. At least, the joint use of this basin seems to be unproblematic (Nickum, 2008). But might there be a risk of future water conflict? The following section describes the trend in terms of interactions and events in the basin.

The existing literature often consists of case studies of conflict or cooperation in the Heilongjiang River basin. Little evidence has been provided as to whether the current status of the basin is cooperative or conflictive. We consider what kind of role water plays in this basin, whether it is a focus of conflict or cooperation, and its historic trends.

The Basins at Risk (BAR) database, created by Oregon State University, compiles reported water-related instances of conflict and cooperation that occur in an international river basin, involve the nations riparian to that basin, and concern freshwater as a scarce or consumable resource (Yoffe, Wolf, & Giordano, 2003). BAR mentions 263 current basins and two historical basins. It lists 37 water-related events in the Heilongjiang River basin, but it is not complete or up to date.

Starting with the BAR data, we created a new database of instances of conflict or cooperation in the Heilongjiang River basin from 1951 to 2016. We coded each event for intensity of conflict or cooperation using the BAR Scale (Table 1). In addition to event data from BAR, we gathered reported events from major media both at home and abroad, such as the *People's Daily*, *Global Times*, *Izvestia*, *Sputnik*, and *New York Times*, as well as local yearbooks. The events we compiled cover a wider range than BAR. Our database includes interactions and events of conflict and cooperation that concern all aspects of transboundary water, e.g., water quality, water quantity, flood control, flooding, navigation, fishery, and boundary or territorial disputes.

Table 1. Water event intensity scale (adapted from the Basins at Risk scale of De Stefano, Edwards, De Silva, & Wolf, 2010, and Yoffe et al., 2003).

Recentred BAR scale	Event description
−7	Formal declaration of war
−6	Extensive war acts causing deaths, dislocation, or high strategic cost
−5	Small-scale military acts
−4	Political/military hostile actions
−3	Diplomatic/economic hostile actions
−2	Strong verbal expressions displaying hostility in interaction
−1	Mild verbal expressions displaying discord in interaction
0	Neutral or nonsignificant acts for the international situation
1	Minor official exchanges, talks, or policy expressions – mild verbal support
2	Official verbal support of goals, values, or regime
3	Cultural or scientific agreement or support (nonstrategic)
4	Non-military economic, technological, or industrial agreement
5	Military economic or strategic support
6	Major strategic alliance (regional or international), international freshwater treaty
7	Voluntary unification into one nation

For 1951 to 2016, 216 water-related events in the Heilongjiang River basin have been retrieved (Table 2). Of the 216 events, 89.72% were cooperative, 10.65% were conflictive (Figure 1). Clearly, in this basin cooperative events over water have far outnumbered conflictive events over water. The overall average BAR value was positive (1.75). No events that registered −7 are listed in the database. The most conflictive events are small-scale military acts (−5), concerning boundary disputes. There are only five, and they all occurred in the 1960s and 1970s. All the conflictive events in the twenty-first century are in the least conflictive category (−1) and concern water quality. The most cooperative events (6 on the BAR scale) are regional fresh-water agreements.

Overall, during the entire period studied, issues of territory and of water quality dominated the conflictive events (Table 3). Cooperative events concerned a much wider range of issues – navigation, water quality, fishery, hydropower, infrastructure and flood control, as well as boundary issues. From a comparative perspective, during the 1951–1999 period, the majority of events concerned navigation, fishery, hydropower or territory. In 2000–2016, hydropower has nearly disappeared, while events regarding water quality and joint management have increased in number.

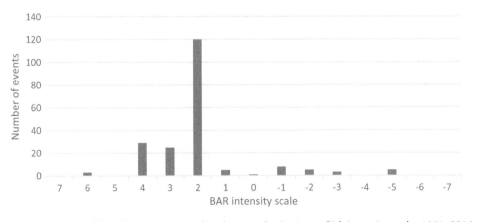

Figure 1. Number of Sino-Russian water-related events by Basins at Risk intensity scale, 1951–2016.

Table 2. Sino–Russian (Soviet) water-related events concerning the Heilongjiang River.

Date	Type of event	Description of event	Basins at Risk scale †	Source
1951.01.02	Navigation	China and the Soviet Union agree to establish navigation rules for the Heilongjiang River, Ussuri River, Argun River, Songacha River and Khanka Lake. This agreement formulated the rules for the navigation in the boundary rivers of the two countries and established the Sino-Soviet Joint Committee on the Navigation in Boundary Rivers.	4	Treaties on Boundary Affairs of the People's Republic of China: Sino–Russian Volume.
1951.01	Navigation	First meeting of the Sino-Soviet Joint Committee on the Navigation in Boundary Rivers	2	Website of the Sino–Russian Joint Committee on the Navigation in Boundary Rivers in China
1952	Navigation	Second meeting of the Sino-Soviet Joint Committee on the Navigation in Boundary Rivers	2	
1953.04.05	Flood control	At the request of the Soviet Union, China provides hydrologic information from hydrometric stations in the mainstream and tributaries of Heilongjiang River.	3	Chorography of Heilongjiang Province, Hydraulic Engineering, Volume 9
1953	Navigation	Third meeting of the Sino-Soviet Joint Committee on the Navigation in Boundary Rivers	2	
1954	Navigation	Fourth meeting of the Sino-Soviet Joint Committee on the Navigation in Boundary Rivers	2	Website of the Sino–Russian Joint Committee on the Navigation in Boundary Rivers in China
1955	Development/ hydropower	China and the Soviet Union sign the Agreement on Joint Investigation of Natural Resources and Potential Productivity of Heilongjiang River Basin and Planning of Comprehensive Utilization of Argun River and Upper Heilongjiang River.	4	Chorography of Heilongjiang Province, The Events, Volume 2
1955	Navigation	Fifth meeting of the Sino-Soviet Joint Committee on the Navigation in Boundary Rivers	2	
1956	Navigation	Sixth meeting of the Sino-Soviet Joint Committee on the Navigation in Boundary Rivers	2	
1956	Flood control	At the request of China, the Soviet Union provides hydrologic information from hydrometric stations in the mainstream and tributaries of the Heilongjiang River.	3	Chorography of Heilongjiang Province, Hydraulic Engineering, Volume 9
1956.08.18	Development/ hydropower	China and the Soviet government approve the agreement signed in 1955 on investigation and planning; the investigation starts in 1956.	4	Keesing's Contemporary Archives; Chorography of Heilongjiang Province, The Events, Volume 2
1957	Navigation	Seventh meeting of the Sino-Soviet Joint Committee on the Navigation in Boundary Rivers	2	Website of the Sino–Russian Joint Committee on the Navigation in Boundary Rivers in China
1957.05	Flood control	First coordination meeting of the Working Group on Hydrology of the Heilongjiang River Basin, which aims to study the Heilongjiang, Argun and Songacha Rivers for their development plans, observational methods, etc.	3	Chorography of Heilongjiang Province, Hydraulic Engineering, Volume 9

(Continued)

Table 2. (Continued).

Date	Type of event	Description of event	Basins at Risk scale †	Source
1957.12.21	Navigation	China and the Soviet sign an agreement on navigation for business in border rivers as well as rivers/lakes related to them.	4	Treaties on Boundary Affairs of the People's Republic of China: Sino–Russian Volume
1958.01	Flood control	Second coordination meeting of the Working Group on Hydrology of the Heilongjiang River Basin. The two countries agreed to exchange hydrologic information, involving 20 hydrometric stations and 19 hydrometric stations.	3	Chorography of Heilongjiang Province, Hydraulic Engineering, Volume 9
1959.01	Flood control	Third coordination meeting of the Working Group on Hydrology of the Heilongjiang River Basin. The two countries exchanged hydrological research results discussed exchange visits of experts.	3	Chorography of Heilongjiang Province, Hydraulic Engineering, Volume 9
1962	Navigation	Eleventh meeting of the Sino-Soviet Joint Committee on the Navigation in Boundary Rivers	2	Website of the Sino–Russian Joint Committee on the Navigation in Boundary Rivers in China
1963.03	Boundary/ territorial dispute	China openly raises the question of the territorial right of the Far East Area, which used to belong to China but was annexed by Tsarist Russia through unequal treaties imposed on China.	-2	The Deterioration of Sino-Soviet Relations: 1956–1966
1963	Boundary/ territorial dispute	The Soviet Union declares that some Chinese residents living in the border area have conducted cross-border fishing in disputed areas of the Heilongjiang and Ussuri Rivers.	-2	The Sino-Soviet Border Problems of 1969
1963.11.02	Navigation	Twelfth meeting of the Sino-Soviet Joint Committee on the Navigation in Boundary Rivers	2	Historical Events of Harbin – 1963
1966.04.19	Navigation	The Chinese government issues rules for managing foreign ships in border rivers, which are much more strict than the Sino–Soviet navigation rules for border rivers signed in 1957.	-3	The Sino-Soviet Border Problems of 1969
1966	Boundary/ territorial dispute	The Soviet army expells Chinese residents of the border area who entered the disputed area near Khabarovsk.	-3	The Sino-Soviet Territorial Dispute
1967	Flood control	China and the Soviet Union suspend the supply of hydrological information to each other.	-3	Water Yearbook
1967.08.17	Boundary/ territorial dispute	14 Chinese residents of the border area land on Wubalao Island, a Chinese-administered island; 40 Soviet soldiers beat up the Chinese residents and forced them off the island.	-5	Chorography of Heilongjiang Province, The Military, Volume 66
1967.09.22	Boundary/ territorial dispute	18 Chinese residents of the border area land on Wubalao Island, a Chinese-administered island; 12 Soviet armed men land on the island and beat up those Chinese.	-5	Chorography of Heilongjiang Province, The Military, Volume 66

(Continued)

Table 2. (Continued).

Date	Type of event	Description of event	Basins at Risk scale †	Source
1969.05.12	Boundary/ territorial dispute	The Soviet military uses weapons to stop Chinese from landing on Wubalao Island and regularly patrols on it. On 15 May 15, four Chinese military come under fire from the Soviet when they go on patrol on Wubalao Island, and one Chinese is killed.	−5	Chorography of Heilongjiang Province, The Military, Volume 66
1969.06.18	Navigation	Fifteenth meeting of the Sino-Soviet Joint Committee on the Navigation in Boundary Rivers	2	XinhuaNet
1969.07.08	Boundary/ territorial dispute	The Soviet army and Chinese local militia exchange fire; one Soviet motorboat is sunk.	−5	Chorography of Heilongjiang Province, The Military, Volume 66
1970.09	Navigation	Sixteenth meeting of the Sino-Soviet Joint Committee on the Navigation in Boundary Rivers	2	The Sino-Soviet Border Problems of 1969
1970.09.08	Boundary/ territorial dispute	Chinese patrol boats are hit by a Soviet naval ship and sunk.	−5	Chorography of Heilongjiang Province, The Military, Volume 66
1973.01.05	Navigation	Eighteenth meeting of the Sino-Soviet Joint Committee on the Navigation in Boundary Rivers	2	People's Daily
1974.03.21	Navigation	Nineteenth meeting of the Sino-Soviet Joint Committee on the Navigation in Boundary Rivers	2	People's Daily
1974.05.22	Boundary/ territorial dispute	The Soviet Union requests the right to Heixiazi Island from China; China refuses.	−2	New York Times
1974.05.30	Boundary/ territorial dispute	The Soviet Union proposes that with its permission China may use two channels near Heixiazi Island; China refuses.	−2	A Calendar of Soviet Treaties: 1974–1980
1977.07.27	Navigation	Twentieth meeting of the Sino-Soviet Joint Committee on the Navigation in Boundary Rivers; the two countries agree on navigation in the Heilongjiang and Ussuri Rivers.	4	A Calendar of Soviet Treaties: 1974–1980
1979.03.27	Navigation	Twenty-first meeting of the Sino-Soviet Joint Committee on the Navigation in Boundary Rivers	2	A Calendar of Soviet Treaties: 1974–1980
1980.03.19	Navigation	Twenty-second meeting of the Sino-Soviet Joint Committee on the Navigation in Boundary Rivers*	2	A Calendar of Soviet Treaties: 1974–1980
1981	Navigation	Twenty-third meeting of the Sino-Soviet Joint Committee on the Navigation in Boundary Rivers*	2	
1982	Navigation	Twenty-fourth meeting of the Sino-Soviet Joint Committee on the Navigation in Boundary Rivers*	2	

(Continued)

Table 2. (Continued).

Date	Type of event	Description of event	Basins at Risk scale †	Source
1983	Navigation	Twenty-fifth meeting of the Sino-Soviet Joint Committee on the Navigation in Boundary Rivers*	2	
1984	Navigation	Twenty-sixth meeting of the Sino-Soviet Joint Committee on the Navigation in Boundary Rivers*	2	
1985	Navigation	Twenty-seventh meeting of the Sino-Soviet Joint Committee on the Navigation in Boundary Rivers*	2	
1986.03	Flood control	A memorandum on hydrological information exchange is signed by China and Russia.	3	Hydrologists from China and Russia hold a meeting in Beijing.
1986.03.05	Navigation	Twenty-eighth meeting of the Sino-Soviet Joint Committee on the Navigation in Boundary Rivers	2	Chorography of Heilongjiang
1986.09.25	Development/ hydropower	18–25 September, experts from China and Russia discuss establishing a committee for comprehensive utilization and planning of border areas in Argun River and Heilongjiang River.	2	Moscow TASS International Service
1986.10.23	Development/ hydropower	China and Russia sign an agreement to establish a committee for comprehensive utilization and planning of border areas in the Argun and Heilongjiang Rivers, including aspects of hydropower development, flood control, navigation, water supply, etc.	4	Treaties on Boundary Affairs of the People's Republic of China: Sino–Russian Volume
1986.12.12	Joint management	First meeting of the Sino–Russian Committee for Comprehensive Utilization and Planning of Border Areas in Argun River and Heilongjiang River	2	XinhuaNet
1987.02.23	Development/ hydropower	9–23 February, Sino–Russian boundary negotiation is held. Both agree to consult on joint development of the Heilongjiang River. The Soviets want to build dams for hydropower and irrigation.	3	Tokyo Kyodo
1987.03	Navigation	Rules on navigation of Sino–Russian border rivers are revised	3	Treaties on Boundary Affairs of the People's Republic of China: Sino–Russian Volume
1987.03.16	Joint management	Scheme of Comprehensive Utilization and Planning of Border Areas in Argun River and Heilongjiang River is formulated and signed. China and Russia plann to conduct investigation of water in the Heilongjiang River in June and July.	4	Moscow Radio Peace & Progress
1987.08.06	Joint management	According to the agreement signed in 1986, experts from China and Russia investigate of water resources in the Heilongjiang and Argun Rivers.	3	Moscow TASS International Service

(Continued)

Table 2. (Continued).

Date	Type of event	Description of event	Basins at Risk scale †	Source
1987.10.23	Joint management	Zade, chairman of the Russian side of the China-Russia Joint Committee for Comprehensive Utilization and Planning of Border Areas in Argun River and Heilongjiang River, speaks highly of this committee's efforts since establishment.	2	Moscow Radio Peace & Progress
1987.11.28	Joint management	Second meeting of the Sino–Russian Committee for Comprehensive Utilization and Planning of Border Areas in Argun River and Heilongjiang River; it is decided to finish the scheme of comprehensive utilization of border rivers before 1992. The major missions include hydropower, flood control and water supply.	4	Moscow TASS International Service
1987	Navigation	Twenty-ninth meeting of the Sino-Soviet Joint Committee on the Navigation in Boundary Rivers*	2	
1988	Navigation	Thirtieth meeting of the Sino-Soviet Joint Committee on the Navigation in Boundary Rivers.*	2	
1988.10.04	Fishery	Agreement on fisheries signed by China and Russia.	4	Treaties on Boundary Affairs of the People's Republic of China: Sino–Russian Volume China Fisheries
1989.01	Fishery	First meeting of the Sino-Soviet Joint Committee on Fishery Cooperation	2	Moscow TASS International Service
1989.01.31	Joint management	Third meeting of the Sino–Russian Committee for Comprehensive Utilization and Planning of Border Areas in Argun River and Heilongjiang River; report concerning hydraulic engineering, flood control, dam site, fishery protection, navigation condition improvement, etc.	3	
1989	Navigation	Thirty-first meeting of the Sino-Soviet Joint Committee on the Navigation in Boundary Rivers*	2	
1989.11.29	Flood control	Water experts from China and the Soviet meet; the Soviets propose a plan for cooperating with China on comprehensive utilization and planning of border water resources, in particular in flood control.	2	Moscow TASS International Service
1990	Fishery	Second meeting of the Sino–Russian Joint Committee on Fishery Cooperation*	2	
1990	Navigation	Thirty-second meeting of the Sino–Russian Joint Committee on the Navigation in Boundary Rivers*	2	
1991.04.20	Fishery	China and Russia meet on fishery resources management in the Heilongjiang and Ussuri Rivers.	2	Fishery Yearbook

(Continued)

Table 2. (Continued).

Date	Type of event	Description of event	Basins at Risk scale †	Source
1991	Fishery	The third meeting of Sino–Russian Joint Committee on Fishery Cooperation was held.*	2	
1991	Navigation	Thirty-third meeting of the Sino–Russian Joint Committee on the Navigation in Boundary Rivers*	2	
1991.05.16	Boundary/ territorial dispute	Agreement on eastern Sino–Russian border is signed, confirming a large part of the Sino–Russian border line.	4	Treaties on Boundary Affairs of the People's Republic of China: Sino–Russian Volume
1992.01.16	Navigation	Agreement on Using Chinese and Russian Ships for Foreign Trade Transportation in Heilongjiang River and Sungari River	4	Treaties on Boundary Affairs of the People's Republic of China: Sino–Russian Volume
1992.02	Navigation	Thirty-fourth meeting of the Sino–Russian Joint Committee on the Navigation in Boundary Rivers	2	Heilongjiang Navigation for Century
1992	Fishery	Fourth meeting of the Sino–Russian Joint Committee on Fishery Cooperation*	2	
1993.03.08	Navigation	Thirty-fifth meeting of the Sino–Russian Joint Committee on the Navigation in Boundary Rivers; rules of navigation mark management in Sino–Russian border rivers are approved.	3	Heilongjiang Navigation for Century
1993.08.13	Fishery	China and Russia meet on fishery resources management; the two countries initial an agreement on Sino–Russian cooperation on protection, enhancement of fishery resources in Heilongjiang River and Ussuri River.	4	Treaties on Boundary Affairs of the People's Republic of China: Sino–Russian Volume
1993.10.11	Fishery	Fifth meeting of the Sino–Russian Joint Committee on Fishery Cooperation	2	Chinese and Russian Experts in Fishery exchange aquaculture technology in Harbin
1994.03.29	Joint management	Agreement on joint nature reserve among China, Mongolia and Russia	4	China Environment News
1994.05.27	Joint management	Agreement on Sino–Russian border management is signed.	4	Treaties on Boundary Affairs of the People's Republic of China: Sino–Russian Volume
1994	Joint management	Thirty-sixth meeting of the Sino–Russian Joint Committee on the Navigation in Boundary Rivers*	2	
1994	Joint management	Agreement on environmental protection cooperation between China and Russia is signed, involving comprehensive utilization of boundary water resources, water protection, and nature reserve management.	4	XinhuaNet
1994.09.03	Navigation	Agreement on Navigation from Ussuri River to Heilongjiang River by way of Khabarovsk is signed.	4	Treaties on Boundary Affairs of the People's Republic of China: Sino–Russian Volume
1994 or 1995	Fishery	Sixth meeting of the Sino–Russian Joint Committee on Fishery Cooperation*	2	

(Continued)

Table 2. (Continued).

Date	Type of event	Description of event	Basins at Risk scale †	Source
1995	Navigation	Thirty-seventh meeting of the Sino–Russian Joint Committee on the Navigation in Boundary Rivers*	2	
1995.06.26	Infrastructure	Agreement on constructing a bridge between Heilongjiang and Blagoveshchensk is signed.	4	
1996	Navigation	Thirty-eighth meeting of the Sino–Russian Joint Committee on the Navigation in Boundary Rivers*	2	
1996.04.25	Joint management	Agreement on Khanka Lake Nature Reserve is signed by China and Russia.	4	XinhuaNet
1996.11.26	Fishery	Seventh meeting of the Sino–Russian Joint Committee on Fishery Cooperation	2	Fishery Yearbook
1997.03.22	Joint management	Russian attorney general says that there have accumulated a lot of problems in natural resources utilization and environmental protection of the Heilongjiang River between China and Russia, which need China's appropriate response.	−1	Moscow ITAR-TASS World Service
1997.03.25	Fishery	China and Russia hold a meeting of the fishery management committee in Khabarovsk.	2	Amur.inform
1997.04.30	Infrastructure	Agreement on Simplifying the Procedure of Customs Related to the Construction of Heilongjiang Bridge is signed.	3	XinhuaNet
1997	Navigation	Thirty-ninth meeting of the Sino–Russian Joint Committee on the Navigation in Boundary Rivers*	2	
1997.11.10	Joint management	Agreement on Guidance for Joint Economic Utilization of Islands and Its Nearby Waters in Sino–Russian Border Rivers is signed.	4	Treaties on Boundary Affairs of the People's Republic of China: Sino–Russian Volume
1997.11.25	Fishery	Eighth meeting of the Sino–Russian Joint Committee on Fishery Cooperation	2	XinhuaNet
1998.03.26	Navigation	Fortieth meeting of the Sino–Russian Joint Committee on the Navigation in Boundary Rivers	2	Website of Chinese office of Sino–Russian Joint Committee on the Navigation
1998.05	Joint management	China and Russia establish a joint working group to provide ways of contracting between them in aspects of ecological resources, fishery and environmental protection in Heilongjiang River and Ussuri River.	2	Amur.inform
1998.07.28	Fishery	Russia's deputy minister of agriculture and food leads a delegation to China and discusses with Chinese agricultural experts about fishery cooperation.	2	Amur.inform
1998	Fishery	Ninth meeting of the Sino–Russian Joint Committee on Fishery Cooperation*	2	

(Continued)

Table 2. (Continued).

Date	Type of event	Description of event	Basins at Risk scale †	Source
1998.12.10	Navigation	Agreement on permissions given to Chinese ships for cargo transportation in the lower Heilongjiang River	4	Treaties on Boundary Affairs of the People's Republic of China: Sino–Russian Volume
1999.03.16	Fishery	China and Russia meet on fishery management in boundary waters of the Heilongjiang and Ussuri Rivers; they agree to conduct joint inspection in these two rivers.	2	Chinese Fishery Yearbook
1999.03.31	Navigation	Forty-first meeting of the Sino–Russian Joint Committee on the Navigation in Boundary Rivers	2	People.cn
1999.11.21	Fishery	Tenth meeting of the Sino–Russian Joint Committee on Fishery Cooperation	2	XinhuaNet
1999.12.09	Joint management	Agreement on Joint Economic Utilization of Islands and Its Nearby Waters in Sino–Russian Border Rivers is signed.	4	Xinhua News Agency
2000.04.08	Navigation	Forty-second meeting of the Sino–Russian Joint Committee on the Navigation in Boundary Rivers	2	XinhuaNet
2000	Fishery	Eleventh meeting of the Sino–Russian Joint Committee on Fishery Cooperation*	2	XinhuaNet
2001.07.16	Joint management	Sino–Russian Treaty of Friendship is signed, including prevention of transboundary pollution, biological resources in boundary waters, protection of natural ecosystems, etc.	4	XinhuaNet
2001	Fishery	Twelfth meeting of the Sino–Russian Joint Committee on Fishery Cooperation	2	Historical Events of Sino–Russian Relations in 2001
2001.04.03	Navigation	Forty-third meeting of the Sino–Russian Joint Committee on the Navigation in Boundary Rivers	2	XinhuaNet
2002.04.09	Navigation	Forty-forth meeting of the Sino–Russian Joint Committee on the Navigation in Boundary Rivers	2	People's Daily
2002	Water quality	Memorandum on Sino–Russian joint monitoring on border rivers is signed; eight joint inspections of the Heilongjiang and Ussuri Rivers.	3	XinhuaNet
2002.12.16	Fishery	Thirteenth meeting of the Sino–Russian Joint Committee on Fishery Cooperation	2	XinhuaNet
2003	Navigation	Forty-fifth meeting of the Sino–Russian Joint Committee on the Navigation in Boundary Rivers*	2	
2003.10.14	Joint management	Some Russian environmentalists and scientists say that Chinese companies are the major cause of Heilongjiang River pollution and there is still no governmental agreement between China and Russia on protecting the Heilongjiang River.	−1	Amur.inform

(Continued)

Table 2. (Continued).

Date	Type of event	Description of event	Basins at Risk scale †	Source
2003	Fishery	Fourteenth meeting of the Sino–Russian Joint Committee on Fishery Cooperation*	2	
2004	Water quality	China and Russia renew the Memorandum of Sino–Russian Joint Monitor on Border Rivers for three years.	3	XinhuaNet
2004	Navigation	Forty-sixth meeting of the Sino–Russian Joint Committee on the Navigation in Boundary Rivers	2	Website of the Department of Transportation of Heilongjiang Province
2004.10.14	Boundary/ territorial dispute	Supplemental agreement on Sino–Russian eastern border is signed, resolving the status of Heixiazi Island and Abagaitu Islet.	4	XinhuaNet
2004.12.07	Fishery	Fifteenth meeting of the Sino–Russian Joint Committee on Fishery Cooperation	2	Amur.inform
2005.07.12	Navigation	Sino–Russian Joint Committee on the Navigation in Boundary Rivers conducts inspection of navigation in boundary rivers.	2	Website of the Department of Transportation of Heilongjiang Province
2005.07.29	Flood control	China and Russia cooperate against floods occurring in their borders.	3	Xinhua Economic News Service
2005.10.12	Navigation	Forty-seventh meeting of the Sino–Russian Joint Committee on the Navigation in Boundary Rivers	2	Website of the Department of Transportation of Heilongjiang Province
2005.11.24	Water quality	The director of China's Environmental Protection Administration meets with the Russian ambassador to China, and reports a pollution accident in the Sungari River. They agree to set up a hotline between environmental protection administrations in Heilongjiang Province and Jewish Autonomous Oblast / Khabarovsk Krai.	2	XinhuaNet; Russia Izvestia
2005.11.25	Water quality	A Russian expert says that there is still no cooperation between China and Russia on eliminating the negative effects on ecosystems of the water pollution accident in the Sungari River.	−1	Amur.inform
2005.12.03	Water quality	Representatives of the Russian Motherland Party protest outside the Chinese embassy in Khabarovsk, asking for more information on the pollution accident in the Sungari River.	−1	BBC
2006.01.02	Water quality	Alexander Alexeyev, deputy foreign minister of Russia, says that China and Russia should establish environmental interaction to prevent accident like the water pollution in the Sungari River in 2005.	1	Russia & CIS General Newswire
2006.03.10	Water quality	China and Russia begin joint monitoring of the Heilongjiang River.	2	Website of the Department of Environmental Protection of Heilongjiang Province

(Continued)

Table 2. (Continued).

Date	Type of event	Description of event	Basins at Risk scale †	Source
2006.03.21	Joint Management	A Sino–Russian Joint Statement is signed, in which the two countries support establishing a subcommittee on environmental cooperation under the framework of the China-Russia Prime Minister's Regular Meeting.	2	Website of the Ministry of Foreign Affairs, People's Republic of China
2006.04.18	Water quality	The governor of Khabarovsk Krai says that the deteriorated water quality is caused by untreated industrial wastewater from China; the water monitoring agreement is not enough; Russia needs to sign an agreement on common responsibility with China.	−1	Russia Izvestia
2006.06.01	Water quality	Sino–Russian Joint Transboundary Water Monitoring Plan is signed, involving five tranboundary water bodies.	3	XinhuaNet
2006.07.15	Navigation	The Sino–Russian Joint Committee on the Navigation in Boundary Rivers conducts inspection of navigation in boundary rivers.	2	Website of the Department of Transportation of Heilongjiang Province.
2006.08.31	Water quality	Alexei Yablokov of Green Russia says that the environmental protection agreement between China and Russia does not cover sanction or inspection, so it needs revision.	−1	Russia & CIS General Newswire
2006.09.12	Water quality	First meeting of the China-Russia Prime Ministers' Regular Meeting Sub-Committee on Environmental Cooperation; China reports progress on joint water monitoring, and Russia speaks highly of China's plan for preventing and controlling water pollution in the Sungari River basin.	2	People.cn
2006.09.29	Boundary/ territorial dispute	A Russian scientist says that China's behaviour of building dams will lead the to border line in the Heilongjiang River drifting towards the Russian side, which is regarded as engineering war.	−1	BBC Monitoring Former Soviet Union – Political Supplied by BBC Worldwide Monitoring
2006.01	Navigation	Forty-eighth meeting of the Sino-Russian Joint Committee on the Navigation in Boundary Rivers	2	Website of the Department of Transportation of Heilongjiang Province
2006.11.09	Joint management	Sino–Russian Agreement on National Border Management is signed, involving rules for border river waters.	4	Website of the Ministry of Foreign Affairs, People's Republic of China
2006.11.22	Water quality	As a response to Sungari water pollution accident, the Russian Emergency Situations Ministry suggests establishing an Asian disaster response centre based on SCO.	4	Sputnik
2006.12.02	Water quality	Iskhakov says that to prevent contamination from happening again, Russia needs to sign an agreement with China to impose sanctions on polluting the Heilongjiang River.	−1	Russia & CIS General Newswire

(Continued)

Table 2. (Continued).

Date	Type of event	Description of event	Basins at Risk scale †	Source
2007.02.27	Technical cooperation	China and Russia conduct joint research on the Heilingjiang and Ussuri riverbeds.	2	BBC Worldwide Monitoring
2007.03.07	Development/ hydropower	A Chinese company plans to implement a hydropower project on the Heilongjiang River; the negotiation process has begun.	1	China Energy Weekly
2007.04.06	Water quality	A group of 400 Russian young men and a local trade association protest outside the Chinese embassy in Khabarovsk against pollution of the Heilongjiang River.	−1	Khabarovsk Territory Media
2007.05.29	Technical cooperation	China and Russia agree to conduct joint investigation of water resources recovery costs after the chemical contaminants leak in 2005. China is highly praised as a strategic partner by Russian representatives.	2	Russian Newspaper, 29 May 2007, p. 5.
2007.06.14	Flood control	Russian state TV reports that China's dam in Plotoka Kazakevicheva Canal might cause flooding on the Russian side of Heixiazi Island (Bolshoy Ussuriysky Island).	0	Financial Times Information
2007.08.30	Joint management	Second meeting of China-Russia Prime Ministers' Regular Meeting Sub-Committee on Environmental Cooperation	2	XinhuaNet
2008	Navigation	Forty-ninth meeting of the Sino–Russian Joint Committee on the Navigation in Boundary Rivers	2	Website of the Department of Transportation of Heilongjiang Province
2008.06.06	Water quality	On the chemical contaminants leak in Heilongjiang Province on 5 June 5, Russia's natural resources minister, Yuri Trutnev, says that Russia has not received any official information, which is thought to be a violation of the related agreement signed by China and Russia.	−1	Sputnik
2008.06.25	Joint management	Third meeting of China-Russia Prime Ministers' Regular Meeting Sub-Committee on Environmental Cooperation	2	China Environment News
2008.07.21	Navigation	The Sino–Russian Joint Committee on the Navigation in Boundary Rivers conducts joint inspection of border river channels and navigation.	2	Website of the Department of Transportation of Heilongjiang Province
2008.09.15	Fishery	Eighteenth meeting of the Sino–Russian Joint Committee on Fishery Cooperation	2	Sputnik
2008	Joint management	Agreement on Rational Utilization and Protection of Transboundary Waters is signed by China and Russia.	6	XinhuaNet
2008	Technical cooperation	Memorandum of Reporting and Information Exchange Mechanism for Transboundary Emergent Environmental Incidents is signed by China and Russia.	3	XinhuaNet

(Continued)

Table 2. (Continued).

Date	Type of event	Description of event	Basins at Risk scale †	Source
2008	Infrastructure	China and Russia sign the Agreement on Construction, Use, Management and Maintainance of Railway Bridge over the Heilongjiang River in the Region of the City of Tongjiang in China and the City of Nizhneleninskoye in Russia.	4	XinhuaNet
2008.12.26	Joint management	First meeting of the Sino–Russian Rational Utilization and Protection of Transboundary Water Committee; both speak highly of the establishment of this commission.	2	Website of Ministry of Foreign Affairs, People's Republic of China.
2009	Navigation	Fiftieth meeting of the Sino–Russian Joint Committee on the Navigation in Boundary Rivers; the two countries officially sign the revised version of the Rules for Navigation of Border Rivers between China and Russia.	3	Website of the Department of Transportation of Heilongjiang Province
2009.05.06	Navigation	The Sino–Russian Joint Committee on the Navigation in Boundary Rivers holds a special meeting to discuss navigation involved with the railway over the Heilongjiang River.	2	Website of the Department of Transportation of Heilongjiang Province
2009.06.03	Joint management	Fourth meeting of China-Russia Prime Ministers' Regular Meeting Sub-Committee on Environmental Cooperation	2	XinhuaNet
2009	Joint management	China and Russia develop the Construction Scheme for Transboundary Nature Reserves in Heilongjiang River Basin.	3	XinhuaNet
2009.07.26	Navigation	The Sino–Russian Joint Committee on the Navigation in Boundary Rivers conducts joint inspection of border river channels and navigation.	2	Website of the Department of Transportation of Heilongjiang Province
2009.08.18	Joint management	China and Russia conduct the first emergency joint exercise on the Heilongjiang River, including navigation and dealing with water pollution accidents.	2	XinhuaNet
2009.10.20	Infrastructure	China and Russia plan to build a pontoon bridge between the city of Heihe and the city of Blagoveshchensk to deal with transportation in winter.	2	Sputnik
2009.10.29	Joint management	Second meeting of the Sino–Russian Rational Utilization and Protection of Transboundary Water Committee; both countries agree to exchange joint monitoring information on transboundary water, hydrological information, flood control, etc. Groups on water quality monitoring and water resources management are set up.	2	Website of the Ministry of Foreign Affairs, People's Republic of China

(Continued)

Table 2. (Continued).

Date	Type of event	Description of event	Basins at Risk scale †	Source
2009.12.15	Water Quality	Fifth Joint Monitoring and Coordination Committee on Transboundary Water Quality and Joint Expert Working Group Meeting	2	XinhuaNet
2009.12.16	Fishery	Nineteenth meeting of the Sino–Russian Joint Committee on Fishery Cooperation	2	Discussion on the development of fishery cooperation economic organizations
2010.04.20	Navigation	Fifty-first meeting of the Sino–Russian Joint Committee on the Navigation in Boundary Rivers	2	Website of the Department of Transportation of Heilongjiang Province
2010.06.21	Joint management	Fifth meeting of the China-Russia Prime Ministers' Regular Meeting Sub-Committee on Environmental Cooperation; Russia proposes holding Heilongjiang River Environmental Academic Exchange Conference in 2011, which obtains China's support.	2	Amur.inform
2010.07.29	Water quality	China quickly informs Russia about a chemical accident on 28 July and takes emergency measures to deal with it, which is highly appreciated by Russia's deputy minister of natural resources and ecology.	1	ifeng.com
2010.07.08	Joint management	Third meeting of the Sino–Russian Rational Utilization and Protection of Transboundary Water Committee	2	Website of the Ministry of Foreign Affairs, People's Republic of China
2010.08.05	Navigation	Sino–Russian Joint Committee on the Navigation in Boundary Rivers conducts joint inspection of border river channels and navigation.	2	Website of the Department of Transportation of Heilongjiang Province
2010.11.23	Joint management	The joint communique of the fifteenth China-Russia Prime Ministers' Regular Meeting proposes conducting joint comprehensive development of Heixiazi Island.	2	Website of the Ministry of Foreign Affairs, People's Republic of China
2011.02.22	Joint management	The Amur Region's deputy minister of natural Resources says that China and Russia plan to create nature reserves along the Heilongjiang River on each side and build a unified management system.	2	Sputnik
2011.03.01	Navigation	The Sino–Russian Joint Committee on the Navigation in Boundary Rivers holds a working meeting to discuss the navigation involved with the Tongjiang cross-border railway bridge.	2	Website of the Department of Transportation of Heilongjiang Province
2011.04.11	Fishery	Twentieth meeting of the Sino–Russian Joint Committee on Fishery Cooperation	2	XinhuaNet
2011.05.24	Navigation	Fifty-second meeting of the Sino–Russian Joint Committee on the Navigation in Boundary Rivers	2	XinhuaNet

(Continued)

Table 2. (Continued).

Date	Type of event	Description of event	Basins at Risk scale †	Source
2011.06.02	Joint management	Sixth meeting of the China-Russia Prime Ministers' Regular Meeting Sub-Committee on Environmental Cooperation; the strategy of constructing a Sino–Russian nature reserve network in the Heilongjiang River basin is approved.	3	XinhuaNet
2011.06.09	Fishery	According to the twenty-first meeting of the Sino–Russian Joint Committee on Fishery Cooperation, China and Russia conducted joint fishery inspection of border rivers (Heilongjiang and Ussuri Rivers) in summer.	2	XinhuaNet
2011.08.23	Navigation	The Sino–Russian Joint Committee on the Navigation in Boundary Riversconducted joint inspection of border river channels and navigation.	2	The website of Department of Transportation of Heilongjiang Province.
2011.09.03	Joint Management	China and Russia sign meeting minute about environmental ecological cooperation development within border regions between Heilongjiang Province and Khabarovski Krai.	2	Website of the Department of Environmental Protection of Heilongjiang Province
2011.09.08	Joint management	China and Russia conduct a second large-scale emergency joint exercise on the Heilongjiang River, on navigation accident response.	2	XinhuaNet
2011.09.17	Fishery	According to the twentieth meeting of the Sino–Russian Joint Committee on Fishery Cooperation, China and Russia conducted joint fishery inspection of border rivers (the Heilongjiang and Ussuri) in autumn.	2	Amur.inform
2011.10.27	Joint management	Fourth meeting of the Sino–Russian Rational Utilization and Protection of Transboundary Water Committee	2	ebsite of the Ministry of Foreign Affairs, People's Republic of China
2012.03.05	Fishery	Twenty-first meeting of the Sino–Russian Joint Committee on Fishery Cooperation	2	China Fishery News
2012.04.18	Joint management	Sixth meeting of the Working Group on Pollution Prevention and Environmental Disaster Response	2	Website of the Department of Environmental Protection of Heilongjiang Province
2012.04.24	Navigation	Fifty-third meeting of the Sino–Russian Joint Committee on the Navigation in Boundary Rivers; the long-standing problem of navigation in border rivers is resolved.	2	Website of the Department of Transportation of Heilongjiang Province
2012.08.08	Fishery	After coordination, four fishing boats and 65 fishermen from China, who were arrested in July by Russia, are returned to China.	2	
2012.08.13	Navigation	The Sino–Russian Joint Committee on the Navigation in Boundary Rivers conducts joint inspection of border river channels and navigation.	2	Website of the Department of Transportation of Heilongjiang Province

(Continued)

Table 2. (Continued).

Date	Type of event	Description of event	Basins at Risk scale †	Source
2012.11.19	Joint management	Seventh meeting of the China–Russia Prime Ministers' Regular Meeting Sub-Committee on Environmental Cooperation	2	XinhuaNet
2012.12.15	Joint management	Fifth meeting of the Sino–Russian Rational Utilization and Protection of Transboundary Water Committee; memorandum is signed.	2	Website of the Ministry of Foreign Affairs, People's Republic of China.
2013.06.25	Fishery	Twenty-second meeting of the Sino–Russian Joint Committee on Fishery Cooperation	2	China Fishery News
2013.07.23	Joint management	The frontier force of China and Russia organizes a joint service exercise in the water area in Heixiazi Island, specifically in jointly arresting those who cross the border.	2	Heilongjiang Yearbook
2013.08.10	Navigation	The Sino–Russian Joint Committee on the Navigation in Boundary Rivers conducts joint inspection of border river channels and navigation.	2	Website of the Department of Transportation of Heilongjiang Province
2013.08.18	Flood control	After the Heilongjiang flood, at the request of China, Russia uses the reservoir in its territory to block the flood, reducing the pressure of flood control for the Heilongjiang mainstream.	2	XinhuaNet
2013.08.20	Flood control	The chairman of the Chinese side of the China–Russia Prime Ministers' Regular Meeting Committee says that Sino–Russian aid agencies played an important role during the Heilongjiang River flood. These two countries might make a joint operations command to prevent climate and man-made disasters.	2	Sputnik
2013.09.04	Joint management	Eighth meeting of the China–Russia Prime Ministers' Regular Meeting Sub-Committee on Environmental Cooperation	2	XinhuaNet
2013.09.17	Joint management	A Sino–Russian emergency joint exercise is conducted in the city of Heihe, aimed to improve response to water emergencies in border rivers.	2	www.huanqiu.com
2013.12.09	Fishery	Twenty-third meeting of the Sino–Russian Joint Committee on Fishery Cooperation	2	China Fishery News
2014.01.23	Joint management	Sixth meeting of the Sino–Russian Rational Utilization and Protection of Transboundary Water Committee; the two countries speak highly of the important role of the committee.	2	Website of the Ministry of Foreign Affairs, People's Republic of China.
2014.04.29	Navigation	Fifty-fifth meeting of the Sino–Russian Joint Committee on the Navigation in Boundary Rivers; the long-standing problem of navigation in border rivers is resolved	2	Website of the Department of Transportation of Heilongjiang Province

(Continued)

Table 2. (Continued).

Date	Type of event	Description of event	Basins at Risk scale †	Source
2014.06.20	Infrastructure	At the invitation of governor of the Russian Jewish Autonomous Region, a delegation from Heilongjiang Province visits Russia. They discuss the construction of the border river railway bridge between Tongjiang and Nizhneleninskoye and sign a memorandum of understanding.	2	Heilongjiang Yearbook
2014.08	Navigation	Memorandum on Navigation Safety Cooperation in Heilongjiang River Basin is signed.	3	Website of the Transport Ministry
2014.09.10	Joint management	Ninth meeting of the China-Russia Prime Ministers' Regular Meeting Sub-Committee on Environmental Cooperation	2	XinhuaNet
2014	Flood control	Memorandum on Sino–Russian Transboundary Flood Control is signed.	3	XinhuaNet
2014	Boundary/ territorial dispute	China and Russia conduct the first joint inspection of the border river line.	2	People.cn
2014.08.21	Navigation	The Sino–Russian Joint Committee on the Navigation in Boundary Rivers conducts joint inspection of certain channels in the Heilongjiang and Ussuri Rivers.	2	Website of the Department of Transportation of Heilongjiang Province
2014.12.18	Infrastructure	Delegations of the transportation departments of the two countries discuss the Heihe border bridge and reach agreements.	3	Website of the Department of Transportation of Heilongjiang Province
2015	Infrastructure	China and Russia sign a revised agreement on co-constructing the bridge between Heihe and Blagoveshchensk.	4	People.cn
2015.02.03	Joint management	Seventh meeting of the Sino–Russian Rational Utilization and Protection of Transboundary Water Committee; results achieved last year are reviewed and plans for next year are made.	2	Website of the Ministry of Foreign Affairs, People's Republic of China.
2015.04.20	Navigation	Fifty-sixth meeting of the Sino–Russian Joint Committee on the Navigation in Boundary Rivers	2	Website of the Department of Transportation of Heilongjiang Province
2015	Infrastructure	China and Russia sign the Agreement on Constructing, Using, Managing and Maintaining Cableway over Heilongjiang River between Heihe and Blagoveshchensk.	4	People.cn

(Continued)

Table 2. (Continued).

Date	Type of event	Description of event	Basins at Risk scale †	Source
2015.06.03	Boundary/ territorial dispute	Chinese delegation of water experts visited Blagoveshchensk, where they inspected newly-built embankments along Heilongjiang River, in order to learn the construction information about new embankments in Russia and their potential influence on China. Experts from China and Russia both considered that there was need to maintain the border river. They discussed how to minimize the influence of newly-built embankments.	2	[Amur.info]
2015.08.25	Water quality	China and Russia conducted joint monitoring of water quality in Heilongjiang River.	2	Website of the Department of Environmental Protection of Heilongjiang Province
2015.08.26	Joint management	Fourth emergency joint exercise in Heilongjiang River at Heihe, aiming to improve water accident response and water pollution prevention.	2	China News
2015.09.21	Joint management	The department of Environmental Protection of Heilongjiang Province and the Ministry of Natural Resources of Khabarovsk Krai hold a third meeting on environmental protection cooperation.	2	Website of the Department of Environmental Protection of Heilongjiang Province
2015.10.12	Navigation	The Sino–Russian Joint Committee on the Navigation in Boundary Rivers conducted navigation inspection in 2015, during which 12 substantive agreements were achieved.	2	Website of the Department of Transportation of Heilongjiang Province
2015.10.20	Joint management	Tenth meeting of the China-Russia Prime Ministers' Regular Meeting Sub-Committee on Environmental Cooperation	2	Website of the Department of Environmental Protection of Heilongjiang Province
2015.12.17	Joint management	Eighth meeting of the Sino–Russian Rational Utilization and Protection of Transboundary Water Committee is held; the results achieved last year are reviewed and plans for next year are made.	2	Website of the Ministry of Foreign Affairs, People's Republic of China.
2016.03.15	Fishery	Twenty-fifth meeting of the Sino–Russian Joint Committee on Fishery Cooperation	2	China Fishery Yearbook
2016.04.23	Navigation	Fifty-seventh meeting of the Sino–Russian Joint Committee on the Navigation in Boundary Rivers; 159 agreements are made.	2	Website of the Department of Transportation of Heilongjiang Province
2016.06.15	Infrastructure	The licence contract for constructing Heilongjiang bridge is signed.	3	Sputnik
2016.07.20	Navigation	The Sino–Russian Joint Committee on the Navigation in Boundary Rivers conducts joint inspection of the upper and middle reaches of Heilongjiang River.	2	Website of the Department of Transportation of Heilongjiang Province

(Continued)

Table 2. (Continued).

Date	Type of event	Description of event	Basins at Risk scale †	Source
2016.08.05	Joint management	Ninth meeting of the Sino–Russian Rational Utilization and Protection of Transboundary Water Committee was held; the two countries spoke highly of the important role of the committee.	2	Website of the Ministry of Foreign Affairs, People's Republic of China
2016.09.14	Infrastructure	Russia's transport minister says that the Heilongjiang bridge project has obtained permission of Russian Direct Investment Fund and Far East Development Fund.	2	Sputnik
2016.09.27	Navigation	The Sino–Russian Joint Committee on the Navigation in Boundary Rivers holds a specific working meeting, at which the agreement on navigation involved with the cross-border cableway between Heihe and Blagoveshchensk was achieved. The vice-president of the Amur Region government paid close attention to cross-border ropeway and border river navigation.	2	Website of the Department of Transportation of Heilongjiang Province
2016.10.12	Water quality	Eleventh meeting of the China-Russia Prime Ministers' Regular Meeting Sub-Committee on Environmental Cooperation; Russia's minister of natural resources and ecology says that transboundary water quality has obviously improved.	2	Website of the Department of Environmental Protection, PRC; Russia Izvestia

*Data are lacking, so the date of this event is estimated.
† For the BAR scale, see Yoffe et al. (2003).

Table 3. Number of events, percentage of cooperation and conflict events, by issue type and time period.

Issue	Total	1951-2016	
		Cooperation	Conflict
Water quality	19	12	7
Joint management	49	47	2
Development/hydropower	6	6	0
Flood control	13*	11	1
Boundary/territorial disputes	15	3	12
Navigation	69	68	1
Fishery	30	30	0
Infrastructure/bridge	12	12	0
Technical cooperation	3	3	0
Total	216	192	23

*Within flood control, one event is neutral.

Dual cooperation at both central and local government levels

Regarding the Heilongjiang, China and Russia have established dual cooperation between their central governments and local governments. Cooperation over the Heilongjiang is jointly managed by the central government and the local governments of the two countries. The central government departments of the two countries are responsible for signing cooperation agreements and formulating cooperation plans and objectives. Implementation is mainly carried out by the relevant departments of local governments. Leading organs at the central government level in China include the Ministry of Ecology and Environment (formerly the Ministry of Environmental Protection), the Ministry of Water Resources, the Ministry of Agriculture and Rural Affairs (formerly the Ministry of Agriculture), and the Ministry of Transport; the main bodies responsible at the local level are the local governments and related departments in Heilongjiang Province. The leading organs at the central government level in Russia include the Ministry of Transport, the Ministry of Agriculture, the Ministry of Natural Resources and Environment, and the Ministry of Civil Defence, Emergencies and Disaster Relief; the main agencies responsible at the local level are the local governments and related departments of Amurskaya Oblast, Jewish Autonomous Oblast, and Khabarovsk Krai.

Under the framework of the transboundary water agreements between the central governments of China and Russia, corresponding regional cooperation has been established between the Heilongjiang provincial government and the governments of Amurskaya Oblast, Jewish Autonomous Oblast, and Khabarovsk Krai. The cooperation covers such areas as agriculture, environmental protection, and hydrometeorology in the border regions. Forms of cooperation include regular meetings to hold dialogues, establish relations and discuss common issues in related fields; joint seminars to exchange information and technologies in related fields; and visits and exchanges to learn about each other's work and draw on each other's experience in related fields.

In the regional cooperation between China and Russia concerning the Heilongjiang River, the cities along it play a key role. These include Heihe and Fuyuan in China, and Khabarovsk and Blagoveshchensk in Russia. These riverside cities are important nodes in the border regions of China and Russia. Their special geographical location and strategic significance make them vital to the cooperation between China and Russia around the river. Since 2007, the Environmental Protection Bureau of Heihe City, Heilongjiang Province, and the Department of Natural Resources of Amurskaya Oblast, Russia, have cooperated in such areas as transboundary waters and environmental protection, and hold regular work meetings. The two sides have successively signed the Minutes of the Talks on Establishing Friendly Relations between the Environmental Protection Bureau of Heihe City and the Natural Resources Department of the Amurskaya Oblast, the 2009–2010 Proposal on International Cooperation between the Natural Resources Department of the Amurskaya Oblast and the Environmental Protection Bureau of Heihe City, and other agreements to conduct in-depth cooperation on transboundary water protection technology, response plans for environmental emergencies, and the establishment of nature reserves. In 2015, the environmental protection departments of Fuyuan County of Heilongjiang Province and the city of Khabarovsk of Russia held a regional environmental protection cooperation conference and signed the Framework Agreement on Regional Environmental Protection between Fuyuan County and Khabarovsk City, establishing cooperation between the two sides on boundary river protection and environmental protection.

How does the Heilongjiang River cooperation affect Sino–Russian relations?

The cross-border function of the Heilongjiang River does not necessarily play a decisive role in Sino–Russian relations, but the cooperation between the two sides over the river can provide a deepening breakthrough point for a strategic cooperative partnership. First, one section of the river forms the border between China and Russia. The common interests of the two sides are to maintain the sustainable development of the border areas, avoid the emergence of cross-border disputes and provide a peaceful environment for the respective development of these border areas. Second, there is room to enhance the two countries' cooperation on the environmental protection of the river. Because China and Russia have made great achievements in political, diplomatic, military and energy cooperation, environmental cooperation, especially concerning the Heilongjiang River, is conducive to comprehensively deepening the Sino–Russian strategic partnership. Third, the current plight of Sino–Russian relations that must be overcome is political heat and economic chill. The cooperation over the river involves more than cooperation water resources; it also includes closely linked regional development. For the joint development of the Silk Road Economic Zone by the North-East and the Far East to connect with the Eurasian Economic Union, the Heilongjiang River can provide a transportation and trade 'water channel'. Constructing a bridge across the Heilongjiang River to link the transport network of Russia in the north to that of China in the south, creating joint transportation on the river and sea, transporting goods from the hinterland of Heilongjiang Province through the lower reaches of the Heilongjiang River to China's coastal areas, Japan and South Korea, and opening up the closed pattern of the Heilongjiang River system will not only provide an important supplement to

North-East Asian international shipping but also offer strong support for the formation of a trans-Eurasian transport trade network. The river can also provide both hydrodynamic and electric power for development. The Heilongjiang River is rich in water resources but has not yet been effectively developed. If development and construction are introduced, they will benefit China and Russia and vigorously promote regional economic development. The economic and social benefits are essentially related to local government and the public.

Current cooperation problems

The current cooperation between China and Russia regarding the Heilongjiang River is far from satisfactory in either depth or breadth. First, the two sides lack an overall plan for the river's development and protection. Although the two countries are cooperating on navigation, fisheries, water quality monitoring and other aspects, they have not agreed on the long-term planning of the comprehensive coordination of all areas, and thus the two sides are passively forced to take emergency remediation measures when problems occur. Potential causes of border disputes, such as the shifting of the channel centreline because of soil erosion, have not been taken seriously, although the issue of Sino–Russian boundary demarcation has been resolved. As a boundary river, the Heilongjiang affects the national border and territorial security of China and Russia. As seen in the aforementioned historical data, the border issue was once a key factor in tension or even conflict between China and the Soviet Union. If ignored, it could have a negative impact on Sino–Russian relations in the future.

Second, the legal basis for cooperation is imperfect. Compared to international water law (like the UN Watercourses Convention), the Heilongjiang River agreement is too simple, with only several clumsy and brief terms that are still missing content (Chen, Rieu-Clarke, & Wouters, 2013). For example, the 2008 agreement stipulates 'fair and reasonable use and protection of transboundary water after economic, social and demographic factors are taken into account', but there are no clearly defined criteria for judging what is a fair and reasonable use of transboundary water.

Third, the basic work on the operational level of cooperation is weak, especially in hydrological information exchange. The two sides have not established a comprehensive information sharing system, and hydrological stations do not cover the entire river. The flood in 2013 clearly exposed this defect.

Fourth, the cooperation model was led by government and signed by the two governments. This model neglected the participation of the public and the market, which has affected the implementation of cooperation.

Cooperation prospects

With the improvement of bilateral relations and the complete settlement of the border issue, China and Russia have shifted from conflict to cooperation concerning the Heilongjiang River. This shift is reflected in the cooperation agreements and cooperative practices between the two countries since the Soviet Union era. However, the breadth and depth of Sino–Russian cooperation involving the river are still inadequate and have not yet been realized at the level of governance.

Because China and Russia have different needs from it, their cooperation over the river is not a simple problem that can be solved by science and technology but a complex issue that involves politics, security and philosophy. Sino–Russian cooperation in navigation, fisheries and environmental protection of the river could be further enhanced through improvement at the technical level. However, the most significant controversy concerning the cooperation between the two sides lies in the development of the boundary river water, which is the major cause of disputes between China and its southern neighbours, although China has demonstrated willingness to cooperate to protect transboundary water ecosystems (Wouters & Chen, 2013). The two countries have different priorities in their focus of cooperation. China focuses on environmental protection as well as water exploitation, while Russia focuses more on flood control and ecological protection. For example, Russia lacks motivation to build hydropower facilities in the main stream of the river because Russia currently has sufficient water resources and excess electricity in the areas near the river. Therefore, in this case, Russia is more concerned about the damage to the ecological environment caused by the development of the boundary river.

Besides, Russia's concern regarding China's water situation in the basin reflects the rapid growth of China's comprehensive strength. Facing a rapidly growing neighbouring country whose power is becoming increasingly superior, Russia might be vigilant and take precautions to avoid becoming China's partner and vassal for raw material. At the government level, the agreements and communiques between China and Russia in the past 10 years have not mentioned the hydropower development of this boundary river. From the public point of view, the Russian people are against the development of hydropower stations. The Russian domestic environmental groups that arose in the 1990s strongly oppose the construction of the supposedly dangerous hydropower projects in the Heilongjiang River's main stream in Russian territory.

Whether China and Russia will cooperate over the river depends on the positioning of the river's function. Borders have two functions: a barrier function, which emphasizes national territorial division and establishes national identity; and an intermediary function, which promotes economic cooperation in border areas. With the development of sub-regional cooperation, the barrier function weakens, while the intermediary function strengthens (Hu, Luo, Li, & Zhang, 2012). Although the emergence of this trend is objective, the extent of the relation between the two functions is rather subjective. A state's interpretation of a border's function depends on the situation (Forsberg, 1995). Since 2001, Sino–Russian trade has been growing, and trade between Heilongjiang Province and Russia has increased significantly; this increase in trade exhibits the continuing intermediary function of the border river. However, the barrier function also has a fundamental interest for both China and Russia, which is reflected in the border negotiations between the two countries, which lasted 40 years, with tremendous disputes, complex circumstances, and difficulties. Particularly as non-traditional border security continues to increase, the barrier function of the border will receive more attention. Because the vast Far East regions are sparsely populated and distant from their European administrative centre, Russia pays more attention than China to the barrier function of the river, concerned to protect its territory in the Far East that was captured from China. Therefore, when it cooperates with China

concerning the Heilongjiang River, Russia attaches more importance to the maintenance rather than the development of the border river.

Conclusions

History suggests that bilateral relations are the factor that most affects Sino–Russian interactions over the shared Heilongjiang River. Thus, the boundary dispute issue, closely related to
national security, once hindered cooperation regarding the river. Fortunately, the territorial
dispute was successfully resolved in 2004. Currently, the controversial issue between China
and Russia in terms of the river is water quality and environmental protection. It is in the
common interest of China and Russia to cooperate over the river, but the two countries have
different interests in the content of this cooperation. How to reduce the different interests and
increase the identical interests is the subject that we are working on. We want to combine the
cooperation over the river's transboundary water resources with regional development and
make it a characteristic emphasis and priority to connect the Silk Road Economic Zone and
the Eurasian Economic Union to constitute an important part of the emerging Asia-Pacific
Economic Circle. These parts should share the same destiny and development, so that
cooperation over the Heilongjiang River helps the common development of China and
Russia. This prospect is positive but faces many obstacles. The continuous improvement of
the two countries' cooperation over the river requires a process, time and unremitting efforts
from both sides. China and Russia have different priorities in terms of this cooperation. Joint
efforts are needed from both sides to reduce the differences in their interests. Russia's lack of
motivation for hydropower development in the mainstream of the Heilongjiang River in fact
causes us to consider whether we must re-examine our own focus on the Heilongjiang River
cooperation – by solving the problem of ecological damage first and then cooperatively
developing boundary river hydropower while protecting the environment. We cannot evade
responsibility for water pollution, and we must continuously strengthen our awareness of
ecological and environmental protection. Therefore, at a certain time in the future, the
cooperation between China and Russia over this river will focus on the maintenance of
boundary rivers and the protection of the environment. The Heilongjiang River can be an
active factor in Sino–Russian relations, but it also faces unavoidable problems. This is a test of
the wisdom of the Chinese government.

Notes

1. The Heilongjiang River and the Amur River are the same river. The former name is used in
    China, and the latter in Russia. For brevity, we use only the former in this article.
2. Rivers that pass through different countries, forming an upstream–downstream relationship.
3. Rivers that divide two countries, forming a border relationship.
4. This article focuses on studying China and Russia on both sides of the border section of the
    Heilongjiang. In the context of the Heilongjiang River basin, it analyzes the interaction
    between China and Russia around the border river of Heilongjiang. Mongolia and North
    Korea will not be further discussed.
5. According to data from China: 48% of the total area of the Heilongjiang river basin is in
    China and 50.4% is in Russia. The figures for Mongolia and North Korea are 1.59% and
    0.01%, respectively; according to data from Russia: 54.1% is in Russia, and 44.2% is in
    China.

6. In January 1951, China and the Soviet Union agreed to establish navigation rules for their border rivers and lakes, specifically the Heilongjiang River, Ussuri River, Argun River, Songacha River and Khanka Lake. In 1992 and 1994, the two countries agreed to open certain ports on each side of the Heilongjiang and Sungari Rivers for mutual trade. And since 1992 Chinese ships have been granted access to the lower Heilongjiang River, which is in Russia's territory. In 1998 and 1994, two agreements initiated cooperation between the two countries over fisheries in the Heilongjiang River, including fishing, increase and protection of fishery resources, and trade of aquatic products.
7. While a tentative hydropower development project was proposed in 1990, no agreement has been reached by the two countries. Until now there is no sign of their considering this proposal.

Disclosure statement

No potential conflict of interest was reported by the author.

Funding

This work was supported by the State Key Laboratory of Hydroscience and Engineering Tsinghua University [sklhse-2014-A-03]; Fundamental Research Funds for the Central Universities [SKZZB2015045]; and China's National Social Science Fund for major projects [16ZDA040].

References

Bolgov, M. V., Demin, A. P., & Shatalova, K. Y. (2016). Russia-Chinese cooperation in utilization and protection of transboundary water: Experience and problems. *Utilization and Protection of Natural Resources in Russia, 2*, 146.

Chen, H., Rieu-Clarke, A., & Wouters, P. (2013). Exploring China's transboundary water treaty practice through the prism of the UN Watercourses Convention. *Water International, 38*(2), 217–230.

Darman, Y., Dikarev, A., Karakin, V., Lomakina, N., Pusenkova, N., & Shvarts, E. (2011). *Environmental risks to Sino-Russian transboundary cooperation: From brown plans to a green strategy. WWF's Trade and Investment Programme report*.

De Stefano, L., Edwards, P., De Silva, L., & Wolf, A. T. (2010). Tracking cooperation and conflict in international basins: Historic and recent trends. *Water Policy, 12*(6), 871–884.

Dinar, A., Dinar, S., McCaffrey, S., & Mckinney, D. (2007). *Bridges over water: Understanding transboundary water conflict, negotiation and cooperation*. Singapore: World Scientific.

Espey, M., & Towfique, B. (2004). International bilateral water treaty formation. *Water Resources Research, 40*, 5.

Forsberg, T. (1995). *Tumas, Forsberg*. Aldershot, U. K.: Edward Elgar Publishing.

Gleick, P. H. (1993). Water and conflict: Fresh water resources and international security. *International Security, 18*(3), 79–112.

Gorbatenko, L. (2012). Russian far east in ATP: Water resources and problems of water utilization. *Paper presented at the 7th international science practical conference "The Rivers of Siberia and the Far East"*, 89.

Gotvanskii, V. (2007). *Amur basin: Development-protection* (pp. 17). Khabarovsk: Arhpelag Fain Print.

Hensel, P. R., Mitchell, M. L., & Ii, T. E. S. (2006). Conflict management of riparian disputes. *Political Geography, 25*(4), 383–411.

Hu, Z. D., Luo, H. S., Li, C. S., & Zhang, W. (2012). Triple functions of country border and its concerning optimized combination under the perspective of geopolitical security. *Human Geography, 3*, 73–77.

Jia, D. X., Bai, J. H., & Liang, F. C. (2010). Perspective analysis of cooperative development of hydropower resources of border rivers between China and Russia. *Energy Technology & Economics, 2* , 5–7.

Jia, S. (2002). Eco-environmental protection of Heilongjiang (Amur River) at Sino-Russian boundary and its sustainable exploitation. *Journal of Environmental Management College of China, 2*, 31–35.

LeMarquand, D. G. (1977). *International rivers: The politics of cooperation.* Vancouver: Westwater Research Centre, University of British Columbia.

Li, C. (2006). A probe on non-traditional security co-operation between neighboring regions of China and Russia. *Russian Central Asian & East European Studies, 6,* 49–57.

Lowi, M. R. (1995). *Water and power : The politics of a scarce resource in the Jordan River basin.* Cambridge, UK: Cambridge University Press.

Nickum, J. E. (2008). The upstream superpower: China's international relations. In: O. Varis, A. K. Biswas, & C. Tortajada, (Eds). *Management of Transboundary Rivers and Lakes. Water Resources Development and Management* (pp. 227–244). Berlin: Springer.

Nosova, S. (2007). Russia-China: Legal relations coordination in utilization of Amur River basin. *Management in Eastern Russia, 3,* 133–140.

Pervushina, N. (2012). Water management and use in the Amur-Heilong River basin: challenges and prospects. In *Environmental security in watersheds: The sea of Azov* (pp. 223–240). Dordrecht: Springer.

Prohorova, N. V. (2011). The development of Russo-Chinese relations in background of utilizing the Amur River basin). In *China in world and regional politics* (pp. 230–244) [in Russian].

Teng, R. (2007). *China-Russia cooperation in border water resources* security. (Master's Degree), Heilongjiang University, Harbin.

Vinogradov, S. (2013). *Sino-Russian transboundary waters: A legal perspective on cooperation.* Social Science Electronic Publishing.

Wang, Z. J., & Zhai, X. M. (2007). Northeast international rivers of China and Northeast Asia' Security. *Northeast Asia Forum, 4,* 15.

Wouters, P., & Chen, H. (2013). China's 'soft-path' to transboundary water cooperation examined in the light of two UN global water conventions–Exploring the 'Chinese way'.*Journal of Water Law, 22,* 229–247.

Yoffe, S., Wolf, A. T., & Giordano, M. (2003). Conflict and cooperation over international freshwater resources: Indicators of basins at RISK. *Journal of the American Water Resources Association, 5,* 1109–1126.

Zhou, G. W., Ying, Z., & Qian, J. (2013). Multilayer cooperation mechanism of cross-border water pollution prevention in Heilongjiang Basin. *China Population, Resources and Environment, 23,* 122–127.

River activism, policy entrepreneurship and transboundary water disputes in Asia

Pichamon Yeophantong ⓘD

ABSTRACT
This article examines the role of non-state actors – namely, 'river activists' – in the management of major transboundary rivers in Asia. Focusing on unresolved disputes over the utilization of the water resources of the Mekong, Nu-Salween and Brahmaputra Rivers, it argues that aside from riparian governments, these activists have contributed considerably to shaping the nature of socio-political contestation in these cases. Drawing upon a 'policy entrepreneurship' framework for analysis, civil society actors are revealed to play an important, if not leading, role in catalyzing and framing water disputes at the national and transnational levels, with cascading consequences for regional water governance.

Introduction

The purpose of this article is to uncover the sources of transboundary water disputes in Asia – that is, the processes of issue emergence that account for the framing of these disputes as such – and examine the role of non-state actors in instigating policy entrepreneurship and change. Contrary to conventional wisdom that often associates such disputes with geostrategic competition or the like, the article posits that apart from riparian governments, 'river activists' have considerably influenced the nature of water-related contestation in the Mekong, Nu-Salween and Brahmaputra River basins through strategic framing dynamics – albeit with varying policy outcomes.[1]

Although the past several decades have witnessed far more incidences of cooperation than of acute conflict over international water resources (Kramer, Wolf, Carius, & Dabelko, 2013, pp. 4–12), one should not overlook the prevalence of socio-political contestation in transboundary river basins, and the implications that protracted water-related disputes hold for the development of effective water-sharing mechanisms and regional institutions. Despite the complex interdependencies that bind riparians together, instability resulting from "civil disobedience, acts of sabotage and violent protest" (p. 8) can still emerge from competition over access to scarce water resources, especially in a transboundary setting. Examples include the clashes between the Maasai and Kikuyu communities in Kenya, which led to the displacement of around 2000 Maasai villagers, over a water diversion project, and the violence that broke out in

Karnataka State over the use and appropriation of the Cauvery River's water resources with neighbouring Tamil Nadu (BBC, 2005, 2016).

Thus, while water treaties, hydrological data-sharing arrangements and technical exchanges – as manifest in the mandates and activities of such basin organizations as the Mekong River Commission (MRC) and the Indus River Commission – may indicate a degree of concord at the interstate level, this does not necessarily translate into a lack of discord at other levels of governance. Water-related disputes over hydropower development on the Lao PDR's section of the Mekong River's mainstream, among other examples, demonstrate how the relatively tempered responses on the part of the MRC and other Mekong governments to Laos's ambitious Xayaburi hydropower project rest uneasily with the sustained opposition mobilized by a broad-based network of river activists. With traditional development discourses espousing the necessity of hydropower dams for electricity production readily accepted by many governments in developing Asia (McCormack, 2001), many of the 'high-profile' disputes over transboundary rivers in the region have largely been driven from the ground up.

The significance of (trans)local forces in resisting large hydro-development projects on Asia's major transboundary rivers thus needs to be critically examined. The tendency to centre analyses of transboundary water disputes on interstate rivalries and power asymmetries between upstream and downstream states, while insightful, risks glossing over the instrumental role played by activists in fomenting contestation from below. As evinced by the Nu-Salween and Brahmaputra cases, the regional and national debates that have emerged over water security can be traced back to more localized concerns over equitable use, water quality, environmental degradation and livelihood loss. Assuming an especially prominent role in issue creation – that is, in catalyzing and framing the political nature of water disputes – river activism can contribute to policy entrepreneurship at the national and transnational levels, with cascading consequences for regional water governance.

In this light, contestation becomes desirable for policy change and, by extension, potentially constructive for transboundary water cooperation in the long term. By defining hydropower development as a 'problem' worthy of public concern, and one that impinges on governing legitimacy and national interests, policy entrepreneurship arising from the contestation stoked by river activists can trigger procedural changes that promise to enhance the reach and effectiveness of transboundary water governance efforts through the inclusion of a wider array of stakeholders. Crucially, as this article demonstrates, this can occur in countries irrespective of the nature of their political system (i.e. democratic or authoritarian).

The article proceeds in three parts. The first section explores the role of river activists in policy entrepreneurship. An analytical framework is advanced here to account for the dynamics of issue emergence and policy change, as well as the importance of contestation for better governance outcomes.

The second section considers, in turn, how policy entrepreneurship unfolded in the Mekong, Nu-Salween and Brahmaputra cases, and with what implications for official policy. These three cases were selected for the following reasons. First, each involves a major international river in the region that is shared by two or more developing states. This adds a complex transboundary dimension to the high-profile disputes over water use in these basins, which should render cooperation and policy change more difficult to attain given the contending interests at stake. Second, all three cases feature large-scale hydropower development schemes that reflect the vested interests of governments and the private sector. These projects have received high-level support from the relevant governments, as they align with prevailing development aspirations (Magee, 2006). With dam developers having invested heavily in these projects, there is also strong opposition on the part of these actors to any prospect of project derailment, such that they could be regarded as 'anti'-policy entrepreneurs themselves. The Lao government and Ch. Karnchang have, for example, pursued the Xayaburi's development despite public censure, whereas China's Huadian Yunnan Corporation and the Yunnan provincial government have continued to lobby for the restart of the Nu dam cascade (Li, 2013; Wong, 2016). Third, sustained civil society–led activism against these projects is ongoing and evident across the cases. Activism in the Nu-Salween case surfaced in late 2003; in the Brahmaputra in 2009; and in the Mekong/Xayaburi in late 2010. These are not cases where activism is the short-lived afterthought of a fashionable conscience. Taken together, these three river basins serve as 'hard test cases' for analyzing how river activism can result in policy entrepreneurship, and the conditions under which policy change may occur.

The concluding section discusses the implications of greater civil society agency for regional water governance.

Caveats

Three caveats warrant note here. First, this article does not seek to evaluate the veracity or accuracy of the claims put forth by the civil society actors examined. Instead, it seeks to understand how such claims have contributed to issue emergence and, consequently, to bringing about a change in policy behaviour. It also focuses more on assessing the processes involved than on absolute outcomes. Central to this inquiry is the question of how river activists are able to act as policy entrepreneurs, framing issues as 'problems' in ways that align with their stated objectives, and that provide impetus for policy change. River activism need not result in the cancellation of a dam or the complete revision of a country's development strategy to be deemed effective. It only needs to generate a perceptible shift in target actors' policy rhetoric and behaviours.

Second, with a vast body of scholarship analyzing the nature of water cooperation and conflict in shared river basins (Giordano, Giordano, & Wolf, 2002; Grey & Sadoff, 2007; Turton, Patrick, & Julien, 2006; Wolf, 1998), the past decade has witnessed a shift in the debate towards more nuanced perspectives that acknowledge how cooperation and conflict constitute "two sides of the same coin of social interaction" (Gulliver, 1971, p. 189). Plotted along the same spectrum, they may coexist at any given time and place (Eidem, Fesler, & Wolf, 2012; Zawahri & Gerlak, 2009; Yoffe, Wolf, & Giordano, 2003).

Even so, the dichotomous language of conflict and cooperation still retains political resonance today. Whether cooperation or conflict prevails over the major rivers that cut across developing Asia (i.e. China and South and South-East Asia) is viewed as an exceedingly important question – one which impinges on regional security and governance outcomes. Accordingly, the myriad challenges posed by Asia's emerging freshwater scarcity and tension over the use of such resources can be seen as either an opportunity for further regional integration, or an imminent source of long-term insecurity. The problem with such views, however, is that they do not sufficiently unpack the socio-political sources of these tensions and consequently neglect how these disputes might have been 'engineered' by certain actors.

Third, this article acknowledges that the private sector (i.e. hydropower companies and financiers) can play a pivotal role as non-state actors in facilitating the pursuit of certain policies and the blockage of others. However, due to space constraints and the fact that the hydropower projects discussed here have been mainly driven by South-East Asian governments and their relevant agencies, the article will focus primarily on river activists as potential agents of change, and state actors as key target actors.

How issues emerge: the role of river activists

A multiplicity of actors exist in transboundary water governance, representing the diverse interests at stake and the plethora of roles to be claimed. In most instances, it is precisely this multiplicity that has contributed to the difficulty of formulating a cohesive framework for the governance of international rivers. The long journey towards ratification of the 1997 UN Watercourses Convention constitutes a potent reminder of this.[2]

Part of the problem arguably lies with the predominant focus in both practitioner and scholarly circles on the role of the state in water governance. Although the principles of sovereignty and territorial integrity imbue governments with primary authority when it comes to river management, there is a need to recognize the inherently fragmented nature of governments and the so-called hydrocracies, especially those in developing Asia, that drive water policy formulation. Indeed, this complex institutional landscape is aptly captured by the idea of "powershed politics" (Magee, 2006). Conflicts of interest, together with contending mandates between rival agencies, have often meant that even in authoritarian and semi-authoritarian regimes such as China, Laos and Cambodia, policy making does not always occur as a top-down, linear process (Yeophantong, 2016). Nor is it the case that decision making on water-related issues takes place in a concealed vacuum, isolated from external influence and other social, political and environmental considerations.

Non-state actors – understood as actors that exist and operate beyond the immediate aegis of the state – have an important, if not always complementary, role to play here. Corporate actors have contributed to the ongoing dam-building boom in South-East Asia, with Chinese financiers (namely, major hydropower state-owned enterprises and policy banks like the China Export-Import Bank) adopting an especially active role in promoting hydropower development in the region and back home (Yeophantong, 2013).

Crucially, in the absence of robust institutions in Asia to coordinate the management of the region's transboundary rivers, coupled with the limited political will and capacity of relevant government agencies to reform the status quo, civil society actors are gaining greater agency within the policy process (Mertha, 2009; Roberts & King, 1991; Schneider & Teske, 1992). As explored in the next section, this is in large part due to their policy entrepreneurship, which has enabled them to take advantage of windows of opportunity to initiate policy change.

Policy entrepreneurship and actor relationships

Under what conditions can weaker actors influence the policies and actions of more powerful actors? The concept of policy entrepreneurship, as developed in the policy studies literature (Crowley, 2003; Lindblom, 1968; Sabatier, 1988), offers a useful framework for analyzing the complex dynamics underlying Asia's transboundary water disputes. That civil society actors can play a key role in pushing forward new thinking to improve water governance outcomes is not a new observation. However, less attention has been directed to their role as policy entrepreneurs – a role usually accorded to water managers (Golan-Nadir & Cohen, 2017) – with the potential to modify or change the rules of the game. Exceptions include Andrew Mertha's (2009) seminal study, which illustrated how the internal aspects of political opportunity (e.g. agency slack or disgruntled officials) can provide a point of entry for policy entrepreneurs. Guobin Yang and Craig Calhoun's (2008) study on the emergence of a 'green public sphere' in China similarly shows how critical media attention – traditional and alternative – can be effectively mobilized by civil society to mount and disseminate opposition.

Because both studies focus on China, this might seem to suggest that policy entrepreneurship is 'unique' to certain country conditions. Mertha (2009), for one, posits that the fragmented nature of China's evolving political system is what allows for increased political pluralization. Yet effective policy entrepreneurship, especially by river activists, should not be construed as a phenomenon unique to any single country or political system. Policy entrepreneurship can – and does – occur across different country and socio-political contexts, as well as on different scales. As the Nu-Salween, Xayaburi and Brahmaputra cases demonstrate, both the internal and external dimensions of political opportunity matter, and river activists in each case can be seen acting either as the 'first movers' in generating new policy contestation, or as creative opportunists in seizing policy windows that enable them to take advantage of emerging areas of public contention (Kingdon, 1995 [1984]).

Contextual factors (e.g. the level of regime repressiveness or political openness) invariably impinge on the freedom of policy entrepreneurs to manoeuvre. However, the fact that river activism has taken root in all three cases suggests that having either a democratic or 'fragmented' authoritarian political system is not a necessary condition for such activism to emerge, but that much may depend on the level of political pluralization as a structural factor and the attributes of the change-agent. As noted by Mintrom and Norman (2009), social acuity and the ability to 'lead by example' are core characteristics of successful policy entrepreneurs. Indeed, to address the thorny question of transboundary environmental harm caused by hydropower development, activists can be seen adopting a range of 'inside-out' and 'outside-in' strategies, which are crucial to targeting diverse stakeholders that frequently operate at overlapping levels of governance.[3] This can likewise be observed in the cases examined in this article, whereby river activists may seek to exert direct pressure on governments or appeal to those within government – usually by way of pre-existing concerns or security interests – to impact policy behaviour.

How state, non-state and international actors interact with one another can, therefore, affect governance outcomes. Focusing on the vocabulary of "interactions" is useful for directing attention to how the language of cooperation and conflict need not be polarizing (Zeitoun & Mirumachi, 2008, pp. 297–316). It also serves to highlight the significance of the "character of relationships" (Avant, Finnemore, & Sell, 2005, p. 3) which exist among these actors, to determining water governance outcomes on multiple scales. In this regard, civil society actors contesting hydropower development on the Mekong and Nu-Salween are able to effect change not by virtue of being 'civil society actors' per se, but more as a result of the position they hold *in relation to* other stakeholders and the interactions that ensue.

The relative authority attributed to activists thus constitutes an important factor that influences their ability to bring about issue emergence and set the agenda. To garner authority in the policy realm, not only do they need to demonstrate leadership and competence in relation to "less capable" others, but they must also convince the general public and governments of the moral exigency and legality of their principled ideas (Avant et al., 2005, p. 13). For river activists in Asia, this usually involves aligning themselves with the interests of 'disadvantaged' local communities and positioning themselves against 'irresponsible' businesses partaking in 'illicit' gains, as well as 'corrupt' governments that disregard the adverse impacts of large dams on local livelihoods and the natural environment. It is on such claims that the disputes over the transboundary waters of the Mekong, Nu-Salween and Brahmaputra are predicated.

Another critical factor that can also contribute to an issue's successfully emerging or disappearing into obscurity is the level of politicization. River activists and their support networks (Keck & Sikkink, 1998) managed to spark widespread contestation in the cases studied here not just by challenging the rationale behind dam development, or informing the public of the negative repercussions of large-scale dams. But they had also succeeded, to varying degrees, in problematizing the inadequacies of extant governing arrangements and acquiring policy relevance through the politicization of their respective causes. Campaigns to stop damming on the Nu-Salween, Mekong and Brahmaputra explicitly appeal to the protection of 'national' and 'strategic' interests (e.g. food security and the stability of ethnically fragile areas). In so doing, activists brought into question the government's capacity and legitimacy to govern by enhancing the reputational costs of nonaction. A 'weak' government, for instance, is depicted as one that remains silent on upstream dam-building by another state, forgoes the country's water-use rights and, by implication, jeopardizes the country's territorial integrity.

The desirability of contestation for policy change

Incrementalism is a fundamental characteristic of the policy-making process (Colebatch, 2002), such that major policy change is unlikely to happen within a day. Indeed, the risk aversion and cautious political posturing of most governments is reflected in the early responses provided by the Cambodian, Vietnamese and Thai governments to the MRC in 2011 in regard to the proposal to construct the Xayaburi Dam (Mekong River Commission, 2011a, 2011b, 2011c). Although Cambodia and Vietnam stand to be the hardest-hit in terms of the Xayaburi's ecological repercussions, the official Cambodian and Vietnamese responses were visibly worded in 'diplomatic speak' (e.g. Vietnam "expressed appreciation" to Laos for its "strong commitment" to the MRC's Procedures for Notification, Prior Consultation and Agreement), presumably so as to not overly antagonize their close neighbour (Mekong River Commission, 2011a, 2011c). The role of policy entrepreneurs – or rather, river activists in this particular case – therefore becomes all the more critical under such conditions.

While conflict is traditionally viewed as a major barrier that undercuts the integrity of governing arrangements and reduces opportunities for cooperation (Avant et al., 2005), contestation can be desirable when it generates new policy ideas and the opportunity for policy change. Indeed, a lack of contestation does not always indicate cooperation or concord, and a false sense of consensus can be counterproductive to inclusive water governance in the long term. Under repressive regimes, for example, little or no contention may emerge due to 'gate-keepers' that prevent certain issues from entering the public policy agenda (Carpenter, 2007). The situation in Laos is instructive here, as internal dissidence is often muted as a result of the country's punitive legal system, which effectively deters whistleblowing behaviour. Even so, debates over sensitive 'water issues' in Laos have not completely ceased, but are taking place 'underground' or discreetly through the subversion of government discourses on development and national interests (interview, Vientiane, Lao PDR, 8 May 2013).

Furthermore, despite the tendency to treat cooperation as a means to better water governance, if not an end in itself, cooperation might not always be desirable or even the best way forward. Given the 'anarchic' nature of transboundary water governance at the international and regional levels, cooperative efforts are likely to take place under less-than-favourable conditions. For example, under conditions of hydro-hegemony, cooperation could result in the entrenchment of asymmetrical power structures that only work to breed further problems (Zeitoun & Warner, 2006). In the case of hydropower development, interstate cooperation could serve as a guise for collusion, in which converging interests lead governments and businesses to prioritize dam-building in spite of social and environmental risks. In other circumstances, cooperation in the form of agreements and treaties may amount to symbolic gestures aimed more at mollifying an irate public, with little practical value in terms of actual enforcement.

On this view, riparians should focus more on producing optimized outcomes than on achieving absolute cooperative gains. They need to appreciate that disputes can give rise to (unintended) positive ramifications. A degree of contestation and competition between actors could thus be desirable for the sake of instigating policy change, if it does not lead to inaction or protracted violence. This is evident in the three cases considered here. In the Xayaburi case, public concern over the social and environmental consequences of Laos's mainstream dam provided a firm basis for the relevant South-East Asian governments and the MRC to exert pressure on the Lao government to reconsider the project. In the Nu-Salween example, pressure from activists eventually persuaded the Chinese government to suspend, in an unprecedented manner, the construction of the proposed dam cascade. Especially under restrictive-state conditions, opposition and contestation may well constitute a 'healthy' response that enables more inclusive stakeholder participation in the management of shared water resources.

Hydropower development and water disputes in Asia

Over 75% of Asia-Pacific countries are estimated to suffer from some degree of water insecurity (ADB, 2013). International rivers have been some of the more prominent sources of such insecurity. With at least 53 transboundary river basins in the region, Asia has been considered a likely site of the next decade's water wars; there appears to be no end to such grim predictions of imminent conflict and severe instability (Bhalla, 2012; Chellaney, 2013; *The Economist*, 2012).

Hydropower development is a major source of water disputes in the region. The ecological and social impacts of large dams, in particular, tend to be prime sites of public contention, especially at the local level. Dam-building often brings into relief the stark power asymmetries between upstream and downstream riparians. It highlights the contending interests that can impede collective water-sharing efforts and exacerbate governance gaps. Across developing Asia, large-scale dam construction has proliferated in tandem with economic modernization discourses, and the desire of the region's governments to capitalize on hydroelectricity – purportedly a cheaper and cleaner energy alternative to oil and fossil fuels. China's cascade of upstream dams on the Lancang-Mekong, for example, has been the subject of long-standing controversy. The cascade has been variously accused of lowering the Mekong's water quality by reducing sediment flows downstream, and causing erratic fluctuations in water levels, resulting in severe droughts and flooding. A similar list of grievances characterizes downstream reactions to upstream dam construction on other transboundary rivers in the region.

It deserves note, nevertheless, that these grievances do not 'naturally' emerge as a result of a large dam project. Put differently, the announcement of a dam project does not necessarily ensure that contestation, issue emergence and policy change will follow. As the following analysis reveals, the socio-political contestation and policy entrepreneurship that have come to characterize plans to dam the Mekong, Nu-Salween and Brahmaputra rivers are largely driven by river activists and their supporters, whose entrepreneurial flair for policy change has shaped the course of these disputes.

The Nu-Salween River and dam cascades in China and Myanmar

The Nu-Salween River is known as one of the world's last free-flowing rivers, coursing through the territories of China, Myanmar and Thailand. The livelihoods of approximately six million people in three countries rely on the ebb and flow of this major river system (Magee & Kelley, 2009). Its watershed also nourishes some of the world's most ecologically and culturally diverse areas, as it passes through China's pristine Three Parallel Rivers World Heritage site.

Plans to dam the upstream reaches of the river (i.e. the Nujiang) in China were proposed as early as 1995. Similarly, downstream in Myanmar, talk of damming the lower stretches of the Salween had taken place since the 1990s. However, it was only in the early 2000s, when proposals to dam both the upstream and downstream sections of the river became public, that domestic and transnational advocacy campaigns were galvanized across the region to protect the Nu-Salween from hydropower encroachment.

Several accounts exist of how the issue of damming the Nu-Salween River entered the Chinese public sphere and became framed as a national problem. Some place emphasis on the role of Chinese civil society or the media (Mertha, 2009; Yang & Calhoun, 2008); others on transnational civil society (Litzinger, 2007) or political wrangling behind the scenes in Beijing and Yunnan (Magee & McDonald, 2011). The present account seeks to reconcile these different versions of the same story, so as to better spotlight how policy entrepreneurship emerged in this case.

In China, the Nu River hydropower project was first approved by the National Development and Reform Commission in 2003. The scheme was slated to involve the construction of a 13-dam cascade with an estimated total power generation capacity of 21 GW. The proposal soon ran into opposition, however. Although local officials had supported the project in light of the central government's Great Western Development (*xibuda kaifa*) plan, Chinese intellectuals, along with civil society and the State Environmental Protection Agency (SEPA),[4] had expressed early on their reservations over the dam's negative ramifications for the surrounding environment. These reservations also emerged at a time when then Premier Wen Jiabao was becoming concerned with the growing political power of large state-owned enterprises within the country's energy sector.

In fact, civil society actors first learnt of the project via concerned officials within SEPA who, fearing that the agency was less 'powerful' than the National Development and Reform Commission, had wanted to raise public scrutiny on the issue in order to slow down the cascade's development. Environmental journalist and activist Wang Yongchen, who led the Beijing-based Green Earth Volunteers, was reportedly asked by a "friend" within SEPA to gather "reinforcements" (*yuanjun*) (Cao & Zhang, 2004). To do so, she reached out to other activists and supporters of China's nascent environmental movement, leveraging an expanding advocacy network that included water experts and supported by popular media outlets such as the *China Youth Daily* (Yang & Calhoun, 2008). This outreach strategy would subsequently culminate in the SEPA-sponsored Nujiang Hydropower Development Expert Forum in Beijing, where intellectual figures like He Daming were invited to give talks on the Nu dams. Crucially, expert appraisal of the project's damaging social and ecological repercussions not only attracted further media attention, but also helped catalyze social opposition to the Nu hydropower project (*Xinjing Bao*, 2003).

Other civil society organizations in China, including well-regarded groups like Friends of Nature and Green SOS, were soon prompted to join the growing anti-Nu-dam campaign. Yu Xiaogang of the Kunming-based Green Watershed, for one, proceeded in late 2003 to conduct a survey of the dam site to assess the project's anticipated impacts. Having previously studied the range of problems associated with the Manwan Dam,[5] he also helped raise awareness among villagers exposed to the Manwan's cascade's social and ecological risks by arranging a trip to the dam's site, where official neglect had led some communities to resort to scavenging in order to make a living (Yardley, 2005; see also Yu, 2004). These developments were further bolstered by the organization of the third NGO Forum on International Environmental Cooperation in November 2003 (http://www.ifce.org/about), where the subject of the Nu River dams coalesced around discussions over other exigent environmental issues.

It is in this sense that civil society was responsible for firing the "first shot" (*di yi qiang*) (Cao & Zhang, 2004), which would subsequently trigger a groundswell of resistance to the scheme among the Chinese public, as well as in certain quarters of the Chinese bureaucracy (e.g. SEPA officials). That said, this opposition was not wholly confined to China but spilled over into neighbouring countries as well. Civil society groups in Thailand, Myanmar and Cambodia such as the Southeast Asia Rivers Network, alongside transnational NGOs such as International Rivers and the World

Wildlife Fund (WWF), had contributed to issue emergence at the regional level through the use of a range of advocacy tools. These included the organization of a series of international conferences in Thailand, as well as the inauguration of joint signature campaigns to 'save' the Nu River – one of which saw signatures from a total of 140 NGOs being delivered to UNESCO and the Chinese Embassy in Bangkok.[6] Notably paralleling these transnational efforts were those of Green Earth Volunteers and other partner organizations, which had collectively spearheaded a successful celebrity petition campaign that garnered national media coverage (Chinese Ministry of Environmental Protection, 2007).[7]

Not only did these campaigns feature prominently in the Chinese media, but they also served to expand the anti-dam movement's support base. Of particular interest is how through the support of well-known public figures, intellectuals and external organizations (e.g. UNESCO), the credibility of Chinese river activists working on the Nu River dams came to be based upon their demonstrated expertise, capacity to lead and principles. Findings from in-depth studies conducted by Chinese scientists on the repercussions of the dam cascade, for example, contrasted sharply with the rhetoric of local Yunnan officials, many of whom had demonstrated little knowledge of the project beyond its basic components and rationale (i.e. being for the sake of poverty reduction and energy production) in public interviews (Mertha, 2009). Cultivating such relative authority is central to policy entrepreneurship, and to engendering compelling problem definitions and instigating successful issue emergence in the public sphere.

At a time when the anti-dam movement was gaining momentum, the country's inaugural Environmental Impact Assessment Law was released in December 2003. This provided additional grounds for anti-dam activists to contest the scheme not only on moral and environmental grounds but on legal terms as well. As such, while the Chinese state-owned enterprise responsible for building the cascade – Huadian Nujiang Hydropower Development Corporation Ltd. – actively denied allegations of the scheme's serious social and environmental costs, efforts on the part of activists like Liang Congjie, who submitted petitions directly to the National People's Congress and the Chinese People's Political Consultative Conference, helped ensure that social dis-approbation of the project would gradually filter into China's elite policy-making circles. These efforts ultimately culminated in the unprecedented announcement in February 2004 of the project's suspension by then Premier Wen Jiabao (Deng, 2013).[8]

Just as Chinese environmental activists were reportedly influenced by the displays of anti-dam advocacy elsewhere in the region,[9] activism against the Nu dam cascade in China would also contribute to stoking grass-roots dam contestation downstream. Here, river activists in Myanmar became involved in contesting a series of planned dams on the Salween River.[10] The scheme itself is a joint undertaking between Myanmar and Thailand, with at least 60% of the hydropower destined for the Thai and Chinese markets. With news outlets such as the *Bangkok Post* reporting on the successes of anti-Nu dam activists in China, this contributed to emboldening grass-roots civil society groups within and outside Myanmar,[11] who likewise sought to draw public attention to this seven-dam cascade slated to be constructed on portions of the Salween in the ethnically fragile Shan and Karen States. Although information about the project has remained tightly guarded, local protests near the dam sites have been consistently staged since 2005 (Solomon, 2014).

These protests – supplemented by a public-awareness campaign that involved, for instance, the delivery of an open letter and petition to the Thai prime minister – were further emboldened by support from a network of transnational NGOs and journalists, as well as ongoing opposition from the Karen National Union (*Democratic Voice of Burma*, 2014) and the Thailand Human Rights Council, the latter having attempted to pressure the Electricity Generating Authority of Thailand (EGAT), one of the project's investors, into abandoning construction of the Hatgyi Dam back in 2006. Crucially, in 2007, as a result of attacks on the Hatgyi Dam site (*The Nation*, 2007), a temporary suspension was placed on the scheme (Naing, 2013; see also Noreen, 2011).

In light of sustained public censure of the Salween dams, EGAT reportedly suspended two of the planned dams in 2008: the Weigyi and the Dagwin (Solomon, 2014). The Nu-Salween case constitutes an important example of how both 'outside-in' and 'inside-out' policy entrepreneurship can be effective in engendering domestic as well as cross-border policy changes. Even now, activism against dams on the Salween continues. But while the accomplishments of activists have at times been eclipsed by the progress made by dam developers, their advocacy efforts have managed to bring under scrutiny a series of large and potentially devastating hydropower dams – schemes which could have otherwise fallen under the public radar. The new focus in China on setting up a national park and turning the Nu River into an international eco-tourist destination further suggests the lasting impression river activists have made on Yunnan's provincial leaders (Leavenworth, 2016).

The Mekong River and the Xayaburi Dam in Laos

The Mekong River flows through the territories of six countries – Cambodia, Laos, Myanmar, Vietnam, Thailand and China (Yunnan Province) – and is frequently dubbed the region's "lifeblood" (Tacon, 2013), as its waters and fisheries sustain the livelihoods of the 60 million people inhabiting the river basin. However, reminiscent of the story of the Nu-Salween's development, a slate of hydropower development schemes is proposed for both its upstream and downstream sections.

In 2007, the Lao government signed a memorandum of understanding with Ch. Karnchang, a large Thai infrastructure developer, for the construction of the 1285 MW Xayaburi Dam in the northern Lao province of Xayaboury.[12] By late 2010, project development by Ch. Karnchang was well underway, with financial backing from a syndicate of major Thai commercial banks that included Bangkok Bank, Siam Commercial Bank and Kasikorn Bank. EGAT had reportedly agreed to purchase approximately 95% of the power generated, with some also destined for the Chinese market (International Rivers, 2016). But although the project is intended to feed into the Lao government's broader strategy of transforming the country into the 'battery' (*mo fai*) of South-East Asia, the expected perverse consequences of the large dam have since undermined these aspirations, serving also as cause for considerable alarm among local communities and civil society actors both upstream and downstream on the river.

While other member states of the MRC were aware of the project by early 2010, it would only be later in the year that Laos's plans became more widely known to the general public. The MRC's prior consultation process on the Xayaburi served as the initial 'entry point' for activists (Walker, 2011). By early 2011, the Xayaburi Dam had become the site of intense bottom-up contestation, with a regional network of activists coming into existence to dispute the dam and its rationale. Environmental activists noted that most of the electricity was to be exported to Thailand; that the large dam would upset fish migration patterns, including those of the endangered Mekong giant catfish; and that the Mekong Delta in Vietnam and Tonle Sap in Cambodia stood to be adversely affected by reductions in water and sediment flow (WWF, 2014). The publication of an investigative report by the *Bangkok Post* in mid-2011 fueled further "outrage" among activists (International Rivers, 2011), as it revealed that preliminary construction activities were being undertaken despite the lack of MRC approval, with villagers also being resettled and given poor compensation (Wongruang, 2011). Here, attention was directed to how Laos, as a member of the MRC, was contravening the 1995 Mekong Agreement's procedural rules on notification, prior consultation and agreement, having failed to undertake a proper environmental impact assessment prior to commencing work on the dam as well (International Center for Environmental Management, 2011).

Given the highly restrictive nature of Laos's political space, river activism on a transnational scale in Cambodia, Thailand and Vietnam has contributed significantly to disputing the legitimacy of the Xayaburi Dam. As in the Nu-Salween case, the public opposition to the dam not only contributed to projecting the social and environmental concerns associated with the dam into the regional public sphere, but it also reinforced the issue's saliency in the policy agendas of the Thai, Vietnamese and Cambodian governments. Whereas early responses to the dam project were fairly tempered (Mekong River Commission, 2011a, 2011b, 2011c), official criticism from these governments of the Xayaburi would become more strongly worded and explicit as local resistance in these countries concomitantly grew (Chen, 2013; Choonhavan, 2014a).

Collectively, Thai-based NGOs and local civil society groups such as Towards Ecological Recovery and Restoration, Living Rivers Siam, EarthRights International and the newly established Mekong Community Institute were instrumental in the galvanization of mass opposition to the dam project. In addition to petitioning Ch. Karnchang and Thai financiers to reconsider their involvement in the project, river activists raised questions over the nature of the power purchase agreement between EGAT and the Lao government, with the Senate Standing Committee on Good Governance Promotion and Corruption Investigation calling for a transboundary impact assessment of the dam (EarthRights International et al., 2014). In late 2012, 30 Thai villagers, through the leadership of the Chiang Khong Mekong Conservation Group and the Community Rights Information Center, and aided by EarthRights International's legal advice, filed a civil lawsuit against EGAT and the Thai Cabinet questioning the legality of the power purchase agreement (Radio Free Asia 2012; 2014).[13] This development received considerable media coverage and support from political figures, including Kraisak Choonhavan, not least because their lawsuit was accepted by the Thai court (interview with Teerapong Pomun, Bangkok, Thailand, 21 January 2016).[14]

In prior months, activists had also come together to demonstrate outside the annual MRC summit, and subsequently met with the MRC's then-CEO Hans Guttman. This was followed by a series of protests in Nong Khai, at the Ch. Karnchang office, and at the Lao embassy, among other venues. In 2014, a joint Declaration of Solidarity was published and signed by 140 NGOs in the region, requesting that the Thai government cancel its power purchase agreement with Laos. The sense of solidarity that arose from this campaign notably helped pave the way for the establishment of the Network of Mekong-Loving Communities and the Council of Mekong Community Organizations, both of which continue to be involved in mobilizing popular opposition to the Xayaburi dam.

Across the Thai border, prominent Cambodian grass-roots organizations like the River Coalitions of Cambodia and Phnom Penh–based NGO Forum played a key role in pushing for the dam's cancellation. Aside from spearheading training workshops to enhance civil society capacity, they also helped organize information sessions and awareness campaigns to educate communities about the risks posed by large dam construction. These initiatives attracted media interest, as well as support from the Cambodian government (Chhith Sam Ath & Marks, 2013), with Prime Minister Hun Sen issuing a formal protest of the Xayaburi project in 2012. Not only did the Cambodian government demand that construction of the dam be suspended until an environmental impact assessment is completed, it also threatened international legal action should the Lao government fail to follow the Mekong Agreement's prior con-sultation and data-sharing protocols (*Asia Times*, 2012).

Similarly, since mid-2012 Vietnamese scientists and organizations, including the Vietnam Union of Science and Technology Association and Vietnam Rivers Network, have been a notable force in prompting official scrutiny of the Xayaburi's anticipated adverse impacts. They had urged the Vietnamese government and the National Mekong Committee to join Cambodia in protesting directly to the Lao government against the dam (Choonhavan, 2014b). A joint letter, signed by Prime Ministers Nguyen Tan Dung and Hun Sen, was later sent to Laos and Thailand, exhorting Laos to extend the consultation time and await further review of the dam's anticipated impacts on the Mekong's mainstream, in accordance with the MRC's strategic environmental assessment report (International Center for Environmental Management, 2010). Moreover, shortly before the MRC summit in 2014, a day-long, multi-stakeholder workshop on the Xayaburi's effects was organized by the People's Aid Coordinating Committee in Ho Chi Minh City, and attended by scientists and National Assembly agencies, among others (Choonhavan, 2014b).[15]

With Thai river activists focusing on the legal aspects of the issue, Cambodian and Vietnamese activists emphasized the dam's threat to regional stability. Appeals were made mainly in light of food security concerns – specifically, how the Xayaburi Dam places Cambodia's fisheries and the livelihoods of at least 20 million people in the Mekong Delta at risk (Henderson, 2013). Framing the issue this way gives it political resonance, as food security concerns have intensified in recent years due to extreme climate conditions, having also become linked to possible social instability. Cambodian commentators have also highlighted that the Xayaburi stands to impact the country's own prospective dams on the Mekong mainstream – a point which has resonated strongly with the Cambodian government (interview, Phnom Penh, 29 January 2016). In this way, the 'outside-in' policy entrepreneurship of river activists has contributed to the creation of politicized issue frames that spotlight the Xayaburi Dam as a grave national and regional security concern, and which position these activists as stewards of the Mekong River and its resources.

In response to external pressure, in May 2012, the Lao government officially con-firmed the Xayaburi's (short-lived) suspension, having also contracted the Compagnie Nationale du Rhone to conduct independent reviews of the project's initial impact assessment report. It also invited for the first time in August Cambodian, Vietnamese and Thai officials to visit the dam site. But despite the Lao government's promising to build a transparent and modern dam, it has proceeded with the Xayaburi's construction without approval from the other MRC member countries. And despite activists' objections to Poyry's design and environmental impact assessment of the dam,[16] the company continues to be responsible for checking the project's compliance with the MRC's recommendations. That said, although activists have had a more limited impact on the Lao government and Poyry, it is interesting to note that the recent announcement of the MRC's impending decentralization and donor funding cut to its budget has been linked to discontent with the organization's inability to speak out against controversial dam projects like the Xayaburi and enforce member states' compliance with its procedural rules (Kossov & Samean, 2016). Here, policy entrepreneurship has evidently resulted in indirect policy outcomes, and in the perceptible shift in the discursive behaviour of the Cambodian and Vietnamese governments.

The Brahmaputra River and the dam cascade in China

The importance of the Brahmaputra River – or Yarlung Tsangpo, as it is known in China – lies not only with its cultural and religious significance, but equally with the vital ecosystems and downstream livelihoods it sustains (South Asia Network on Dams, Rivers and People, 2013). Yet, with disputes over Chinese upstream dam-building on the Yarlung Tsangpo–Brahmaputra River focusing on the security implications of such activities for Sino–Indian relations, the key role assumed by non-state actors in India in mobilizing popular opposition and disputing China's hydropower schemes has largely been ignored.

One cannot, of course, overstate the importance of security concerns in this instance: media reports of Chinese plans to divert 40 km^3 of water from the Brahmaputra stirred up considerable alarm in Indian policy-making circles back in 2003 (Nandy, 2013). However, according to a leaked US cable dated November 2006, although the Brahmaputra issue was raised during a 'readout roundtable', hopes of expanded Sino–Indian engagement following then-President Hu Jintao's visit to India over the issue had been dashed, with one scholar noting that the visit was "a disappointment for Northeast India" (Wikileaks, 2006). During this time public awareness of, and opposition to, Chinese hydro-development on the Brahmaputra was relatively limited. High levels of public disapprobation of China's six-dam cascade on the river's upper reaches would only emerge in more recent years due to the activism of dam opponents based primarily in north-eastern India and their coalition of supporters, which includes political figures, journalists and public intellectuals.

Official discourses frequently emphasize that the government of India (GoI) has been cooperating with its Chinese counterparts on the management of the Brahmaputra through hydrological data-sharing,[17] as well as dialogue via the Expert Level Mechanism (*Times of India*, 2009). In fact, shortly after Chinese plans to dam the Yarlung Tsangpo–Brahmaputra had first come to light in 2009 as a result of a report produced by the National Remote Sensing Agency, then Minister of External Affairs S.M. Krishna sought to allay mounting public concern by saying that the Chinese project was "small", with no impact on the flow in north-eastern India (*Indian Express*, 2010). Moreover, once public concerns over Chinese dam-building evolved into fears of a bigger Chinese diversion project on the Brahmaputra, Krishna swiftly responded that no diversion scheme was found to have been planned by China (ibid.).

But with a growing number of Indian scientists, scholars and policy analysts, such as Sreemati Chakrabati of Delhi University and Ramaswamy Iyer of the Centre for Policy Research[18] (Ramachandran, 2015), speaking out against the ecological ramifications of

Chinese dams, together with national media outlets like the *Times of India* criticizing the government for putting the livelihoods of people in north-eastern India at risk,[19] the GoI and the minister soon came under intense fire for their perceived inaction and willingness to accept Beijing's assurances that its planned upstream dams would cause no adverse impacts downstream. Allegedly turning a "Nelson's eye" to Chinese plans (*New Indian Express*, 2011), the GoI was depicted as having failed to press Beijing for additional information, compromising India's national interests and its water-use rights as a downstream riparian in the process (*Times Now*, 2011).

Hence, contrary to depictions of Sino–Indian tensions over the Brahmaputra as derived from interstate rivalry, much of the contestation over China's upstream dams on the Brahmaputra has emerged from prominent activist organizations in north-eastern India. By 2010, the All Assam Students Union and the influential Krishak Mukti Sangram Samiti, a farmers' rights movement led by Akhil Gogoi, had set in motion a broad-based resistance movement.[20] This fed into the broader regional politics of Arunchal Pradesh, which since 2006 has seen the staging of a series of campaigns against a proposed mega-dam network to be built in the area (Sharma, 2012). Both organizations were especially active in contesting the 2000 MW Lower Subansiri hydroelectricity project, which is expected to have deleterious social and ecological repercussions on surrounding communities and on the Brahmaputra River (*Indian Express*, 2014). The fact that their resistance discourses were framed in the language of social injustice and political exploitation helped further galvanize public support for their cause. The student union Asom Jatiyatabadi Yuba Chatra Parishad, for instance, called upon the GoI to establish a neutral area in Arunachal Pradesh, lest the central government be seen as neglecting the country's north-eastern areas (Talukdar, 2014).

It would be this same language, supplemented by a recurring narrative of safeguarding national security, which was subsequently used to draw out public ire and opposition to China's upstream dams. This approach speaks to a key characteristic of effective policy entrepreneurs: their ability to convince target policy makers to pursue a certain course of action by reducing perceptions of associated risk (Golan-Nadir & Cohen, 2017). The Brahmaputra case illustrates that the reverse can also be true, with activists seeking to heighten the perception of risk from *inaction* (i.e. jeopardizing national interests and internal cohesion) as a means to pressure the GoI into modifying their political posture vis-à-vis China on the issue.

The announcement in 2013 of Chinese plans to construct three more hydropower dams on the middle sections of the Yarlung Tsangpo–Brahmaputra as part of the State Council's new Energy Development Plan (Krishnan, 2013; Walker, 2013) triggered scathing responses from both the All Assam Students Union and the Krishak Mukti Sangram Samiti, the latter in particular demanding that "bold steps" be taken by India's central and state governments to "protect Assam interests" (*Assam Tribune*, 2013). The president of the Bharatiya Janata Party, Rajnath Singh, has similarly urged India to strongly protest China's plan and not "soft-pedal" the matter (John, 2013). Notably, these calls to action were quickly followed by a strongly worded statement from the Minister of External Affairs to "remind" Beijing of India's "established user rights to the waters of the river" (qtd. in *Economic Times*, 2014). In late 2014, protests were held in Assam against the new Chinese proposal and the operation of the cascade's first hydropower station (Zangmu, 540 MW), with demonstrations also organized in front of the Chinese Embassy in New Delhi to call upon the Chinese government to suspend the newly built dam (Karmakar, 2014; *Telegraph India*, October 16, 2014; November 24, 2014).

Just before the Zangmu's operationalization in November 2015, Assam Chief Minister Tarun Gogoi accused Prime Minister Narendra Modi of committing a "grave injustice" to the state and its people by failing to raise the issue of the Brahmaputra during his visit to China (*Economic Times*, May 19 2015). This matter was later raised by the Rajya Sabha (Council of States), to the effect that the central government has been placed under constant pressure and public scrutiny to ensure that "India's interests are not harmed by any Chinese activities" upstream (*Economic Times*, December 7 2015).

Support for the grass-roots anti-dam campaign in Assam has also come from the outside. International Rivers has helped build the capacity of Assam civil society groups, while Chinese activist Wang Yongchen has given interviews with the Chinese media warning the Chinese government of the impact of extensive dam-building on the surrounding area's fauna and flora (Gupta, 2010). The Dhaka-based Asia-Pacific Forum of Environmental Journalists has likewise raised concerns over Chinese upstream dam-building, asking Beijing to disclose more information about its dams' ramifications – though not much attention has, as yet, been given by the Bangladeshi government to this issue (*Times of Assam*, 2012).

Despite eliciting a tempered response from the Chinese government, which has reiterated its "responsible attitude towards cross-border river development" (*China Daily*, 2013), river activists – notably assisted by high-profile political figures and parties in the country – have been central to heightening the public saliency of the Brahmaputra issue and to its strategic framing as a national problem. Highlighting the potentially adverse effects of Chinese dams on the Brahmaputra's hydrological flows and the water security of local communities, together with the 'illegality' of Chinese upstream dam-building (i.e. it undermines India's downstream water-use rights), they have not only sought to occupy the moral and legal 'high ground', but have also contributed to the dispute's politicization since 2006. Even now, the issue attracts public speculation regarding an inevitable water war between the two regional powers (Ramachandran, 2015). How the GoI responds to Chinese hydropower plans has, therefore, become increasingly linked to its domestic governing legitimacy, particularly over the country's politically and ethnically fragile north-eastern areas, as well as to its ability to act as a regional counterweight to a rising China.

That said, it warrants note that the GoI continues to assert that the Chinese dams are run-of-the-river dams, and that they are expected to have limited effects on the Brahmaputra's hydrological regime (Denyer, 2013; Dikshit, 2013; *The Hindu*, 2013). Why the GoI appears to harbour little political will to change its policies towards China on this matter is closely related to why it has been so keen to assuage discontent over Chinese hydropower plans, given the considerable criticism it attracts in the process. There are two possible explanations. First, the GoI remains wary of the potential for nationalistic outbursts to result in domestic instability, especially in the country's north-eastern region, and increased tensions with its northern neighbour. Indeed, New Delhi remains cautious of overly upsetting Beijing (Borah, 2016), and while this is unlikely to be the GoI's only consideration, China remains one of India's largest trading partners. Between 2000 and 2012, bilateral trade in goods between the two countries increased more than 20-fold (Madan, 2013).

Second, the GoI has its own controversial plans to dam the Brahmaputra (Vidal, 2013). Despite being widely criticized, the GoI remains committed to completing the construction of the Lower Subansiri project (*Times of India*, 2012). This places the GoI in a difficult position, as over-criticism of China's hydropower plans could lead to its own "rhetorical entrapment" (Schimmelfennig, 2001) by vindicating the anti-dam discourse (interview, Hong Kong, 27 May 2016) and providing Bangladesh with a similar basis to oppose its plans (Khadka, 2014). What this suggests, in effect, is that while policy entrepreneurs in the Brahmaputra case have succeeded in securitizing the problem within public and parliamentary circles, these framing efforts have yet to provide sufficient impetus to New Delhi (and Beijing) to substantively change its policy stance.

Discussion and conclusion

Although the central role played by non-state actors in managing transboundary waters is now widely acknowledged, with emphasis often placed on the necessity of mainstreaming participatory and inclusive mechanisms, the importance of these non-state actors as policy entrepreneurs and their impacts on prevailing water governance dynamics remain under-theorized. Focusing on the role of river activists, this article has illustrated how these evolving policy entrepreneurs manoeuvre through complex and often restrictive political terrain to strategically frame hydropower development on the Mekong, Nu-Salween and Brahmaputra Rivers as a source of dispute, and consequently bring into question the viability of existing governance arrangements. Seizing windows of opportunity at different intervals in the policy process, these actors have employed methods of contestation to create issue frames that captured public interest and, in so doing, helped set the policy agenda by exerting pressure on target actors (i.e. governments and relevant companies). By building coalitions of diverse supporters, they were able to share resources and expertise, as well as gain credibility by presenting a unified front to the general public (Mintrom & Norman, 2009). Activism against China's upstream dam-building on the Brahmaputra constitutes an instructive example of how organizations in north-eastern India worked to politicize upstream Chinese dam-building as a national problem, enhancing its policy resonance at home and across the border.

Neither the Nu-Salween, the Brahmaputra nor the Xayaburi example reveals an absolute 'victory' for civil society actors – that is, the termination of the dam project. Still, their procedural achievements remain noteworthy. Not only did river activism in the Nu-Salween case bring into relief the tenuous link between official discourses on hydropower development and socio-economic growth, it also prompted the Chinese government to direct more attention to the consequences of large-scale dam construction. Through the policy entrepreneurship that transformed hydro-development on the river into a social as well as transboundary water issue, activists contributed to opening up more political space for civil society (Mertha, 2009), and bringing stronger public scrutiny to dam-building activities in the country – as was later seen in the successful case of Tiger Leaping Gorge (Liu, 2013).

In the Brahmaputra case, opposition to China's hydropower plans was effectively linked by activists in north-eastern India to deeper concerns over socio-economic injustice and the country's own 'unsustainable' dam projects. While engaging in counterfactual hypothesizing is often a precarious enterprise, it can be argued that without the heated contestation from these river activists, China's dam-building on the Brahmaputra is unlikely to have become as prominent an issue as it currently is in Indian policy debates – a notable achievement, even if it has yet to spawn substantive policy changes.

These observations point to another contribution of river activism: its ability to inaugurate new opportunity structures through coalition building. As mentioned, the campaign against the Xayaburi Dam bolstered the creation of community networks, such as the Council of Mekong Community Organizations within Thailand, and the Network of Mekong-Loving Communities. These community structures offer informal venues where, as Leifeld and Schneider (2012, p. 3) note, "actors can communicate without incurring significant costs". In other words, tapping into relevant policy and knowledge networks reduces the transaction costs of information-sharing among members. The formation of such coalitions also helps enhance the social acuity (i.e. public impact and issue salience) of these advocacy campaigns. River activism in the Xayaburi case has helped embolden activism against other projects, including Laos's controversial Don Sahong Dam, which now sees Cambodian youth groups and NGOs working to raise local awareness of the dam's impacts (interviews with Oudom Ham and Samnang Heng, Phnom Penh, 29 January 2016).

Building strong institutions to govern transboundary waters is undoubtedly important; yet it is unlikely to be sufficient to ensure the sustainable use of, and equitable access to, finite water resources. Nuanced appreciation of the roles assumed by river activists is crucial to encouraging a much-needed shift in current debates on transboundary water governance away from state-centric perspectives. Considering how much of the discord over the Mekong, Nu-Salween and Brahmaputra stems from local concerns and anxieties, such a shift in focus is needed for the sake of uncovering more workable solutions to the competition over shared water resources seen in Asia today.

Notes

1. By 'river activists', I refer here to international and local NGOs, as well as civil society groups, which focus their activism and advocacy efforts on the protection of rivers and the livelihoods of riparian communities.
2. Vietnam became the convention's 35th contracting state in May 2014, which resulted in its entering into force on 17 August 2014.
3. I am grateful to workshop participants of the 'Water Politics and Regional Stability' workshop, organized by the Lee Kuan Yew School of Public Policy, for stressing this observation to me.
4. SEPA was the forerunner of the Ministry of Environmental Protection.
5. This was the first dam built as part of the Lancang dam cascade on the upper stretches of the Mekong River. With the dam's construction resulting in the mass displacement of communities as well as extensive environmental degradation, this had prompted Yu Xiaogang to undertake a social impact assessment of the scheme in 2002.
6. The one to UNESCO was reportedly signed by a total of 60 NGOs, while the one delivered to the Chinese Embassy featured 80 signatures from organizations based inside and outside China.
7. A total of 62 celebrities signed the petition. The petition itself built upon the momentum gathered from the Green Forum organized earlier by the China Environmental Culture Promotion Association, the country's only 'national social group' specializing in environmental culture.
8. Even so, reports have now surfaced that plans to dam the Nu River are poised to be restarted under the Xi Jinping administration.

9. Wang Yongchen, for one, has cited Thailand's anti–Pak Mun Dam movement as a source of inspiration that made her realize that dams were a part of environmental protection.
10. The Salween River constitutes the lower part of the Nu River that flows through Myanmar.
11. A number of these transnational organizations are located along the Thai-Myanmar border.
12. The dam is slated to be part of a network of 11 large dams.
13. In July 2014, the Thai Supreme Administrative Court agreed to hear the lawsuit brought by the villagers, overruling the previous decision made by the Administrative Court.
14. Although the court subsequently dismissed the case, as of December 2015 it has been with the Thai Supreme Court.
15. The People's Aid Coordinating Committee is an organization mandated by the Vietnamese government to mobilize and coordinate NGO activity in the country.
16. This Finnish company is the engineering firm for the Xayaburi.
17. A memorandum of understanding was signed in 2002 in which China agreed to provide hydrological data during the flood season to India, as a consequence of the earlier massive flooding and landslide. The agreement was renewed in 2008 and again in 2014.
18. This is one of the country's leading public policy think-tanks.
19. For example, in June 2011, the *Newshour* television show organized a panel debate on Chinese dams on the Brahmaputra, where it brought together a cast of policy makers and scholars.
20. Gogoi had risen to popular fame as a result of his high-profile campaign against corruption.

ORCID

Pichamon Yeophantong ⓘ http://orcid.org/0000-0002-4685-5421

References

Asian Development Bank (ADB). (2013, March 13). 3 in 4 Asia-Pacific nations facing water security threat - study. Retrieved from http://www.adb.org/news/3-4-asia-pacific-nations-facing-water-security-threat-study

Asia Times. (2012, May 3). Cambodia pressures Laos to halt work on Xayaburi damRetrieved from http://www.atimes.com/atimes/Southeast_Asia/NE03Ae01.html.

Assam Tribune. (2013, January 31). China to construct 3 more dams on Brahmaputra. Retrieved from http://www.assamtribune.com/scripts/mdetails.asp?id=jan3113/at08

Avant, D. D., Finnemore, M., & Sell, S. K. (2005). *Who Governs the Globe?* Cambridge: Cambridge University Press.

BBC. (2005, January 24). Thousands flee Kenyan water clash. Retrieved from http://news.bbc.co.uk/2/hi/africa/4201483.stm

BBC. (2016, September 13). Why water war has broken out in India's Silicon Valley. Retrieved from http://www.bbc.com/news/world-asia-india-37346570

Bhalla, N. Thirsty South Asia's river rifts threaten "water wars". (2012, July 23). Reuters. Retrieved from http://www.reuters.com/article/2012/07/23/us-water-southasia-idUSBRE86M0C820120723

Borah, R. (2016, April 29). What of India and China relations? *Al-Jazeera*. Retrieved from http://www.aljazeera.com/indepth/opinion/2016/04/india-china-relations-160427112839371.html

Cao, H., & Zhang, P. (2004). Nujiang dam suddenly shelved [due to] civil society power [Nujiang da be turan gezhi muhou de minjian liliang]. *Jingji* [*Economics*], 5. Retrieved from http://finance.sina.com.cn/g/20040520/1615770147.shtm

Carpenter, R. C. (2007). Studying Issue (Non)-Adoption in Transnational Advocacy Networks. *International Organization*, 61, 643–667. Retrieved from http://www.jstor.org/stable/4498160

Chellaney, B. (2013). *Water: Asia's New Battleground*. Washington, DC: Georgetown University Press.

Chen, D. (2013, January 20). Mekong countries at odds over Xayaburi Dam. *Cambodia Daily*. Retrieved from https://www.cambodiadaily.com/archives/mekong-countries-at-odds-over-xayaburi-dam-7943/

Chhith Sam, A., & Marks, D. (2013, March 22). The NGO Forum's Campaign Against Xayaburi. *Stimson*. Retrieved from http://www.stimson.org/summaries/ngo-forums-campaign-against-xayaburi/

China Daily. (2013, February 4). China talks with India on cross-border river issue. Retrieved from http://europe.chinadaily.com.cn/china/2013-02/04/content_16199818.htm

Chinese Ministry of Environmental Protection. (2007, August 14). China environmental culture promotion association. Retrieved from http://english.sepa.gov.cn/About_SEPA/Social_Organizations/200708/t20070814_107912.htm.

Choonhavan, K. (2014a April 30). Vietnam screams for halt to Mekong dams as delta salts up. *The Nation*. Retrieved from http://www.nationmultimedia.com/opinion/Vietnam-screams-for -halt-to-Mekong-dams-as-delta-s-30232520.html

Choonhavan, K. (2014b, June 10). Vietnam demands halt to Mekong dams. *China Dialogue*. Retrieved from https://www.chinadialogue.net/article/show/single/en/7032-Vietnam-demands-halt-to-Mekong-dams

Colebatch, H. K. (2002). *Policy* (2nd ed.). Buckingham: Open University Press.

Crowley, J. E. (2003). *The Politics of Child Support in America*. New York: Cambridge University Press.

Democratic Voice of Burma. (2014, March 17). KNU wants genuine peace before dam construction resumes. Retrieved from https://www.dvb.no/dvb-video/knu-wants-genuine-peace-before -dam-construction-resumes-burma-myanmar/38564

Deng, Q. (2013, Feburary 11). Campaigners re-ignite Nu River dam debate. *China Dialogue*. Retrieved from https://www.chinadialogue.net/article/show/single/en/5694-Campaigners-re-ignite-Nu-River-dam-debate

Denyer, S. (2013, February 7). Chinese dams in Tibet raise hackles in India. *Washington Post*. Retrieved from https://www.washingtonpost.com/world/asia_pacific/chinese-dams-in-tibet-raise-hackles-in-india/2013/02/07/ee39fc7a-7133-11e2-ac36-3d8d9dcaa2e2_story.html

Dikshit, S. (2013, April 2). One River, two countries, too many dams. *The Hindu*. Retrieved from http://www.thehindu.com/opinion/op-ed/one-river-two-countries-too-many-dams/article4570590.ece

EarthRights International, Community Resource Centre, FACT, Samreth Law Group, ECA Watch Austria, Law and Policy of Sustainable Development […] & International Rivers (2014, April). Specific instance complaint under the oecd guidelines for multinational enterprises regarding the contributions of Andritz AG to human rights abuse and environmental damage in connection with the Xayaburi Hydropower project in Lao PDR. Retrieved from https://www.earthrights.org/sites/default/files/documents/andritz-oecd-complaint-re-xayaburi-4.9.2014.pdf

Eidem, N. T., Fesler, K. J., & Wolf, A. T. (2012). Intranational Cooperation and Conflict over Freshwater: Examples from the Western United States. *Journal of Contemporary Water Research & Education*, 147, 63–71. Retrieved from http://onlinelibrary.wiley.com/store/10.1111/j.1936-704X.2012.03103.x/asset/j.1936-704X.2012.03103.x.pdf?v=1&t=ijmrsa69&s= 647287c20bbefdbec1d58596f946e7b6a9980425

Economic Times. (2014, December 11). Taking up concerns on Brahmaputra river with China: Government to Rajya Sabha. Retrieved from http://articles.economictimes.indiatimes.com/ 2014-12-11/news/56955585_1_considerable-established-user-rights-brahmaputra-river-downstream-states

Economic Times. (2015, May 19). PM Narendra Modi did grave injustice to people of Assam: Tarun Gogoi. Retrieved from http://articles.economictimes.indiatimes.com/2015-05-19/news/ 62369131_1_tarun-gogoi-pm-narendra-modi-brahmaputra-river

Economic Times. (2015, December 7). China urged to be mindful of Indian interests downstream of Brahmaputra: Government. Retrieved from http://articles.economictimes.india

times.com/2015-12-07/news/68835790_1_brahmaputra-river-minister-uma-bharati-trans-border-rivers

Giordano, M., Giordano, M., & Wolf, A. (2002). The geography of water conflict and coopera-tion: Internal pressures and international manifestations. *The Geographical Journal, 168*, 293–312. Retrieved from http://www.jstor.org/stable/3451473

Golan-Nadir, N. and Cohen, N. (2017). The role of individual agents in promoting peace processes: business people and policy entrepreneurship in the Israeli-Palestinian conflict. *Policy Studies*, 38, 21–38.

Grey, D., & Sadoff, C. W. (2007). Sink or Swim? Water security for growth and development. *Water Policy*, 9, 545–571. Retrieved from http://cip.management.dal.ca/publications/Water%20security%20for%20growth%20and%20development.pdf

Gulliver, P. H. (1971). *Neighbours and Networks: The Idiom of Kinship in Social Action among the Ndendeuli of Tanzania*. Berkeley: University of California Press.

Gupta, J. (2010, November 24). Nervous neighbours. *China Dialogue*. Retrieved from https://www.chinadialogue.net/article/show/single/en/3959-Nervous-neighbour

Henderson, S. (2013, December 3). Mekong dams a long-term risk to food security. *Cambodian Daily*. Retrieved from https://www.cambodiadaily.com/archives/mekong-dams-a-long-term%E2%80%88risk-to-food-security-48415/

Indian Express. (2010, April 22). Chinese dam will not impact flow of Brahmaputra: Krishna. Retrieved from http://archive.indianexpress.com/news/chinese-dam-will-not-impact-flow-ofbrahmaputra-krishna/609953/

Indian Express. (2014, December 9). Assam groups say no to resumption of Lower Subansiri hydel [sic] project. Retrieved from http://indianexpress.com/article/india/india-others/assam-groups-say-no-to-resumption-of-lower-subansiri-hydel-project/

International Center for Environmental Management. (2011, November 23). Xayaburi HPP: Gains and losses for the LMB: Comments on the Xayaburi HPP Compliance Report: With a focus on the Mekong Delta. Retrieved from http://www.icem.com.au/documents/envassessment/mrc_sea_hp/VUSTA%20poyry%20review.pdf

International Center for Environmental Management. (2010). *Strategic environmental assessment of hydropower on the Mekong mainstream*. Vientiane: Mekong River Commission Secretariat.

International Rivers. (2016, January 25). Media kit on Xayaburi Dam lawsuit. Retrieved from https://www.internationalrivers.org/resources/9230

International Rivers. (2011, April 18). Outrage over secret Xayaburi Dam construction. Retrieved from https://www.internationalrivers.org/resources/outrage-over-secret-xayaburi-dam-construction-3723

John, J. (2013, February 13). Protest China's plan to build dams on Brahmaputra river, Rajnath Singh says. *Times of India*. Retrieved from http://timesofindia.indiatimes.com/india/Protest-Chinas-plan-to-build-dams-on-Brahmaputra-river-Rajnath-Singh-says/articleshow/18480639.cms

Karmakar, R. (2014, November 25). Assam protests China dam on upper Brahmaputra. *Hindustan Times*. Retrieved from http://www.hindustantimes.com/india/assam-protests-china-dam-on-upper-brahmaputra/story-Ar1t9f1ciAxUn14Z7ZOxKL.html

Keck, M. E., & Sikkink, K. (1998). *Activists beyond borders: Advocacy networks in international politics*. Ithaca, NY: Cornell University Press.

Khadka, N. S. (2014, March 20). Megadams: Battle on the Brahmaputra. *BBC*. Retrieved from http://www.bbc.com/news/world-asia-india-26663820

Kingdon, J. W. (1995 [1984]). *Agendas, alternatives, and public policies* (2nd ed.). Boston: Little, Brown & Company.

Kossov, I., & Samean, L. (2016, January 14). Donors slash funding for MRC. *Phnom Penh Post*. Retrieved from http://www.phnompenhpost.com/national/donors-slash-funding-mrc

Kramer, A., Wolf, A., Carius, A., & Dabelko, G. (2013). The Key to Managing Conflict and Cooperation over Water. *A World of Science*. *11*, 4–12. Retrieved from http://unesdoc.unesco.org/images/0021/002191/219156E.pdf#page=4

Krishnan, A. (2013, January 30). China gives go-ahead for three new Brahmaputra dams. *The Hindu.* Retrieved from http://www.thehindu.com/news/international/china-gives-goahead-for -three-new-brahmaputra-dams/article4358195.ece

Leavenworth, S. (2016, May 12). China may shelve plans to build dams on its last wild river. *National Geographic.* Retrieved from http://news.nationalgeographic.com/2016/05/160512- china-nu-river-dams-environment/

Leifeld, P. and Schneider, V. (2012). Information exchange in policy networks. *American Journal of Political Science,* 56, 731–744. doi:10.1111/j.1540-5907.2011.00580.x

Li, J. (2013, January 25). Ban lifted on controversial Nu River dam projects. *South China Morning Post.* Retrieved from http://www.scmp.com/news/china/article/1135463/ban-lifted- controversial-nu-river-dam-projects

Lindblom, C. E. (1968). *The policymaking process.* New Jersey: Prentice-Hall.

Liu, J. (2013, April 19). How Tiger Leaping Gorge was saved. *China Dialogue.* Retrieved from https://www.chinadialogue.net/books/5923-How-Tiger-Leaping-Gorge-was-saved/en

Litzinger, R. (2007). In search of the grassroots: Hydroelectric politics in Northwest Yunnan. In E. J. Perry & M. Goldman (Ed.), *Grassroots Political Reform in Contemporary China.* Cambridge: Harvard University Press.

Madan, T. (2013, October 8). India's relations with China: The good, the bad and the (Potentially) ugly. *Brookings East Asia Commentary.* Retrieved from https://www.brookings. edu/opinions/indias-relations-with-china-the-good-the-bad-and-the-potentially-ugly/

Magee, D. (2006). Powershed politics: Yunnan Hydropower under Great Western development. *The China Quarterly,* 185, 23–41. doi:10.1017/S0305741006000038

Magee, D., & Kelley, S. (2009). Damming the Salween River. In F. Molle, T. Foran, & M. Kakonen (Ed.), *Contested Waterscapes in the Mekong Region: Hydropower, Livelihoods and Governance.* London: Earthscan.

Magee, D. & McDonald, K. (2011). Beyond Three Gorges: Nu Over Hydropower and energy decision politics in China. *Asian Geographer,* 25, 39–60.

McCormack, G. (2001). Water Margins: Competing Paradigms in China. *Critical Asian Studies,* 33, 5–30. doi:10.1080/14672710122114

Mekong River Commission. (2011a). Mekong river commission procedures for notification, prior consultation and agreement form for reply to prior consultation: The socialist Republic of Viet Nam. Retrieved from http://www.mrcmekong.org/assets/Consultations/2010- Xayaburi/Viet-Nam-Reply-Form.pdf

Mekong River Commission. (2011b). Mekong river commission procedures for notification, prior consultation and agreement form for reply to prior consultation: The Kingdom of Thailand. Retrieved from http://www.mrcmekong.org/assets/Consultations/2010-Xayaburi /Thailand-Reply-Form.pdf

Mekong River Commission. (2011c). Mekong river commission procedures for notification, prior consultation and agreement form for reply to prior consultation: The Kingdom of Cambodia. Retrieved from http://www.mrcmekong.org/assets/Consultations/2010-Xayaburi /Cambodia-Reply-Form.pdf

Mertha, A. (2009). "Fragmented Authoritarianism 2.0": Political Pluralization in the Chinese Policy Process. *The China Quarterly,* 200, 995–1012. doi:10.1017/S0305741009990592

Mintrom, M & Norman, P. (2009). Policy entrepreneurship and policy change. *The Policy Studies Journal,* 37, 649–667.

Naing, S. Y. (2013, May 22). On Salween River, growing signs that work on hat Gyi dam resumes. *The Irrawaddy.* Retrieved from http://www.irrawaddy.org/conflict/on-salween-river growing-signs-that-work-on-hat-gyi-dam-resumes.html

New Indian Express. (2011, June 16). Satellite image shows 'no change' in Brahmaputra. Retrieved from http://www.newindianexpress.com/nation/article438509.ece

Noreen, N. (2011, March 15). Dam investors 'fuelling violence'. *Democratic Voice of Burma.* Retrieved from http://www.dvb.no/news/dam-investors-fuelling-violence/14753

Radio Free Asia. (2014, June 24). Thai court agrees to hear lawsuit over controversial Xayaburi Dam in Laos. Retrieved from http://www.rfa.org/english/news/laos/lawsuit-06242014170128.html

Radio Free Asia. (2012, August 7). Thai villagers sue over dam. Retrieved from http://www.rfa.org/english/news/laos/xayaburi-08072012171723.html

Ramachandran, S. (2015, April 3). Water wars: China, India and the great dam rush. *The Diplomat*. Retrieved from http://thediplomat.com/2015/04/water-wars-china-india-and-the-great-dam-rush/

Roberts, N. C., & King, P. J. (1991). Policy entrepreneurs: Their activity structure and function in the policy process. *Journal of Public Administration Research and Theory, 1*, 147–175.

Sabatier, P. A. (1988). An advocacy coalition framework of policy change and the role of policy-oriented learning therein. *Policy Sciences, 21*, 129–168.

Schimmelfennig, F. (2001). The Community trap: Liberal norms, rhetorical action, and the Eastern enlargement of the European Union. *International Organization, 55*, 47–80. doi:10.1162/002081801551414

Schneider, M., & Teske, P. (1992). Toward a theory of the political entrepreneur: Evidence from local government. *American Political Science Review, 86*, 737–747. doi:10.2307/1964135

Sharma, T. (2012, March 9). Fighting India's mega-dams. *China Dialogue*. Retrieved from https://www.chinadialogue.net/article/show/single/en/4799

Solomon, F. (2014, March 14). Stop Salween dams, say 34,000 people in Burma. *Democratic Voice of Burma*. Retrieved from http://www.dvb.no/news/stop-salween-dams-say-34000-people-in-burma-myanmar/38461

South Asia Network on Dams, Rivers and People. (2013, July 17). Brahmaputra — the beautiful river or the battleground? Retrieved from https://sandrp.wordpress.com/2013/07/17/brahmaputra-the-beautiful-river-or-the-battleground/

Tacon, D. (2013, June 14). River be damned. *Sydney Morning Herald*. Retrieved from http://www.smh.com.au/world/river-be-damned-20130613-2o6r4.html

Talukdar, S. (2014, July 4). Assam bodies oppose move to restart work on mega dam. *The Hindu*. Retrieved from http://www.thehindu.com/todays-article/tp-national/tp-otherstates/assambodies-oppose-move-to-restart-work-on-mega-dam/article6175552.ece

The Economist. (2012, September). Damned if they do. Retrieved from http://www.economist.com/node/21563764

The Hindu. (2013, October 23). China allays fears over Brahmaputra dam. Retrieved from http://www.thehindu.com/news/national/china-allays-fears-over-brahmaputra-dam/article5264861

The Nation. (2007, September 4). Thai killed in attack on Hat Gyi. Retrieved from http://www.nationmultimedia.com/national/Thai-killed-in-attack-on-Hat-Gyi-30047654.html

Telegraph India. (2014, October 16). Protest against China dams - AJYCP to demonstrate in Delhi on November 4. Retrieved from http://www.telegraphindia.com/1141017/jsp/northeast/story_18933789.jsp

Telegraph India. (2014, November 24). Protests in Assam on dam in China - Project on Brahmaputra ruffles feathers. Retrieved from http://www.telegraphindia.com/1141125/jsp/frontpage/story_19082164.jsp

Times of Assam. (2012, January 7). APFEJ expresses concern at China Dams. Retrieved from https://www.timesofassam.com/headlines/apfej-expresses-concern-at-china-dams/

Times of India. (2009, October 16). India to check if China is building Brahmaputra dam. Retrieved from http://timesofindia.indiatimes.com/india/India-to-check-if-China-is-building-Brahmaputra-dam/articleshow/5129067.cms

Times of India. (2012, February 16). Construction of Subansiri dam not to be stopped. Retrieved from http://timesofindia.indiatimes.com/city/guwahati/Construction-of-Subansiri-dam-not-to-be-stopped/articleshow/11917447.cms?referral=PM

Times Now. (2011, June 14). Debate: China diverts Brahmaputra. Retrieved from http://www.timesnow.tv/Debate-China-diverts-Brahmaputra-1/videoshow/4375862.cms

Turton, A. R., Patrick, M. J., & Julien, F. (2006). Transboundary Water Resources in Southern Africa: Conflict or cooperation? *Development, 49*, 22–31. doi:10.1057/palgrave.development.1100269

Vidal, J. (2013, August 10). China and India 'water grab' dams put ecology of Himalayas in danger. *The Guardian*. Retrieved from https://www.theguardian.com/global-development /2013/aug/10/china-india-water-grab-dams-himalayas-danger

Walker, B. (2013, February 1). China gives green-light to new era of mega-dams. *China Dialogue*. Retrieved from https://www.chinadialogue.net/blog/5678-China-gives-green-light-to-new-era-of-mega-dams/e

Walker, S. (2011, February 11). NGOs weigh in on Laos dam. *Phnom Penh Post*. Retrieved from http://www.phnompenhpost.com/national/ngos-weigh-laos-dam

Wikileaks. (2006). Hu Jintao fails to commit to Indian Relationship. Retrieved from https:// wikileaks.org/plusd/cables/06NEWDELHI8067_a.html

Wolf, A. T. (1998). Conflict and cooperation along international waterways. *Water Policy, 1*, 251–265. Retrieved from http://www.ce.utexas.edu/prof/mckinney/ce397/Topics/conflict/ Conflictandcooperation.pdf

Wong, E. (2016, June 18). China's Last Wild River Carries Conflicting Environmental Hopes. *New York Times*. Retrieved from http://www.nytimes.com/2016/06/19/world/asia/china-climate-change-nu-river-greenhouse-gases.html?_r=0

Wongruang, P. (2011, September 18). No stopping flow of construction at "suspended" dam. *Bangkok Post*. Retrieved from http://www.terraper.org/web/en/node/1194

World Wildlife Foundation (WWF). (2014, March 30). NGOs set one-year deadline to stop Xayaburi dam. Retrieved from http://wwf.panda.org/?218410/NGOs-set-one-year-deadline-to-stop-Xayaburi-dam

Xinjing Bao [Beijing News]. (2003, November 25). Experts oppose Nujiang dams [Nujiang jian shuiba zhuanjia qi fandui]. Retrieved from http://www.china.com.cn/chinese/2003/Nov/ 448837.htm

Yang, G., & Calhoun, C. (2008). Media, civil society, and the rise of a green public sphere in China. In P. Ho & R. L. Edmonds (Ed.), *China's Embedded Activism: Opportunities and constraints of a social movement*. Oxon: Routledge.

Yardley, J. (2005, December 26). Seeking a public voice on China's 'Angry River'. *International New York Times*. Retrieved from http://www.nytimes.com/2005/12/26/world/asia/seeking-a-public-voice-on-chinas-angry-river.html?_r=0

Yeophantong, P. (2013). *China, Corporate Responsibility and the Contentious Politics of Hydropower Development: Transnational Activism in the Mekong Region* (Global Economic Governance Working Paper 2013/82). Retrieved from Global Economic Governance Programme: http://www.globaleconomicgovernance.org/sites/geg/files/Yeophantong_GEG% 20WP%202013_82.pdf

Yeophantong, P. (2016). China's Hydropower Expansion and Influence over Environmental Governance in Mainland Southeast Asia. In E. Goh (Ed.), *Rising China's Influence in Developing Asia*. Oxford: Oxford University Press.

Yoffe, S., Wolf, A. T., & Giordano, M. (2003). Conflict and Cooperation over International Freshwater Resources: Indicators of Basins at Risk. *Journal of the American Water Resources Association, 39*, 1109–1126. Article No. 02036 http://www.transboundarywaters.orst.edu/pub lications/abst_docs/Yoffe_Wolf_Giordano.pdf

Yu, X. (2004). The new development view is calling for participatory social impact evaluation – case study of Manwan Hydropower Station. Presented at the UN Symposium on Hydropower and Sustainable Development, 19-27 October 2004. Retrieved from http://www.un.org/esa/ sustdev/sdissues/energy/op/hydro_yu.pdf

Zeitoun, M., & Mirumachi, N. (2008). Transboundary water interaction I: Reconsidering conflict and cooperation. *International Environmental Agreements: Politics, Law and Economics, 8*, 297–316. doi:10.1007/s10784-008-9083-5

Zeitoun, M., & Warner, J. (2006). Hydro-hegemony - a framework for analysis of trans-boundary water conflicts. *Water Policy, 8*, 435–460. doi:10.2166/wp.2006.054

Zawahri, N. A., & Gerlak, A. K. (2009). Navigating International River Disputes to Avert Conflict. *International Negotiation, 14*, 211–227. doi:10.1163/157180609X432806

Dam Diplomacy? China's new neighbourhood policy and Chinese dam-building

Carla P. Freeman

ABSTRACT

This analysis examines whether the Chinese state has transformed state dam-construction firms engaged abroad into agents of 'hydro-diplomacy' to reflect larger diplomatic initiatives to improve relations with neighbouring countries. It concludes that there is little evidence of strengthened direct oversight over Chinese dam-building companies in the region, which remain principally profit-seeking actors. In implementing projects, they prioritize the standards of host countries over the strategic concerns of the Chinese state. This suggests that either the 'agency costs' of hydropower firms' behaviour remain acceptable to Beijing or there are other impediments to policy change.

Introduction

China has become the face of controversial hydropower construction along its borders. Its massive dam-construction firms have carried out some of the largest dam-building projects in neighbouring countries, drawing protests for their often damaging social and environmental impacts from local communities and international observers. At the same time, China's more assertive conduct in its maritime disputes with neighbouring states has introduced new strains into its relationships with countries on its periphery. Beijing has sought to counter these tensions through an enlivened neighbourhood diplomacy, which, among other stated goals, includes a sensitivity to the natural environment it shares with its neighbours (China Council for International Cooperation on Environment and Development, 2013; Mu, 2013).

This analysis draws on the principal–agent framework in examining whether Beijing has engaged in 'hydro-diplomacy', making hydropower development an object of its efforts to improve relations with countries along its periphery. This study focuses on whether China has sought to strengthen its oversight over its state-owned firms that are engaged in often controversial regional hydropower projects to ensure that their behaviour aligns with China's foreign policy goals. This question engages the debate over the extent to which China's state-owned enterprises (SOEs) are used to serve Beijing's strategic goals abroad. One view is that China's SOEs are among the tools deployed by Beijing to serve its international strategic interests; the other is that China's state-owned firms are fundamentally economic actors, driven first and foremost by profit motives rather than national strategic objectives. The former point of view emphasizes the state ownership of these enterprises as agents of China's "state capital-ism" (Milhaupt & Zheng, 2015, p. 668). Linked to the latter perspective is that the Chinese state lacks the capacity (is too weak, uncoordinated, and without the technical sophistication) to use its state firms as its agents (see the discussion in Norris, 2016). Evidence that Beijing has changed the practices of its state-owned hydropower firms in line with its neighbourhood diplomatic campaign would support the view that Beijing has the capacity to direct the behaviour of state-owned companies to serve particular foreign policy aims.

A growing body of work evaluating the social and environmental impacts of China's hydropower development abroad informs the article. A study by Maurin and Yeophantong (2013), for example, examines how China's regulatory regime aimed at promoting sustainable investing may extend to the investment strategies of Chinese-owned firms, including those engaged in hydropower development, in developing countries. It concludes that, while the Chinese government appears more aware of the financial risks of poor corporate practices on the part of Chinese-owned companies operating abroad, both the Chinese government and Chinese companies remain on a "learning curve" (Maurin & Yeophantong, 2013, p. 303). Research by Frauke Urban, Johan Nodensvard, Giuseppa Siciliano and Bingqian Li (2015) assesses the social sustainability of Chinese hydropower projects overseas, focusing on two Chinese-financed projects, the Bui Dam in Ghana and the Kamchay Dam in Cambodia. Noting that Chinese companies are driving the current renaissance of dam construction around the world, the authors identify numerous weaknesses in the social sustainability practices of the Chinese firms involved. They attribute these weaknesses to both the degree of political accountability in the project's host country and the absence of external regulatory bodies with standardized guidelines.

This analysis focuses more narrowly than these and other existing studies on large Chinese *state-owned* dam-building enterprises, firms over which the Chinese state as owner exercises exceptional influence, and their operations in China's geographic periphery, a region to which China assigns high priority in its foreign policy. It looks for evidence that Beijing has introduced new policies to modify the behaviour of its state-owned hydropower companies as they engage in operations across its borders to serve its strategic foreign policy interests.

The article is organized into four sections. The first section expands on this intro-duction with relevant background material. The second section discusses the Myitsone Dam project as a baseline example of the kinds of practices that have made Chinese dam-construction firms the object of international criticism. The problems the project created also made clear to Chinese authorities how the environmental and social practices of Chinese-owned firms abroad could adversely affect its relationship with a neighbour with which China had enjoyed particularly congenial ties. The third section scrutinizes Chinese policies for hydropower overseas projects since Myitsone for potential changes. The analysis shows that, while there is evidence that the Chinese government has begun assigning a greater weight to the social and environment impacts of international projects carried out by Chinese firms, it has taken few concrete steps to assert greater direct control over the behaviour of its firms abroad. In other words, there is little evidence that Beijing has sought to adjust the behaviour of its state-owned firms to serve its diplomatic goals. In its fourth section, the article concludes with reflections on potential explanations for this finding.

Background

China has long given priority in its security calculations to the geopolitical zone comprising the countries along its borders. Since ancient times, the rulers of China's unifying central state have viewed stability along the country's long territorial frontiers as critical to the country's security. Under the dynastic tribute system, the lure of trade with China was used to quell threats from the barbarians on the edge of the Chinese imperium (Lary, 2008). In the immediate post-Mao era, China pursued a regional foreign policy of 'good neighbourly relations', improving its ties with the countries in its immediate neighbourhood. Its diplomatic efforts included trying to resolve or set aside territorial disputes, where those existed, in order to focus on developing close economic ties in its immediate region supportive of economic development.

However, as noted, China's growing power and assertive pursuit of disputed maritime claims have given rise to new tensions between it and its neighbours. This has occurred as American policy, framed during the Obama Administration after 2011 as the 'pivot' (or 'rebalance') to Asia, has engaged the United States with countries in China's near abroad in ways that for China read as encirclement and containment. Countries bordering China, including Myanmar and Vietnam, have seized the opportunity presented by more active US attention to deepen economic and security ties with Washington to balance their powerful neighbour. In 2013 the Chinese leadership held a widely attended work forum on periphery policy, the first work forum on foreign policy since 2006 and the first such high-level meeting devoted to China's periphery. Chinese President Xi Jinping laid out the goals of China's neighbourhood policy as to "make our neighbors more friendly in politics, economically more closely tied to us, and we must have deeper security cooperation and closer people-to-people ties" ('Xi Jinping: China', 2013). In addition to a shared geography, the natural environment was among the factors Xi stated that made the periphery "extremely strategically significant" for China ('Xi Jinping: Let', 2013).

China is already the leading bilateral trade and investment partner of nearly all of its neighbours. However, some Chinese cross-border economic activities have raised concerns. As noted, hydropower development has been a particularly sensitive transboundary issue. As the upper riparian, China has largely rejected a multilateral approach to the management of rivers that originate in its territory. It has demonstrated a preference for preserving maximum control over all of its waterways as national resources, regardless of the preferences of many of its downstream neighbours for cooperative river management (Ho, 2014; Wouters, 2014). As China has constructed dams with downstream impacts on lower riparian countries, affected states have grown concerned about China's growing control over water flows in their countries. Political leaders in Vietnam, with which China also has maritime disputes, have gone so far as to speculate that China's upstream water management could become a trigger for conflict (Parameswaran, 2012). The concerns of some of China's neighbours about China's upstream river management have been exacerbated by the contrast between China's broad resistance to multilateral water resource management and the growing participation by downstream riparians in international freshwater agreements – marked most recently by Vietnam's May 2014 accession to the United Nations Watercourses Convention (le Clue, 2012).

Given complex and substantial impacts on local ecosystems and human populations, which include potentially destabilizing distributive effects, the development of large-scale hydropower by Chinese firms in neighbouring countries also involves particular political sensitivities. Projects may also have far-reaching transboundary consequences (Bao, 2012). Chinese state-owned hydropower companies are among the world's most prolific dam-building companies, constructing some of largest-scale infrastructure projects in countries bordering on China. Not only has China become the face of mega-dam construction, but China's dam-construction companies have also become the face of China in some countries along its borders. Between 2006 and 2011 in South-East Asia alone, Chinese-financed projects accounted for 46% of total additions to hydro-electricity capacity in Cambodia, Laos and Myanmar. These projects often involve adverse environmental impacts and displacement of riparian communities. The perception that much of the capacity added by Chinese hydropower investments benefits China more than local populations, through energy exports to China or elsewhere, or to supply energy to other Chinese projects, from which Chinese companies will profit, adds to the controversy that often surrounds them (International Hydropower Association, n.d.; U.S. Energy Information Administration, 2013).

As will be described, the limited regulatory oversight of the Chinese state over SOEs operating abroad has enabled Beijing to distance itself from the conduct of its firms (giving it plausible denial, so to speak, in cases of misconduct). To frame this arrangement in terms of the principal–agent model: in effect Beijing has found the agency costs (the costs of an agent – here the state-owned company – whose actions or interests diverge from those of the principal – in this case, the state) resulting from its limited direct control over its agents acceptable. As will be shown in the discussion of the Myitsone case, the question is whether China's foreign policy interests may have changed this calculation. Other studies using the principal–agent model to understand the challenges governments face as principals when they seek to enlist their firms in the cause of public policy, such as Ramamurti (1991) or Dombrowksi (1996), show how difficult this is to make happen given the generally wide gap between the interests of the government and the firm. Inducements or incentives, increasing information flows, and engaging in greater oversight to minimize opportunities for "hidden action" become key to the state's success in enlisting its firms in achieving policy goals (Dombrowski, 1996, p. 187). Applying the principal–agent logic to 'hydro-diplomacy', Beijing should therefore pursue a number of goals in order to make its state-owned dam-building companies better agents of the Chinese state's diplomatic interests. These goals can be expected to include methods to strengthen coordination or promote overlap between the interests of the state and the firm, through such devices as requiring or incentivizing firms to follow international best practices. They might also include changing management or oversight practices to reduce the opportunities of its firms to skirt or ignore these requirements. Raising standards and implementing review processes for project financing and strengthening the enforcement power of Ministry of Commerce (MOFCOM) officials would seem logical targets for policy change.

Although this article touches on several controversial hydropower projects in countries neighbouring China, it centres its analysis on the case of the Myitsone Dam in northern Myanmar, the most potentially catalytic among them from the perspective of China's foreign policy. The project's high costs, limited benefits for the local population, and significant environmental costs drew local and international criticism from the time it was announced. However, the political transition in Myanmar opened the door to unprecedented public outcry over the damage to the riparian environment associated with the dam's construction. Amid US–Myanmar rapprochement, the dam became the focal point for debate within Myanmar over China's growing influence in the country, with its ultimate suspension an unequivocal signal to Beijing from Naypyitaw that its relationship with its south-western neighbour was changing (Gordon, 2012).

The Myitsone lesson for Chinese diplomacy

When China Power Investment Corporation (CPI) and Myanmar's Asia World Group (AWG) established a joint venture in 2005, the partners envisioned constructing a cascade of seven hydropower projects along the Mali and N'Mai rivers in Myanmar. This represented a planned investment of USD 20 billion, with project financing from China's Export-Import Bank (ExIm Bank). State-owned CPI is one of China's largest electricity producers. Its Yunnan subsidiary for Myanmar operations, Upstream Ayeyawady Confluence Basin Hydropower Co. Ltd. (UACHC), was slated to operate the dam for a 50-year period. Asia World is among Myanmar's largest and most diverse conglomerates; it is also on the list of companies sanctioned by the US for alleged ties to drug trafficking (Wijaya, 2012). In 2007, CPI and AWG negotiated an "agreement in principle" to proceed with this ambitious scheme (German Institute of Global and Area Studies, 2012).

Among the keystone projects planned was the Myitsone Dam, a 152-metre-high mega-dam, with the potential to generate 6000 MW of electricity. The site of the dam lies at the headwaters of the Irrawaddy near the confluence of the Mali and N'Mai in Kachin State, Myanmar's northernmost province, where an organized ethnic insurgency continues to challenge Naypyidaw's political authority. When completed, the project would have created a reservoir covering 766 km^2, roughly the area of Singapore (International Rivers, n.d.a). According to an analysis by multiple experts, published by International Rivers (n.d.b), of the environmental impact assessment (EIA) conducted by CPI, the EIA fell far short of best practice: "The analysis of the dams' impacts on terrestrial fauna was 'relatively robust', but ... there were serious flaws in the methodology and structure of the EIA, total neglect of the temporal and spatial scale of the social and environmental impacts of the dams, superficial analysis of the dams' impacts on freshwater biodiversity, [while] ... public participation failed to meet best practice." The EIA downplayed the significant environmental risks the project posed to the Irrawaddy, not only to its fish stocks but also to its role in irrigating Myanmar's rice crops and providing silt to the Irrawaddy Delta. These costs stood in contrast to the limited benefits of the project for local actors, as plans were to export 90% of the power generated by the dam to China in exchange for USD 17 billion over 50 years (Hadfield, 2012; Motlagh, 2012). In addition to the UACHC and Asia World, lead beneficiaries of the project were to have been Yunnan Province, which would have received the bulk of the power, and the Myanmar central government. According to a CPI-published report, the latter would have received more than USD 54 billion from the project (Martov, 2013).

Almost immediately after the venture was announced, a dozen Kachin leaders appealed in writing to Myanmar's then top leader, Senior General Than Shwe, for cancellation of the project. Opposition to the project was also expressed by the Kachin Independence Organization, which controls the territory in which the project is located. Local and international NGOs also volubly criticized the project (German Institute of Global and Area Studies, 2012). Despite this opposition, however, between 2007 and 2010, construction proceeded. Along with Myanmar Electric Power Enterprise, CPI and AWG formed the Irrawaddy Myitsone-Myintnya-MyintWan Hydropower Company Limited to construct and operate the Myitsone Dam and other planned hydropower projects, with several Chinese subcontractors. Subcontractors for aspects of the construction were selected from among China's largest dam builders: China Gezhouba Group Corporation, China Power Investment Corporation Materials and Equipment Co. Ltd., and Sinohydro (International Rivers, n.d.c.). By 2009, the process of resettling thousands of residents from the hundreds of villages affected by the dam had moved ahead, with the initial phase of construction launched soon after (International Rivers, 2015).

As Myanmar's government transitioned to a new, more participatory and quasi-civilian political system in 2010, Beijing proceeded with plans to strengthen its cooperation with Naypyitaw. It announced a "comprehensive strategic cooperative partnership" with the new government and signed a raft of new economic agreements, including one for another hydropower project. At much the same time, Yunnan, the Chinese province bordering on Myanmar, which sees Myanmar as a key economic partner and source of energy, including hydropower, announced a "bridgehead economic strategy" to meet its growing demand. The "bridgehead" envisions logistical interconnections linking Yunnan and the rest of landlocked south-western China to the Greater Mekong Subregion and South-East Asia, as well as to the South Asian Subcontinent and the Indian Ocean. Some analysts see the bridgehead concept as linked to a Two Oceans Strategy of Beijing to expand its naval access westward, signalled by regular stops by the Chinese navy at Myanmar on route to and from the Gulf of Aden (Liu, 2013; Yun, 2012).

The suspension of the Myitsone project in 2011 was a sign that relations between China and Myanmar were changing, with significant economic and strategic implications for China for which Beijing was not prepared (Sun, 2012). In addition to the CPI project manager for Myitsone going on record to express his "shock", Chinese government officials reportedly reacted to the decision by Myanmar assertively, pressing Naypyidaw to honour its commitments (Watts, 2011). The downturn in China–Myanmar relations, accompanied by warming ties with the United States and its allies, jeopardized the bridgehead vision, including its geostrategic dimension. Also significantly, it challenged the expectations Chinese officials had held that it could count on Naypyidaw's support within the Association of Southeast Asian Nations for its position in its disputes with other claimants in the South China Sea, including that these should be resolved through bilateral negotiations. Many Chinese officials assessed the suspension of Myitsone in September 2011 as symptomatic of improving relations between the

US and Myanmar, a decision resulting from American influence rather than from domestic pressure. The same month, Myanmar Foreign Minister Maung Lwin Wunn met with US State Department officials to discuss improved US–Myanmar ties. China had not anticipated that the armed clashes between the Kachin Independence Organization and the Myanmar army that had erupted at the Myitsone Dam site by June 2011 would be a serious barrier to the project. Nor had it seen the growing domestic and international criticism of the project as an intractable impediment to progress – even the publication in August 2011 of a letter by Daw Aung San Suu Kyi urging termination of the project, 'Appeal to Save the Irrawaddy', was not seen as a threat to implementation. Experts familiar with all three governments suggest that Beijing's recognition that it had misread the situation and miscalculated the extent to which the political transition in Myanmar could affect its relations with its neighbour, in which it had considerable stakes, contributed to its decision to give new attention to its relations with its regional neighbours (interviews, 2014).[1]

More like a good neighbour – policy changes after 2013?

A scan of Chinese firms' dam-building activities in countries in China's periphery after Myitstone's suspension shows that these continued unabated. According to the International Rivers database of dams (a list based on open source materials gathered through November 2014), approximately 76% of the dams under construction or proposed in Asia in 2013 and 2014 involved Chinese firms. Through late 2014, in Cambodia, Kazakhstan, Kyrgyzstan, Lao PDR, Myanmar, Nepal and Pakistan alone, there were 26 dams involving Chinese firms still under construction and an additional 45 new projects in the proposal or approval process.[2] Of these, more than 100 were large (over 70 MW) in scale. Others are mega-dams. The two largest Chinese projects proposed for the region are a 7100 MW project by the Three Gorges Corporation on the Indus in Pakistan, a 4500 MW dam on the Indus which could be financed by both China Development Bank and USAID, and a 4000 MW dam planned by Datang on the Salween in Myanmar (International Rivers, 2014).

As these projects proceeded in the wake of Myitsone's suspension and Xi Jinping's announcement of a renewed diplomatic emphasis by Beijing on neighbourhood relations, were there changes to the ways Chinese state-owned dam-building firms were doing business in neighbouring countries? Recalling Dombrowski's observation that narrowing the gap between the interests of the government and the firm (including reducing opportunities for the agents' "hidden action") is key to the state's success in enlisting its firms in achieving policy goals, did Beijing pursue measures to bring Chinese enterprises' goals and related behaviour in line with China's goal of improving its relationships with its neighbours? Did Beijing seek to increase its oversight of SOEs constructing projects in neighbouring countries? Were standards that aligned more closely with international best practices (or China's own domestic regulations for its firms) introduced and enforced? If Beijing took no steps to improve the operations of its state firms in the region, how can this be explained?

There are a number of social and environmental guidelines for hydropower projects, developed by international NGOs, that collectively serve as the basis for international best practices. The World Commission on Dams, the International Finance

Corporation, and the dam industry's International Hydropower Association are among them (International Rivers, 2008). Extrapolating from these sources, best practices reflecting sustainable guidelines for dam construction to mitigate environmental damage and social costs should include several fundamental criteria, with the principles of transparency and social participation at their core. These include decision making with stakeholder participation, EIAs based on precautionary environmental standards, and the establishment of mechanisms for monitoring and compliance (International Energy Agency, 2006; International Financial Corporation, 2012; World Bank, 1999; World Commission on Dams, 2000).

It should be noted, however, that while the criteria distilled above have become standard best practices for most governments in developed countries, they are not widely accepted in developing countries. Most developed countries made EIAs mandatory beginning in the late 1960s – EIA legislation was passed in the United States in 1970, for example. Many developing countries also require EIAs; however, numerous studies of EIA requirements in developing countries find that local conditions often impose significant impediments to such processes as broad stakeholder involvement in the approval and design stage of the project or to project monitoring and compliance (Li, 2008). At the same time, the imperative of economic growth through energy development may lead even those developing countries that assign a high value to the environment to opt to make the environmental and social trade-offs that so often go hand in hand with large-scale hydropower development. (Here, the case of the Coca Codo Sinclair Dam in Ecuador offers an example—see Coca Codo Sinclair Hydroelectric Project, n.d.)[3]

Domestic regulations for dam construction in China do incorporate many international best practices. Since 2002, infrastructure projects, including hydropower, have required an EIA under the Environmental Impact Assessment Law adopted by the State Council that year. Article 5 of the 2006 Chinese Company Law requires that companies "undertake social responsibility" (Lin, 2010, p. 65). The State-Owned Assets Supervision and Administration Commission (SASAC) issued Corporate Social Responsibility Guidelines for State-Owned Enterprises in 2008. Along with these requirements aimed at companies in general, hydropower development in China is also subject to additional industry-specific regulations. These include the Regulations on Environmental Management of Construction Projects, issued in 1998; the Notice on Strengthening Environmental Protection Work in Hydropower Development, jointly promulgated in 2005 by the State Environmental Protection Administration (SEPA) – today the Ministry of Environmental Protection (MEP) – with the National Development and Reform Commission (NDRC); Provisional Measures on Public Participation in Environmental Impact Assessment, issued by SEPA in 2006; the State Council's Regulations on Open Government Information, issued in 2008; and the Regulations on the Environmental Impact Assessment of Planning (known as PIA Regulations), promulgated by the State Council in 2009 (Cameron & Wei, 2012, p. 290; McDonald, Brossard & Brewer, 2009)). In addition, the 2006 Rules of Land Compensation and People Resettlement in Medium and Large Hydraulic and Hydroelectric Projects set strict standards for compensation and resettlement. The Provisional Measures mentioned above also require strategic plans for dam development and environmental management on the basis of basin-wide analysis; this requirement extends to transboundary projects (Chinese Laws, n.d.). In 2008, more-over, SASAC issued a Guideline on Fulfilling Social Responsibility by Central Enterprises, prompting several dam-building companies to independently commit to various corporate social responsibility principles. For example, Sinohydro's Engineering Bureau No. 1 adopted environmental management standard ISO 14001. China Datang and CPI separately adopted the UN Global Compact Agreement on human rights, labour, the environment and anti-corruption. These corporations played a role in crafting guidelines promoted by the Chinese International Contractors Association (CICA), released in December 2010 "against the background of ever-increasing expec-tations from the Chinese government and the international community", as the asso-ciation's vice-chairman put it (Guide on Social Responsibility, 2010; International Rivers, n.d.d).

However, the uneven implementation of environmental regulations and standards has been a long-standing challenge in China. A lack of transparency in the planning and construction process is a key factor in this (Tilt, 2015). For example, EIAs are conducted by assessors who cannot be construed as objective third parties as they are both trained and certified by China's MEP and are often paid a commission by the developers of the project being assessed. In addition, public oversight is constrained by the weakness of public-interest litigation in China. Until amendments in 2015 to China's environmental protection law opened some channels for non-governmental groups and public prosecutors to file public-interest lawsuits, litigation to challenge project design and implementation on behalf of the public interest or on environmental grounds was not an option (de Boer, 2015; Wang & Li, 2014; Zhang, 2011).

Historically, China's domestic regulations have not been extended to operations by Chinese firms outside Chinese territory. Projects involving hydropower installations designed for transboundary exports of electricity to China are no exception. At the time that the Myitsone Dam project was suspended, beyond the broad proviso that Chinese firms should respect local laws, Chinese firms were operating in the absence of any Chinese regulations or investment standards for overseas hydropower projects. The only regulations then in place that applied broadly to overseas investments by Chinese companies were issued in 2007 by MOFCOM with the Ministry of Foreign Affairs and SASAC: the Regulations on Further Regulating the Development of Contracting Foreign Projects. These regulations, with a series of other guidelines issued between 2008 and 2010,[4] exhort Chinese firms to comply with local laws and customs, protect the interests of local workers, participate in local charity work and protect the local environment, as well as to consider social responsibility in doing business overseas (International Rivers, 2012). However, in the absence of details about implementation and enforcement, these 'regulations' lack regulatory teeth. [5]

There are a number of clear junctures of opportunity for the Chinese state to strengthen its hand over the behaviour of its state-owned firms. These exist in project financing and MOFCOM's role in overseeing foreign aid projects. In the area of project finance, lines of credit issued through China's ExIm Bank or the China Development Bank and extended to the Chinese enterprises and the governments of host countries have helped facilitate overseas projects and mitigate the financial risk to SOEs asso-ciated with such substantial investments (Kong, 2010).[6] Chinese policy has been that investments of USD 200 million and over, involving over USD 50 million in foreign exchange, and projects using ExIm Bank loans for more than USD 100 million require approval from China's State Council, the highest level of the Chinese government. MOFCOM approval has been necessary for any overseas project that exceeds an investment of USD 100 million in countries that are deemed risky or in countries with which China lacks formal diplomatic ties. Provincially based MOFCOM bureaus have been permitted to approve projects involving investment from USD 10 million to USD 100 million by local-level SOEs. All central SOEs have been required to obtain a Certificate of Enterprise/Organization Investment from MOFCOM after they have finalized an overseas foreign investment project of any scale (Sauvant & Chen, 2014). Moreover, sign-off by China's economic planning agency, the NDRC, on large overseas projects involving more than USD 100 million in investment has also been required. In addition, the Ministry of Finance has set the quantity of foreign exchange available to Chinese companies (International Rivers, 2012).

In February 2014, however, far from tightening regulations on and regulatory over-sight of firms operating abroad, the Chinese government appeared to relax them. The NDRC promulgated the Administrative Measures for Approval and Filing of Outbound Investment Projects (Order 9), reducing the NDRC's role in project approval (Xiong & Sun, 2014). One law firm characterized the new rules as "liberalizing". Under the revised measures, only projects for which the total investment amount exceeds USD 1 billion require NDRC approval. Projects involving overseas direct investment of less than USD 300 million have devolved to provincial-level approvals. State Council approval is required only for projects involving a sensitive location and industry with investments above USD 2 billion (Ma, McKenzie, & Deng, 2014).[7]

Financing from China's development banks, which serve as the largest sources of financing for Chinese-backed hydropower projects, offers another opportunity for Beijing to impose higher social and environmental standards for international infra-structure development. China's ExIm bank has been exploring setting standards along the lines of those set by the US Export-Import Bank in 1995. The Export-Import Bank of the United States became the first of a growing number of export credit agencies to adopt a set of environmental and social 'due diligence' guidelines for environmental as well as social impact assessments. These requirements include Environmental and Social Impact Assessments (ESIAs) that follow the International Financial Corporation's Performance Standards on Environmental and Social Sustainability, supplemented by the World Bank Group's guidelines, referenced above (Friends of the Earth, n.d.).

China's ExIm Bank notes on its website that it has undertaken "intensive exchanges with the World Bank, International Financial Corporation and others on environmental assessment"; however, it has not yet signed on to either institution's standards. As China's ExIm Bank describes on its website, "During pre-lending investigation, a high threshold is put in place according to the principle of differentiated treatment, i.e. environment-friendly sectors are more likely to get lending support while loans to support highly-polluting or energy-intensive projects or exports of such products are tightly controlled." Current lending guidelines have not changed for several years. They set host-country environmental policies and standards as the "basis for evaluation". Guidelines state that: "When the host country does not have a complete environmental protection mechanism or lacks environmental and social impact assessment policy and standards, we should *refer* [emphasis added] to our country's standards or international practices" (China Export and Import Bank, 2008). The China Development Bank still does not require EIAs as part of its lending criteria. In addition, Sinosure, China's state-owned provider of export credit insurance, makes no mention of sustainable development practices, environmental pollution or social ethics in its public documents, although it is signatory to the Berne Union's Value Statement, which does include such principles ('Working Party on Export Credits', 2015).

Another point of opportunity for strengthening oversight of hydropower firms lies in MOFCOM's responsibility for oversight of overseas foreign aid projects through the Economic and Commercial Counsellor's office within the relevant Chinese embassy or consulate abroad. In principle, this responsibility is buttressed by MOFCOM's power to fine the Chinese companies or even to withdraw its permission for firms to proceed with the project (Powers, Mohan, & Tan-Mullins, 2010). In early 2013, MOFCOM, in tandem with the MEP, issued Guidelines on Environmental Protection for Chinese Companies Investing and Operating Abroad ('MOFCOM Holds Press Conference', 2013). The guidelines are the first by Chinese authorities setting environmental criteria for the behaviour of Chinese companies overseas – and the very first to give the MEP a role in overseeing the environmental behaviour of Chinese firms operating abroad. They underscore that Chinese companies should act according to the environmental laws of the project's host country, conduct environmental assessments, and do their best to mitigate any adverse consequences for local cultural heritage, among other practices. However, as these stand, they are nonbinding appeals to good corporate citizenship, not mandatory requirements. Chinese authorities on the ground seeking to encourage firms to adopt higher standards must thus still rely on exhortation rather than regulation to encourage firms to act according to best practices. Two years after Myitsone was suspended, the Chinese Embassy in Naypyidaw and the Chinese-Myanmar Enterprises Association promised at a press event that Chinese firms would exercise "moral self-discipline" with respect to investments in Myanmar (Sands, 2013).

Finally, most of China's large SOEs are under the oversight of SASAC (n.d.; Fan & Hope, 2013). SASAC does have considerable authority to monitor and regulate the operations of state firms (Sauvant & Chen, 2014). In 2012, SASAC issued the Interim Supervision Measures of SOEs' Investment Overseas. Reports suggest that SASAC is exploring ways to evaluate the corporate social responsibility performance of SOEs and extend its Interim Supervision Measures to environmental issues. However, new supervision measures have yet to be introduced at the time of writing (Liu, 2015).

Concluding analysis

These examples make clear that Beijing has done little to strengthen its regulatory control over the performance of its dam-building companies abroad, whether by tightening regulations on firms operating abroad, by setting more stringent requirements for receiving financing, or through other forms of oversight. There is some evidence that Chinese dam-building companies are approaching operations with a greater awareness of the potential risks of failing to manage community concerns and environmental impacts in their international operations. Some of China's largest firms have independently adopted measures to strengthen their own internal guidelines. A

2015 benchmarking study by International Rivers of the policies and practices of major Chinese dam-building firms reports that most large SOEs have introduced commitments to applying high international standards. Datang, for example, has stated that it will follow the World Bank Environmental and Social Safeguards Policies as its operating standard. Gezhouba now requires that an EIA be carried out for construction projects and that an environmental management system be set up, following the GB/T24001-2004 standard. Gezhouba, Huaneng and Huadian all have an internal policy of endorsing Global Compact standards, which require businesses to adopt a precautionary approach to environmental issues and take initiatives aimed at promoting greater environmental responsibilities. Sinohydro issued broad commitments to sustainable development principles in 2011 – the first publicly articulated commitment by a hydropower construction firm. These principles included biodiversity preservation and ensuring that ESIAs were conducted – pledges that in effect made the World Bank's safeguard policies minimum standards for the company. The company, which had come under intense fire over concerns about human rights violations associated with the Merowe Dam project in Sudan, also made commitments to public consultations, including in planning where resettlement might be needed, providing a grievance and complaints mechanism, and avoiding damage to cultural heritage sites (International Rivers, 2012; Sinohydro Group, 2012). According to an International Rivers study, Sinohydro has promised to comply with Chinese laws as well as local laws, following whichever reflect higher standards. If, as China's largest dam-building company, Sinohydro adheres to this corporate policy, standards could be lifted for all Chinese dam-building companies.

Generally, however, Chinese dam companies operating internationally have continued to implement the standards associated with international best practices situationally rather than as a matter of general practice. Where host governments do not insist on environmental standards, dam construction typically proceeds in ways redolent of the Myitsone project (International Rivers, 2015). For example, Sinohydro had been moving ahead with the Chhay Areng hydropower dam in Cambodia's Koh Kong Province, despite violent local protests and international environmental concerns, but with the strong backing of Cambodia's prime minister, Hun Sen (Hussein, 2015). Only after unrelenting and growing domestic opposition and negative international criticism was construction on the project suspended until 2018. However, many experts believe that this decision will be reversed (Parameswaran, 2015). Similarly, dam construction in Lao PDR appears to be moving forward following business-as-usual practices. New project approvals proceed under highly opaque circumstances, even when there are well-known objections to projects by local citizens. With dam-building companies from Thailand and Malaysia proceeding with projects that could have devastating environmental consequences, there are growing concerns that Laos's push to maximize the development of its hydropower resources is causing an environmental 'race to the bottom' in the country (Hirsh, 2013).

Where host governments set high standards for the environmental and social impact of hydropower projects, however, Chinese dam-building companies demonstrate far better environmental and social performance standards. In Pakistan, for example, the 720 MW Karot Hydropower Project (a project valued at USD 1.4 billion) underwent a rigorous and lengthy approval process. It was selected in 2006 as a potential hydropower site, with studies conducted by a local firm and a German consultancy. Following a 2009 ESIA, China's Three Gorges Corporation, a state firm under SASAC, was awarded the contract for the development of the hydropower generation plant in conjunction with a Pakistani firm. However, following this the project received still further scrutiny, leading to a decision to apply International Finance Corporation standards to the ESIA process. A new assessment was completed in January 2015 by Pakistan Engineering Services (Pvt) Limited, including a stakeholder engagement plan involving a series of public hearings. With land procurement being the investor's responsibility, the government is involved in negotiations over prices of land that must be acquired, to be decided between the landowners and the company ('International Water Power', 2015; Karot Power Ltd, n.d.; Mang, 2012; Pakistan Engineering Services, 2015; Poindexter, 2015).

Similarly, Chinese developers have met the criteria insisted on by the government of Nepal to win the contract and approval for developing Nepal's massive West Seti project. China's Three Gorges (through a subsidiary company) is the Chinese investor and the lead developer in this project as well, with financing from China's ExIm Bank. The project, first envisioned more than three decades ago, had an initial EIA that was approved by the Asian Development Bank (ADB). After criticism of the ADB over the project's resettlement plan for the thousands of people affected by the project – a plan alleged to fall short of the ADB's own standards – a new resettlement process reportedly has been approved. In addition to ADB standards, the World Bank's international safeguard standards are now also being applied (Fast, 2013). The project is expected to be completed by 2022. Originally, much of the power it produced was to be sold to India; now some is to be provided to locals free of charge, with the rest sold to Nepal (Giri, 2014; International Rivers, n.d.e; Pokharel, 2012; World Bulletin News, 2015).

This raises questions about *why* the Chinese state has done so little to strengthen its regulatory hand over its state-owned hydropower firms operating across its borders, given the priority assigned to neighbourhood diplomacy and the capacity of its firms to meet high performance standards when the host country requires them. Returning to the principal–agent model, one explanation is that the agency costs of firms' behaviour remain acceptable to Beijing. Beijing may calculate that strains over dam construction do not damage its regional relationships in significant or lasting ways as regional populations receive the benefits of the energy supplied by these developments.

What is acceptable may also be what is politically viable in Beijing. Efforts to adjust policies to serve foreign policy interests may face political and structural impediments. These may include resistance to new policies from the firms themselves, which may be able to draw on powerful champions in China's political hierarchy. The hydropower industry has had direct ties to China's top leadership. Former Chinese president Hu Jintao worked at Sinohydro from the late 1960s to the mid-1970s. Li Peng, who served as China's premier under Jiang Zemin, was the Chinese leader who in the mid-1990s personally initiated talks with Myanmar on hydropower development. His son, Li Xiaopeng, headed Huaneng, and his daughter, Li Xiaolin, was the CEO of CPI until 2015 (Gan, 2015; Lee, 2013). Subnational actors may also be a factor in the political equation. In the Myitsone case, for example, Yunnan Province holds a direct stake in the project through the UACHC and as a source of power to serve its economic development objectives. Moreover, Yunnan has dense ties to political and economic interests in Myanmar that may give the province a preference for business as usual. Indeed, in interviews with industry and policy experts in Myanmar, some expressed the view that Myanmar's sense of fraternal ties and appreciation for China's historical support would eventually result in the resumption of Myitsone's construction (interviews, Yangon, August 2014).

The fragmented nature of the Chinese system and the competing lines of authority governing state-owned firms may also underlie the evident policy inertia. For example, central SOEs require approval for overseas direct investment from MOFCOM, NDRC and SASAC (Sauvant & Chen, 2014). Modifying regulations would require these powerful bureaucratic players as well as industry and other institutional actors to all agree to changes. Additional answers might also be found in the Chinese state's regulatory capacity. While Chinese firms may have the technical capability and access to state-backed financing to execute large-scale projects overseas, it may also be that the Chinese state has not developed the regulatory capacity to exert disciplinary oversight over its largest firms through the existing system. Indeed, the findings in this article may support the conclusion of Margaret Pearson (2005) and others (see the discussion in Ferchen, 2008) that some economic sectors have become strategic tools of the Chinese state to preserve its own authority, and therefore reforming them exceeds its capabilities.

If these structural issues are key factors, then to make state-owned dam companies operating overseas better agents, the state would have to undertake significant changes not only to current policies but also to institutional arrangements. This is a notoriously challenging task even for a highly motivated Chinese leadership – one certainly made more difficult when the personal networks of senior officials may be involved. There is the possibility that change could be driven by Beijing's new international activism itself, linked to Beijing's grand transnational development vision. China's post-Mao reform history offers numerous examples of how international commitments by Beijing have been used to drive internal modernizing institutional and policy reforms. Beijing has used the language of public goods to describe such ambitious new institutions as the New Development Bank and the Asian Infrastructure Investment Bank (AIIB), both of which focus on infrastructure development in developing and emerging economies. The AIIB charter sets high standards for sustainable development through its Environmental and Social Framework. Given the governance of the AIIB by an international non-resident (and unpaid) board of directors, implementation of these standards could help move the whole coterie of institutions involved in Chinese infrastructure development, including dam building, towards greener and more socially sensitive investment standards.

Notes

1. It should be noted that some experts in China argued for reforming Chinese practices. Li Fusheng, then a deputy general manager at the China ExIm Bank, emphasized the risk to Chinese firms of their failure to engage with stakeholders. (Li edited *Chinese Investment Overseas Case Studies on Environmental and Social Risks* [Zhongguo Jinwai Touzi Huanjing Shehui Fengxian Anli Yanjiu], Peking University Press, 2014). Others urged transparency in China's overseas development projects ('Chinese Reactions to the Suspension', 2011).

Scholar Zhu Feng opined that, "Unless China begins to offer necessary public goods – which in addition to trade must include mature regional governance based on the rule of law, human rights and regional economic growth – its neighbours will not sincerely consider China's interests" ('Chinese Reactions to the Suspension', 2011).

2. Data through November 2014, International Rivers, China Overseas Dams List (10 November 2014).

3. The Coca Codo Sinclair Dam in Ecuador, developed by a consortium led by Sinohydro, will cause the loss of the San Rafael Falls, the tallest waterfall in Ecuador, while road construction to the project site has caused deforestation in the UNESCO Sumaco Biosphere Reserve.

4. Since 2007, for example, China's ExIm Bank, which handles export and import credits as well as concessional loads, has required companies to conduct EIAs for overseas projects and compensate communities for any environmental damage (Herbertson, 2011). The MEP has endorsed the Equator Principles (a set of standards adopted by financial institutions around the world on environmental and social risk) as guidelines for Chinese banks (Hensengerth, 2010).

5. This seems to be a fair assessment by International Rivers, which drew my attention to this point in its 26 November 2012 report 'The New Great Walls: A Guide to China's Overseas Dam Industry'.

6. To capture this relationship, Bo Kong describes China's national oil majors as "co-governed" by the state.

7. Sensitive countries are those without normal diplomatic relations with China, countries that are under international sanctions, and countries that are experiencing wars or social turmoil. Sensitive industries span basic telecommunications, large-scale land development, main electrical grids, news media, and cross-border water resources development and utilization.

Disclosure statement

No potential conflict of interest was reported by the author.

References

Bao, X. (2012, December). Dams and intergovernmental transfer: Are dam projects pareto improving in China? *Job Market Paper.* Retrieved from http://www.columbia.edu/~xb2112/XiaojiaBaoJMP.pdf

Cameron, A., & Wei, L. (2012). An environmental impact assessment for hydropower development in China. *Vermont Journal of Environmental Law, 14,* 275–302. doi:10.2307/vermjenvilaw.14.2.275

China Council for International Cooperation on Environment and Development. (2013, October 30). Important speech of Xi Jinping at peripheral diplomacy work conference. Retrieved from http://www.cciced.net/encciced/newscenter/latestnews/201310/t20131030_262608.html

China Export and Import Bank. (2008). Issuance notice of the "Guidelines for Environmental and Social Impact Assessments of the China Export and Import Bank's (China EXIM Bank) Loan Projects [unofficial translation with original Chinese]. Retrieved from https://www.internationalrivers.org/resources/guidelines-for-environmental-and-social-impact-assessments-of-the-china-export-and-import

Chinese Laws on Domestic Dam Building. (n.d.). Decarbon.ise. Retrieved from http://decarboni.se/publications/new-great-walls-guide-china%E2%80%99s-overseas-dam-industry/chinese-laws-domestic-dam-building

Chinese Reactions to the Suspension of the Myitsone Dam Project. (2011, November 12). Exporting China's development to the world. *Macquarie University and Free University.*

Retrieved from https://mqvu.wordpress.com/2011/11/12/chinese-reactions-to-the-suspension-of-the-myitsone-dam-project/

Coca Codo Sinclair Hydroelectric Project - Power Technology. (n.d.). Power-Technology.com. Retrieved from http://www.power-technology.com/projects/coca-codo-sinclair-hydroelectric-project/

de Boer, D. (2015, March 20). Under the Dome' may be a turning point for China's environment policy. China Dialogue. Retrieved from https://www.chinadialogue.net/article/show/single/en/7795–Under-the-Dome-may-be-a-turning-point-for-China-s-environment-policy

Dombrowski, P. (1996). Policy responses to the globalization of American banking. Pittsburgh: University of Pittsburgh Press.

Fan, G., & Hope, N. (2013). The role of state-owned enterprises in the Chinese economy. China-US 2022,Chapter 16. Retrieved from http://www.chinausfocus.com/2022/wp-content/uploads/Part+02-Chapter+16.pdf

Fast, T. (2013). Another dam development project? Development-forced displacement and resettlement in Nepal. *Centre for East and South-East Asian Studies Lund University.* Retrieved from http://lup.lub.lu.se/luur/download?func=downloadFile&recordOId=3990958&fileOId=3990962

Ferchen, M. (2008, January). Regulating market order in China: Economic ideas, marginal markets and the state (PhD dissertation draft). Cornell University. Retrieved from http://www.mattferchen.com/wp-content/uploads/2013/08/Dissertation-Final-Submission-Draft-January-3-2008.pdf

Friends of the Earth US. (n.d.). *A review of the environmental and social policies of national development banks in Brazil, China, and India.* Advancing sustainable and accountable finance: National Development Banks and their emerging role in the global economy, Conference Hong Kong, China, Hong Kong.

Gan, N. (2015, July 8). Top power industry job for Li Xiaolin, daughter of former Chinese premier. *South China Morning Post.* Retrieved from http://www.scmp.com/news/china/policies-politics/article/1834390/top-power-industry-job-li-xiaolin-daughter-former

German Institute of Global and Area Studies. (2012). Chronology of the Myitsone dam at the confluence of rivers above myitkyina and map – 34. *Journal of Current Southeast Asian Affairs.* Retrieved from http://d-nb.info/1024416801/34

Giri, S. (2014, April 14). IBN okays west seti bid from Chinese firm. *EKantipur.com.* Retrieved from http://www.ekantipur.com/2015/04/14/top-story/ibn-okays-west-seti-bid-from-chinese-firm/404061.html

Gordon, J. (2012). China: A hidden danger in the reform process. *PacNet, 75.* Retrieved from http://csis.org/files/publication/Pac1275.pdf

Guide on Social Responsibility for Chinese International Contractors. (2010, December 22). Sino-German CSR Project. Retrieved from http://www.chinacsrproject.org/Events/Event_Show_EN.asp?ID=110

Hadfield, P. (2012, March 4). Burmese villagers exiled from ancestral home as fate of dam remains unclear. *The Guardian.* Retrieved from http://www.theguardian.com/environment/2014/mar/04/burma-village-myitsone-dam-project-china

Hensengerth, O. (2010). Sustainable dam development in China between global norms and local practices (Discussion Paper). German Development Institute. Retrieved from http://www.die-gdi.de/uploads/media/DP_4.2010.pdf

Herbertson, K. (2011). Leading while catching up?: Emerging standards for China's overseas investments. Sustainable Development, Law & Policy, 11.3, 22–26, 41–43. Retrieved from http://digitalcommons.wcl.american.edu/cgi/viewcontent.cgi?article=1483&context=sdlp

Hirsh, P. (2013). Laos mutes opposition to controversial mekong dam. *China Dialogue.* Retrieved from https://www.chinadialogue.net/article/show/single/en/6509-Laos-mutes-opposition-to-controversial-Mekong-dam

Ho, S. (2014). River politics: China's policies in the Mekong and the Brahmaputra in comparative perspective. *Journal of Contemporary China, 23*(85), 1–20. doi:10.1080/10670564.2013.809974

Hussein, K. (2015, April 23). Courting China with care. *Dawn*. Retrieved from http://www.dawn. com/news/1177577/courting-china-with-care

International Energy Agency. (2006, May). Implementing agreement for hydropower technologies and programmes, Annex VIII, hydropower good practices: Environmental mitigation measures and benefits. Retrieved from http://www.ieahydro.org/reports/Annex_VIII_ Summary_Report.pdf

International Financial Corporation. (2012, January 1). Performance standards on environmental and social sustainability. Retrieved from http://www.ifc.org/wps/wcm/connect/ 115482804a0255db96fbffd1a5d13d27/PS_English_2012_Full-Document.pdf?MOD=AJPERES

International Hydropower Association. (n.d.) Laos Retrieved from http://www.hydropower.org/ country-profiles/laos

International Rivers. (2008, December). Social and environmental standards for large dams. *International Rivers*. Retrieved from http://www.internationalrivers.org/files/attached-files /international_rivers_wcd_ifc_iha_comparison.pdf

International Rivers. (2012, November 26). The New Great Walls: A guide to China's overseas dam industry. Retrieved from http://www.internationalrivers.org/resources/the-new-great- walls-a-guide-to-china%E2%80%99s-overseas-dam-industry-3962

International Rivers. (2014, November 10). China overseas dams list. Retrieved from http://www. internationalrivers.org/resources/china-overseas-dams-list-3611

International Rivers. (2015, June 22). Benchmarking the policies and practices of international hydropower companies. Retrieved from http://www.internationalrivers.org/resources/9065

International Rivers. (n.d. a). Irrawaddy Myitsone dam. Retrieved from http://www.internatio nalrivers.org/campaigns/irrawaddy-myitsone-dam-0

International Rivers. (n.d. b). International expert review of the Myitsone dam EIA. Retrieved from http://www.internationalrivers.org/files/attached-files/independent_expert_review_of_ the_myitsone_dam_eia_0.pdf

International Rivers. (n.d. c). The Myitsone dam on the irrawaddy river: A briefing. Retrieved from http://www.internationalrivers.org/resources/the-myitsone-dam-on-the-irrawaddy-river -a-briefing-3931

International Rivers. (n.d. d). Chinese dam builders going overseas. Retrieved from http://www. internationalrivers.org/campaigns/chinese-dam-builders

International Rivers. (n.d. e.) West Seti Dam, Nepal. Retrieved from http://www.internationalri vers.org/campaigns/west-seti-dam-nepal-0

International Water Power & Dam Construction Industry News Archive, International Water Power. (2015, May). Water power magazine. Retrieved from http://www.waterpowermagazine. com/news/industry_news_archive.html

Karot Power Ltd. (n.d.). Retrieved from http://www.karotpower.com/index.php?action= Environment%20&%20Sociall

Kong, B. (2010). *China's international petroleum policy*. Santa Barbara, CA: Praeger Security International.

Lary, D. (Ed.). (2008). *The Chinese state at the borders*. Vancouver, BC: University of Washington Press.

le Clue, S. (2012, April 12). Water treaties – A question of rights. *China Water Risk*. Retrieved from http://chinawaterrisk.org/resources/analysis-reviews/water-treaties-a-question-of- rights/

Lee, Y. B. (2013, Fall). Water power: The 'Hydropower Discourse' of China in an age of environmental sustainability. *Asianetwork Exchange*, *21*(1), 1–10.

Li, J. C. (2008). Environmental impact assessments in developing countries: An opportunity for greater environmental security? (Foundation for Environmental Security & Sustainability Working Paper No. 4). Retrieved from http://www.fess-global.org/workingpapers/eia.pdf.

Lin, L. (2010). Corporate social responsibility in China: Window dressing or structural change. *Berkeley Journal of International Law*, *28*(1), 64. doi:10.15779/Z38F35Q

Liu, J. (2013, November 16). China's Bridgehead Strategy and Yunnan Province - East by Southeast. *East by Southeast*. Retrieved from http://www.eastbysoutheast.com/chinas-bridgehead-strategy-yunnan-province/

Liu, M. (2015, March 18). Is corporate social responsibility China's Secret weapon? *Eco-Business*. Retrieved from http://www.eco-business.com/opinion/corporate-social-responsibility-chinas-secret-weapon/

Ma, X., McKenzie, P., & Deng, J. (2014, April 24). New NDRC rules set to facilitate China outbound investments, Morrison-Foerster client alert. Retrieved from http://www.mofo.com/~/media/Files/ClientAlert/140423NewNDRCRules.pdf

Mang, G. (2012, January 26). China's Global Quest for Resources and Implications for the United States, Testimony before the U.S.– China Economic and Security Review Commission. Retrieved from http://origin.www.uscc.gov/Hearings/hearing-china%E2%80%99s-global-quest-resources-and-implications-united-states

Martov, S. (2013, November 6). The Irrawaddy. Retrieved from http://www.irrawaddy.org/burma/myitsone-dam-project-hold-far-dead.html

Maurin, C., & Yeophantong, P. (2013, June). Going global responsibly? China's strategies towards "Sustainable" overseas investments. *Pacific Affairs*, 86(2), 281–303. doi:10.5509/2013862281

McDonald, K., Brossard, P., & Brewer, N. (2009). Exporting dams: China's hydropower industry goes global. *Journal of Environmental Management*, XXX, 294–302. doi:10.1016/j.jenvman.2008.07.023

Milhaupt, C. J., & Zheng, W. (2015). Beyond ownership: State capitalism and the Chinese firm. *The Georgetown Law Journal*, 103, 667–722.

MOFCOM Holds Press Conference on CSR of Chinese Companies Operating Abroad. (2013, March 1). Ministry of Commerce, People's Republic of China. Retrieved from http://english.mofcom.gov.cn/article/newsrelease/press/201303/20130300051296.shtml

Motlagh, J. (2012, July 14). Myanmar's Ethnic Kachin fear China may restart Dam project. *Seattle Times*. Retrieved from http://www.seattletimes.com/nation-world/myanmars-ethnic-kachin-fear-china-may-restart-dam-project/

Mu, X. (2013, October 26). Xi Jinping: China to further friendly relations with neighboring countries. *Xinhua*. Retrieved from http://news.xinhuanet.com/english/china/2013-10/26/c_125601680.htm

Nepal Clears $1.6 Bln Hydropower Project. (2015, April 13). Worldbulletin News. Retrieved from http://www.worldbulletin.net/todays-news/157810/nepal-clears-16-bln-hydropower-project' ·

Norris, W. J. (2016). *Chinese economic statecraft: Commercial actors, grand strategy, and state control*. Ithaca, NY: Cornell University Press.

Pakistan Engineering Services (Pvt.) Ltd. (2015, January). Karot Power Company (Pvt.) Limited 720 mw karot hydropower project (巴基斯坦卡洛特电力有限公司) stakeholders engagement plan. Retrieved from http://Www.Karotpower.Com/Cms/Lib/Downloadfiles/Stakeholder.Pdf

Parameswaran P. (2012, July 26). Lao dam pr blitz backfires. Radio Free Asia. Retrieved January 7, 2017, from http://www.rfa.org/english/commentaries/east-asia-beat/xayaburi-07262012050806.html

Parameswaran, P. (2015, February 25). Cambodia's strongman taunts opposition with rocket fire. *The Diplomat*. Retrieved from http://thediplomat.com/2015/02/cambodias-strongman-taunts-opposition-with-rocket-fire/

Pearson, M. (2005). The business of governing business in China: Institutions and norms of the emerging regulatory state. *World Politics*, 57, 296–322. doi:10.1353/wp.2005.0017

Poindexter, G. B. (2015, March 4). Pakistan awards US$1.4 billion contract for 720-MW Karot hydroelectric project on Jhelum River. *Hydroworld.com*. Retrieved from http://www.hydroworld.com/articles/2015/03/pakistan-awards-us-1-4-billion-contract-for-720-mw-karot-hydroelectric-project-on-jhelum-river.html

Pokharel, P. (2012, April 2). Nepal parliament OKs China dam project. *Wall Street Journal*. Retrieved from http://www.wsj.com/articles/SB10001424052702303816504577319 061311691568

Powers, M., Mohan, G., & Tan-Mullins, M. (2010). *China's resource diplomacy in Africa: Powering development?* Basingstoke, Hampshire: Palgrave MacMillan.

Ramamurti, R. (1991). Controlling state-owned enterprises. In R. Ramamurti & R. R. Vernon (Eds.), *Privatization and control of state-owned enterprises* (pp. 206–233). Washington, DC: The World Bank.

Sands, G. (2013, July 22). China learns CSR in Myanmar. *Foreign Policy Association*. Retrieved from http://foreignpolicyblogs.com/2013/07/22/china-learns-csr-in-myanmar/

Sauvant, K. P., & Chen, V. Z. (2014). China's regulatory framework for outward foreign direct investment. *China Economic Journal*, 7(1), 141–163. doi:10.1080/17538963.2013.874072

Sinohydro Group Ltd. (2012, April 20). CSR report 2011. *China Sustainability Reporting Resource Center, sustainabilityreport.cn*. Retrieved from http://www.sustainabilityreport.cn/en/ReportShow.asp?ReportId=3261

State-owned Assets and Supervision Administration Commission. (n.d.) Guidelines to state-owned enterprises directly under the central government on fulfilling corporate social responsibilities. Retrieved from http://en.sasac.gov.cn/n1408035/c1477196/content.html

Sun, Y. (2012). China's strategic misjudgment on Myanmar. *Journal of Current Southeast Asian Affairs*, 31(1), 73–96. ISSN: 1868-1034.

Tilt, B. (2015). *Dams and development in China: The moral economy of water and power*. New York, NY: Columbia University Press.

Urban, F., Nordensvard, J., Siciliano, G., & Li, B. (2015). Chinese overseas hydropower dams and social sustainability: The Bui Dam in Ghana and the Kamchay Dam in Cambodia. *Asia & the Pacific Policy Studies*, 2(3), 573–589. doi:10.1002/app5.103

U.S. Energy Information Administration (EIA). (2013, August 16). Chinese investments play large role in Southeast Asia hydroelectric growth. Retrieved from http://www.eia.gov/todayinenergy/detail.cfm?id=12571

Wang, K., & Li, C. (2014, June 9). China: Roadblocks to effective EIA. *China Water Risk*. Retrieved from http://chinawaterrisk.org/opinions/china-roadblocks-to-effective-eia/

Watts, J. (2011, October 4). China angry over Burma's decision to suspend work on £2.3bn dam. *The Guardian*. Retrieved from http://www.theguardian.com/environment/2011/oct/04/china-angry-burma-suspend-dam

Wijaya, M. (2012, May 20). Myanmar lures Singapore Inc. *AsiaTimes Online*. Retrieved from http://www.atimes.com/atimes/Southeast_Asia/NC20Ae01.html

Working Party on Export Credits and Credit Guarantees. (2015, March 16). Trade Committee, OECD. Retrieved from http://www.oecd.org/officialdocuments/publicdisplaydocumentpdf/?cote=TAD/ECG(2015)3&doclanguage=en

World Commission on Dams. (2000). *Dams and development: A new framework for decision-making*. UK and USA: Earthscan Publications Ltd. Retrieved from http://www.internationalrivers.org/resources/dams-and-development-a-new-framework-for-decision-making-3939

World Bank. (1999, January). OP 4.01, environmental assessment. Retrieved from http://web.worldbank.org/WBSITE/EXTERNAL/PROJECTS/EXTPOLICIES/EXTOPMANUAL/0,contentMDK:20064724~menuPK:4564185~pagePK:64709096~piPK:64709108~theSitePK:502184,00.html

Wouters, P. (2014, November 17). Keeping peace: China's upstream Dilemma. *China Water Risk*. Retrieved from http://chinawaterrisk.org/resources/analysis-reviews/keeping-peace-chinas-upstream-dilema/

Xi Jinping: China to Further Friendly Relations with Neighboring Countries. (2013, October 26). Xinhuanet. Retrieved from http://news.xinhuanet.com/english/china/2013-10/26/c_125601680.htm

"习近平：让命运共同体意识在周边国家落地生根" [Xi Jinping: Let a sense of shared destiny take root among neighboring countries] (2013, October 25). *Xinhuanet*. Retrieved from http://news.xinhuanet.com/politics/2013-10/25/c_117878944.htm

Xiong, J., & Sun, R., (2014, April 21). The other shoe finally falls': NDRC issues 'Administrative Measures for Verification and Registration on Overseas Investment Projects,'" King & Wood Mallesons. *China Law Insight*. Retrieved from http://www.chinalawinsight.com/2014/04/arti cles/corporate/mergers-acquisitions/the-other-shoe-finally-falls-ndrc-issues-administrative-measures-for-verification-and-registration-on-overseas-investment-projects/

Yun S. (2012). China's strategic misjudgment on Myanmar. Journal of Current Southeast Asian Affairs, 31(1),73–94.

Zhang, J. (2011, July 19). Plight of the public: Citizen participation in China's environmental laws. *ChinaDialogue*. Retrieved from https://www.chinadialogue.net/article/show/single/en/ 4414-The-plight-of-the-public-1-

Multi-track diplomacy: current and potential future cooperation over the Brahmaputra River Basin

Yumiko Yasuda, Douglas Hill, Dipankar Aich, Patrick Huntjens and Ashok Swain

ABSTRACT

This article analyzes key factors affecting transboundary water cooperation in the Brahmaputra River basin at multiple scales. The analysis of multi-track diplomacy reaffirms the potential of actor-inclusive approaches, arguing for a need to go beyond purely focusing on formal legal norms and consider the possibilities of cultural norms of informal processes of cooperation. Various 'windows of opportunity' exist in the current phase of the Brahmaputra basin's development, leading to exploration of a Zone of Possible Effective Cooperation, arising from the effort to scale up multi-track initiatives as well as broader geo-political-economic changes happening across and beyond the basin.

Introduction

The Brahmaputra[1] is one of the most significant river basins within Asia, encompassing approximately 400 million people, many of whom depend on the basin for their livelihoods (WWF, 2017), hydropower generation potential, and mega-biodiversity, with a great number of endemic flora and fauna. The river flows from the high Tibetan plateau, at an altitude of 5000 metres, descending rapidly as it enters India, where tributaries from Bhutan join the mainstream, and flowing down towards Bangladesh through diversifying terrain (Sharma, Gorsi, & Paithankar, 2016).

There are international dimensions to the management and distribution of the resources among the riparians, as four countries – China, India, Bhutan and Bangladesh – share various parts of the basin. Several unresolved issues of transboundary water management pose challenges to the institutionalization of an integrated basin management approach, including trust issues between and within the countries of the basin. The management of the Brahmaputra has long been marked by political complexity at a range of scales, including long-standing civil society activism in India over the social and environmental impacts of large-scale hydropower projects and bilateral disputes between India and Bangladesh over the terms of water sharing, which has generated widespread enmity in some quarters in the latter country. The emergence of China as a more significant stakeholder in the basin in recent years has added further dimensions to this complexity, not least because many stakeholders in India have felt aggrieved at the perceived threat to the latter's traditional role as the hegemonic power in the basin. Certainly, there are also cooperative mechanisms in place among the riparian nations, but these presently mostly take the form of bilateral cooperation, and even well-established mechanisms are seen by some commenters as insufficient to create a whole-of-basin approach. Civil society actors are also taking the lead in facilitating dialogue by leading some of the concrete actions of cooperation among riparian actors.

In exploring a better governance of this complex basin, it is crucial to understand the political economy context, including the relationships of various actors. In particular, understanding the key factors that affect current and potential future cooperation can shed light on how further cooperation could be driven. However, most scientific analysis surrounding the Brahmaputra River in the past has focused on its biophysical conditions.[2] And an emerging body of analysis that provides understandings for socio-political-economic perspectives of the river basin mostly provides limited geographic scope rather than a whole basin.[3] Just as significantly, most of the literature does not draw specifically on the work on multi-track diplomacy, though there are many initiatives taking place in the basin at a range of scales that fit into this conceptual approach. At the same time, much of the popular commentary in different parts of the basin is mired in polemical assertions and distrust that discursively construct the basin as a zero-sum game (Hill, 2017).

To fill this gap, this article analyzes the key factors affecting transboundary water cooperation over the Brahmaputra River, drawing on the insights of the bourgeoning literature on transboundary water governance and multi-track diplomacy. In doing so, the analysis encompasses both current and potential future cooperation over the river. Key action situations analyzed in this article include China–India bilateral cooperation as well as civil-society-driven multi-track dialogue processes such as the Brahmaputra Dialogue. In examining these different action situations, the article argues that there are different lessons to be drawn that are of wider applicability. First, current processes of potential water cooperation between India and China are problematized by the lack of transparent information flow between the parties, a situation that only increases the discursive construction in some Indian media outlets of water sharing in the Brahmaputra as a zero-sum game. However, our analysis demonstrates that an actor-inclusive approach that includes sub-national bureaucrats and civil society members can increase the legitimacy and robustness of processes. This demonstrates the need to consider the potential of Track II/III diplomacy to create the conditions by which Track I arrangements can take up proposals that might otherwise gain little traction. More broadly, evolving processes of economic cooperation can enhance the potential for building trust and workable institutions, which could then enhance cooperation on water.

In bringing these factors together, our analysis demonstrates some of the key features of the proposed Zone of Possible Effective Cooperation (ZOPEC) for the Brahmaputra, where basin-wide cooperation among all the riparian countries might occur in con-junction with economic cooperation, allowing cross-sectoral cooperation and benefit sharing. In doing so, we demonstrate the necessity of thinking through the mutual gains of a whole-of-basin approach to developing the Brahmaputra basin.

Methodology

To analyze the key factors that affect water diplomacy over the Brahmaputra River, the research team[4] first developed a multi-track water diplomacy framework (to be discussed in the next section) as a conceptual and analytical framework for this research (Huntjens et al., 2016). Using this framework, eight cases of transboundary water cooperation action situations in the Brahmaputra Basin were analyzed, further refining the framework and subsequent analysis.[5] These were intended to capture the range of different stakeholders involved in this incredibly diverse and complex basin, as well as looking at how the sharing of water is embedded in a broader range of action situations that can both enhance and constrain water cooperation.

Our mixed-methods research involved collecting and triangulating data of three types: literature, interviews, and inputs from a stakeholder workshop.[6] Our 59 informants came from a variety of sectors, including government agencies, research organizations, media, NGOs and civil society in all four basin countries.[7] To validate the research findings and to gain further inputs to the research, a stakeholder workshop was conducted with 27 participants from different parts of the basin and beyond,[8] representing a variety of sectors (Furze, 2016).

Multi-track water diplomacy framework

The conceptual approach used in this article to understand the potential and limitations of cooperation over the Brahmaputra Basin draws on the bourgeoning literature that analyzes the factors affecting water cooperation, including those that work within political economy and institutional approaches.[9] The Multi-Track Water Diplomacy Framework (Figure 1) developed as the analytical framework for this research contains several key components that assist in understanding the potential cooperation, including the action situation, the broader and specific contextual situations, the mix of formal and customary institutions involved, and how these influence the interaction of actor and agency. The unique value of this framework lies in the fact there were no other frameworks that attempted to lay out the comprehensive political economy factors that affect water diplomacy, and that could also be used to analyze cooperation at different temporal scales. In addition, each component of the framework has detailed indicators that were used as guiding points for semi-structured interviews, as well as code for analysis using Max QDA, allowing detailed analysis of various factors affecting cooperation.[10] The framework takes its structure from the theory of structure-agency, where social structure (institutions) shapes actors and agencies, and vice versa (Giddens, 1984; Wendt, 1987).

Action situation

Our conceptual and analytical framework is centred around an *action situation*, which Huntjens et al. (2016) refer to as the social space where participants with diverse preferences

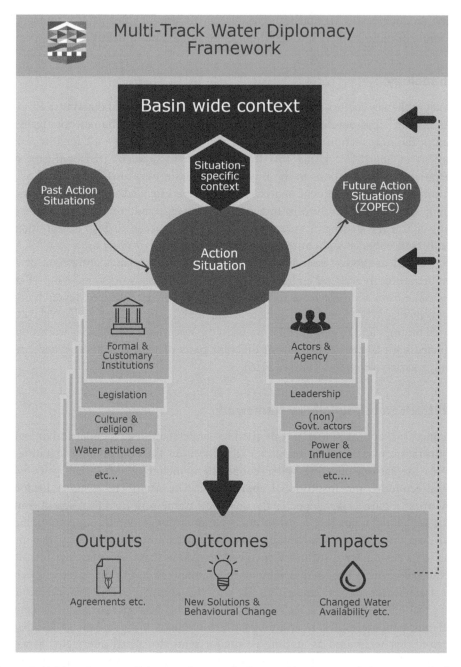

Figure 1. Multi-track water diplomacy framework. Source: Huntjens et al. (2016, p. 25); figure created by the research team, including the authors.

interact, exchange goods and services, solve problems, dominate one another, or fight (among the many things that individuals do in action situations). Although institutions may have some permanency, in our analysis of action situations the institutions are sustained or altered by the actions of the people that reproduce or change them (Huntjens et al., 2016). It is exactly at this juncture (i.e., in the action situation) that institutions are 'renegotiated' and changed. Indeed, as Ostrom (2005, p. 32) defined it, an action situation can be thought of as a 'situation when two or more individuals are faced with a set of potential actions that jointly produce outcomes'. Consequently, each action situation will have a certain set of outputs (e.g., decisions, agreements), outcomes (e.g., new solutions or behavioural change) and impacts (e.g., changed water availability). This, then, is a dynamic and contingent situation, since the outputs, outcomes and impacts may in turn impact, or feed back into, structure and agency. Each particular action situation has a situation-specific context that is multi-scalar in its dimensions, since some aspects of this context might be viewed as basin-wide, while other aspects are far more localized.

To understand the particulars of this multi-scalar context, our analysis needs to include key analytical components such as the biophysical material characteristics of the river; key socio-economic characteristics; the nature and extent of development; and past and ongoing water cooperation. Thus, our analysis of action situations acknowledges that pre-existing conditions will influence the structure and agency of different actors in the Brahmaputra, but also asserts that such situations are laden with the potential for change. In particular, we outline a range of instances that might be thought of as constituting a 'window of opportunity', where policy shifts might occur.

Formal and customary institutions

Both formal and customary institutions were identified as important factors affecting water cooperation. Adopting the definition of Calhoun (2002, p. 233), this framework defines institutions as 'deeply embedded patterns of social practices or norms that play a significant role in the organization of society'.

Formal institutions are institutions that are adopted through a formalized process. They include constitutional rules, codified laws, rules adopted by organizations, and policies (Huntjens et al., 2016). Codified laws and policies are often considered symbols of cooperation and often influence the status of cooperation (Wouters, Vinogrado, Allan, Jones, & Rieu-Clarke, 2005). Formal institutions not only include rules adopted by government entities but also refer to rules adopted by non-governmental organizations or the community through a formalized process.

Customary institutions are institutions that are embedded in organizations or groups without a formalized process. Culture, norms, religion, historical factors, and attitudes to water can all influence water cooperation (Aggestam & Sundell-Eklund, 2014; Creighton, Priscoli, Dunning, & Ayres, 1998; Johnston, 1998; Swain, 2004; Wigfield & Eccles, 2000).

Actors and agency

Water cooperation encompasses various actors, including government, political leaders, non-governmental organizations, civil society actors, religious organizations, academia, researchers and the private sector. In the context of water cooperation, actors' relationships, particularly their power relationships, is one of the key factors that affects cooperation (Zeitoun & Warner, 2006). Agency, which is the ability of an actor to exert influence, is therefore one of the key factors that can influence cooperation (Ali-

Khan & Mulvihill, 2008). Thus, this framework includes the existence of actors, actors' influence, power relationships, and leadership as key criteria for analysis.

Results

There is a well-documented tension in the literature and among practitioners regarding state-led Track I approaches and multi-track diplomacy approaches (Grech-Madin, Döring, Kim, & Swain, 2018). Our research sought to understand the state of cooperation across the basin by analyzing Track I (state-to-state) water diplomacy, Track II/III cooperation (facilitated by non-state actors), and potential future cooperation.

Track I cooperation

Status of cooperation
Although there are long-standing bilateral treaties between several riparian nation-states on the Brahmaputra River,[11] there is currently no Track I cooperation or dialogues involving all the riparian states which is specifically focused on the Brahmaputra.

   While multilateral cooperation clearly holds great potential for the Brahmaputra River basin, current state-to-state cooperation is focused on bilateral cooperation, including China–India, India–Bangladesh, Bhutan–India, Bangladesh–Bhutan and Bangladesh–China. There is no bilateral cooperation related to the Brahmaputra River between China and Bhutan. While our broader research (Yasuda, Aich, Hill, Huntjens, & Swain, 2017) has analyzed the factors affecting these five situations of bilateral cooperation, as well as the factors affecting non-cooperation, this article focuses on analysis of bilateral cooperation between China and India.[12] Arguably, this is the most vital of any of these relationships, since it concerns the two largest and most influential countries in the basin and because the evolving relationship between these two is also the source of a great deal of uncertainty and tension.

   India and China have signed two memoranda of understanding (MoUs) related to the Brahmaputra basin, which constitute the core of their existing cooperation. The first was signed in 2002, and has been renewed twice since then. Through this MoU, China collects flood-season hydrological data in remote locations in the upper reaches of the Brahmaputra River and provides them to India, and India pays China to cover the cost of data collection (Central Water Commission of India and Bureau of Hydrology and Water Resources of China, 2014). Another MoU was signed in 2013 to strengthen cooperation on the transboundary river through the existing 'expert-level mechanism' between the two countries, where member experts discuss technical issues of hydrology information sharing, monitoring and information sharing, and building hydrology models (CH13, 2016). There is also a cooperation on emergency responses to hazards (Ministry of External Affairs, Government of India, 2013).

Factors affecting cooperation
Key factors affecting this cooperation were analyzed using the Multi-Track Water Diplomacy Framework (Figure 2). One of the key basin-wide contexts that affected this cooperation is the recent development of Chinese hydropower dams in the

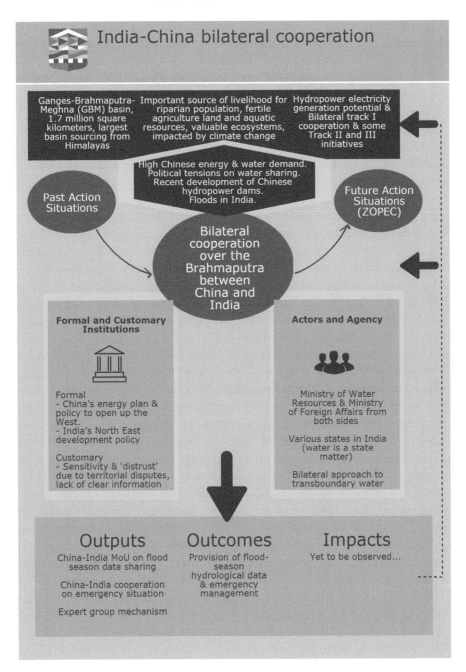

India-China bilateral cooperation

Ganges-Brahmaputra-Meghna (GBM) basin, 1.7 million square kilometers, largest basin sourcing from Himalayas

Important source of livelihood for riparian population, fertile agriculture land and aquatic resources, valuable ecosystems, impacted by climate change

Hydropower electricity generation potential & Bilateral track I cooperation & some Track II and III initiatives

High Chinese energy & water demand. Political tensions on water sharing. Recent development of Chinese hydropower dams. Floods in India.

Past Action Situations

Future Action Situations (ZOPEC)

Bilateral cooperation over the Brahmaputra between China and India

Formal and Customary Institutions

Formal
- China's energy plan & policy to open up the West.
- India's North East development policy

Customary
- Sensitivity & 'distrust' due to territorial disputes, lack of clear information

Actors and Agency

Ministry of Water Resources & Ministry of Foreign Affairs from both sides

Various states in India (water is a state matter)

Bilateral approach to transboundary water

Outputs
China-India MoU on flood season data sharing

China-India cooperation on emergency situation

Expert group mechanism

Outcomes
Provision of flood-season hydrological data & emergency management

Impacts
Yet to be observed...

Figure 2. Factors affecting water cooperation between China and India. Source: Yasuda et al. (2017, p. 31); figure created by the authors.

upstream of the Brahmaputra River. The first of these built in the Chinese part of the Brahmaputra, the Zangmu Dam, was operationalized in 2014, and three more (Da Gu, Jie Xu and Jia Cha) are planned by the government (Samaranayake, Limaye, & Wuthnow, 2016; State Council, 2013). There has also been a debate in China about whether to divert water from the Brahmaputra to water-scarce parts of that country.

Some commenters report that the plan is currently on hold as other inter-basin transfers within the country are undertaken (Amano, 2015; Samaranayake et al., 2016), while some commenters outside China assert that satellite images show possible signs of the construction of water-diversion tunnels (Bhat, 2017; Dasguptal, 2017). What is undoubtedly true is that these relatively new developments in the basin have created concerns downstream in India. These concerns accelerated after 2000, when a naturally formed dam in one of the tributaries of the Brahmaputra in the Chinese part of the basin broke and flooded Arunachal Pradesh and Assam (north-eastern provinces of India), killing 30 people and leaving 50,000 homeless. Samaranayake et al. (2016) argues that this incident was catalytic in initiating the cooperation on data sharing between China and India.

Several formal institutions affect this basin-wide context. Recent development of Chinese hydropower dams is specified in the energy plan under the 12th Five-Year Plan, a socio-economic development plan covering the development period of 2010–2015 (Samaranayake et al., 2016; State Council, 2013). More specifically, the Open Up the West policy, launched in 2000 to encourage development of the impoverished western part of China, means a greater focus on areas within or closely connected to the Chinese part of the Brahmaputra (Lai, 2002; Samaranayake et al., 2016). In the lower riparian parts of the basin, in India, there are also attempts to increase infrastructure development. Indeed, while the North-East region has historically been politically unstable and comparatively neglected, the Modi government's Act East policy has increased Indian government attention to this part of the country and seeks to foster increased interconnectivity with neighbouring countries (Ministry of Development of North Eastern Region, Government of India, n.d.).

While there is evidently a lack of trust between the two nations with regard to competing plans for economic integration, including around hydropower, this distrust has broader and deeper contours, which means that it is a key customary institution that affects the cooperation. Indeed, disagreement over territory between China and India has existed since 1947, when the British colonial government of India drew a border not recognized by China, but this enmity intensified as a consequence of the Indo–Chinese war in 1962 (Lidarev, 2012). A number of interviewees in both China and India indicated this unresolved border dispute as a key source of distrust between the two nations, affecting the status of their transboundary cooperation (CH7, 2016; CH8, 2016; CH11, 2016; CH14, 2016; IN5, 2016). In addition to distrust arising from this long-standing border dispute, the lack of information sharing in the past has contributed to distrust. Particularly in relation to the Brahmaputra River, information about its hydropower plans were not disclosed officially by Chinese government until 2010, and it had denied construction and planning of these dams, raising concern and suspicion in downstream nations (IN8, 2016; Samaranayake et al., 2016).

In addition to traditional government actors engaged in transboundary water cooperation, such as the Ministry of Foreign Affairs and the Ministry of Water Resources, many interviewees indicated the role of media, particularly in India, as raising awareness (and nationalist fervour) over development in upstream parts of the Brahmaputra River (CH1, 2016; CH15, 2016; CH17, 2016; CH18, 2016; IN11, 2016). There was a perception among some of the Chinese interviewees that Indian media were not necessarily communicating the 'true' information (CH1, 2016; CH6, 2016; CH17,

2016), contributing to distrust between the two nations. Such comments fit with the widely noted role of Indian media as too often encouraging hyperbole and public speculation over Chinese actions in the basin (Hill, 2017, 2015)

Track II/III cooperation

Status of cooperation
In addition to bilateral cooperation by state actors, non-governmental actors also take initiative in facilitating cooperation. One of these facilitated processes is the Brahmaputra Dialogue (hereafter, the Dialogue), which was initiated by NGOs and research institutes, including the South Asian Consortium for Interdisciplinary Water Resources Studies (SaciWATERS), the Indian Institute of Technology Guwahati and the Institute of Water and Flood Management Bangladesh. The Dialogue was initiated in 2013 and since then has had three phases. The first phase focused on facilitating dialogues among non-state actors, starting with national dialogue in India and Bangladesh, then moving to a transboundary dialogue between Indian and Bangladeshi stakeholders (Banerjee, Salehin, & Rames, 2014). The second phase of the Dialogue expanded its scope to encompass dialogues between Indian states, including Assam and Arunachal Pradesh, which included some of the state actors from both central and provincial governments (SaciWATERS, 2015). The first multi-country dialogue was also initiated during this phase and included Bhutan and China (SaciWATERS, 2016b). The third and current phase of the dialogue aims to include all the riparian countries of the Brahmaputra basin, to bring political willingness, and to develop a joint mechanism for effective basin management (SaciWATERS, 2016b).

While the Dialogue is still an ongoing process, that this process has shifted its nature from Track III to Track I.5 is one of the key achievements. This shift to Track 1.5 was undoubtedly helped by the fact that a government official facilitated one of the country dialogues during this phase. As a concrete output from the Dialogue, a joint research proposal was developed by research organizations in China, India and Bangladesh to conduct a vulnerability assessment for the entire stretch of the river (BA10, 2016), which would be a useful input to policy development.

Factors affecting cooperation
The Brahmaputra Dialogue process recognizes the existence of competing and complementary stakeholders in each country's water sector. Thus, during the first phase, the processes sought to build confidence and create understanding between the stakeholders in India and Bangladesh. In India, the dialogue initially engaged civil society groups, academics, and, later on, bureaucrats from the upper and lower riparian states of Assam and Arunachal Pradesh, as well as representatives from other parts of the Indian water sector (SaciWATERS, 2015). Considering that effective development of the basin is not under the ambit of a single ministry, but rather of many actors that have overlapping jurisdictions and claims and exist in a variable chain of influence, it is also necessary to have representatives of different bureaucracies, such as the Ministry of Environment, Forest and Climate Change and the Central Water Commission, as well as other state and central authorities. A wide range of governmental actors have a stake in reconciling the interests of different states, and these actors are not necessarily

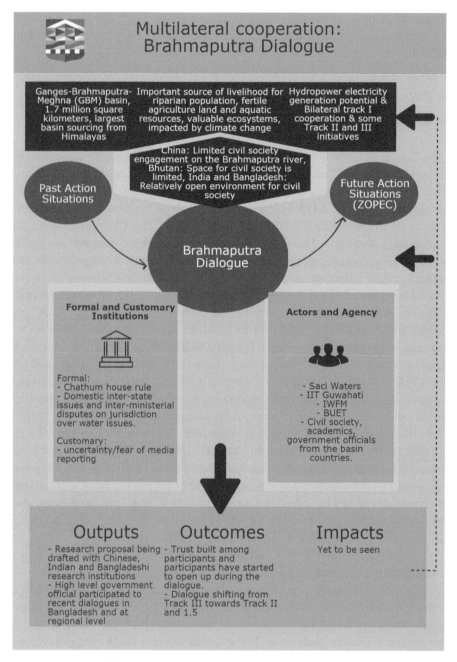

Figure 3. Factors affecting the Brahmaputra Dialogue. Source: Yasuda et al. (2017, p. 96); figure created by the authors.

involved when India is negotiating with another country in the Brahmaputra. The Brahmaputra Dialogue broadened the inclusion of actors to a multilateral level at a later stage, recognizing the issue as multilateral. Recognizing the complexity of the issue, the Brahmaputra Dialogue gradually scaled up the process to encompass various actors who have stake in the basin.

The inclusion of various kinds of stakeholders in the dialogue was also affected by how India's formal institutions shape responsibility over water resources management. The Constitution of India (1949, Seventh Schedule) gives major responsibility over water to the states, which makes it challenging to resolve inter-state water disputes. Within the Brahmaputra basin, there is an ongoing inter-state conflict between Assam and Arunachal Pradesh, particularly related to Assam's opposition to hydropower dam plans in Arunachal Pradesh and fear of the flooding it may create (IN17, 2016). This situation resulted in the need for attention to inter-state dialogue and discussions within India as one of the key dialogue process for the Brahmaputra.

Another formal institution that guided the dialogue process is adherence to the Chatham House Rule, which allows anonymity and openness of discussion (Chatham House, n.d.). The lead author's observation of one of the Dialogue meetings affirms that this rule allows openness in discussion among participants. One of the Bangladeshi interviewees echoed this observation, while also pointing out that some of the participants in their national dialogue were uncomfortable with the situation where they are not able to speak freely about what they had learnt during the Dialogue process (BA10, 2016). While the Dialogue process could be elevated to a wider public discussion, media representatives have not been invited to any of these sessions. This is related to uncertainty and fear of how the media may misreport the dialogue, particularly the regional dialogue processes (BA10, 2016; SaciWATERS, 2016a).

While many of our respondents were supportive of the Chatham House Rule, some believed more engagement with the media could be helpful. For example, some Bhutanese interviewees commented that these kinds of initiatives could foster more discussion in their relatively small and closely knit society, where it is not customary for the public to openly react or to express grievances (BT4, 2016). More broadly, civil society actors in the Brahmaputra basin work in a range of varying social contexts. Of the four countries, civil society in India and Bangladesh acts more openly compared to China and Bhutan. There is also variance in how much attention the Brahmaputra basin is afforded by civil society in each country. In China, a number of interviewees commented on the absence of NGOs working on the Brahmaputra (CH1, 2016; CH3, 2016; CH4, 2016), and one of the Chinese participants in the Dialogue appreciated that their participation allowed them to understand each other and the issues other riparian countries face.[13]

Figure 3 illustrates these factors and how they have influenced the Brahmaputra Dialogue process.

Zone of Possible Effective Cooperation

As well as analyzing the contours of disputes between the different stakeholders in the basin and the range of institutional responses currently devoted to their resolution, the research also analyzed potential areas of future cooperation. The ZOPEC refers to a combination of viable future action situations that allow mutual gains, where parties gain more by cooperating than by not cooperating (Huntjens et al., 2016). This research identified the ZOPEC through three inputs: analysis of existing cooperation action situations along the Brahmaputra River; analysis of current and future factors that potentially influence cooperation; and a stakeholder workshop involving 27 participants from all the riparian countries and regional actors (Furze, 2016).

Emerging factors that could affect the ZOPEC

Several emerging factors in the Brahmaputra basin could affect the ZOPEC (Figure 4). While there is currently no basin-wide Track I cooperation over the Brahmaputra River involving all the riparian countries, the basin encompasses existing and emerging regional economic cooperation mechanisms that could benefit cooperation over

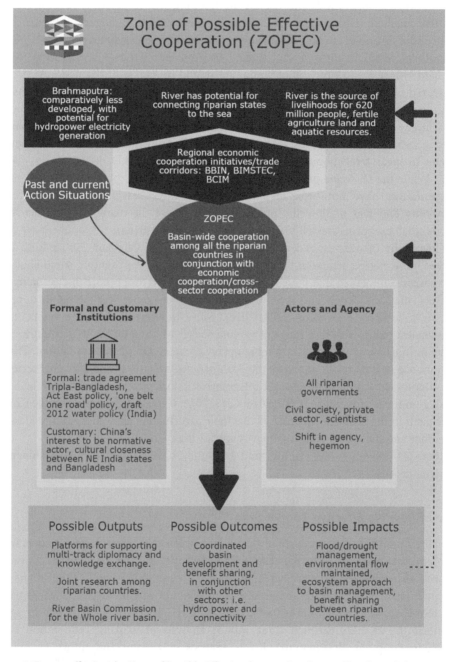

Figure 4. Factors affecting the Zone of Possible Effective Cooperation. Source: Yasuda et al. (2017, p. 104); figure created by the authors.

water. One of the most significant mechanisms is BBIN (Bangladesh Bhutan India Nepal), which hosts a Joint Working Group on Water Resources Management and Power as well as one on connectivity and transit (Energy Bangla, 2015). The terms of reference for the Joint Working Group include exploring hydro-power, undertaken jointly with at least three countries on an equitable basis; exchanging experiences and best practices; and developing grid connectivity (Sachin, 2016). Clearly, BBIN has the potential to directly influence cooperation over the water and energy sectors associated with the Brahmaputra (BA8, 2016; Energy Bangla, 2015; Economic Times, 2015).

While BBIN might be the most directly notable initiative in terms of directly seeking to coordinate the possibilities of water and energy cooperation in eastern South Asia, there are a range of overlapping and competing mechanisms for enhancing economic cooperation that could impact the Brahmaputra basin. For example, the Bangladesh-China-India-Myanmar Forum for Regional Cooperation is a sub-regional organization that aims to promote a multimodal trade corridor connecting the south-western part of China with Bangladesh through Myanmar and India (Aneja, 2015; Sajjanhar, 2016). If this vision is realized, the Brahmaputra River could become a trade route.

At a global scale, China is leading the Belt and Road Initiative, which aims to establish new land-based and ocean-going trade routes (Verlare & van der Putten, 2015). While the enormous roll-out of infrastructure mooted under the Belt and Road Initiative could alter geo-economic realities around the globe, in the Brahmaputra basin it is likely to have an impact through the section of the policy that encompasses the China–Myanmar–Bangladesh–India Corridor. The Belt and Road Initiative sees an increasing role for Bangladesh in particular, as it can offer an ocean trade route through its port facing the Bay of Bengal. Another, somewhat competing, version of regional economic cooperation is the Bay of Bengal Initiative for Multi-Sectoral Technical and Economic Cooperation (BIMSTEC, 2015), which includes Bangladesh, Bhutan, India, Nepal, Sri Lanka, Myanmar and Thailand.

These more general regional economic cooperation initiatives demonstrate a desire among a range of stakeholders across the Brahmaputra basin for greater economic integration, which may enhance the potential for cross-sectoral collaboration, expand and the 'pie' of possible cooperation. Again, however, we need to be conscious that such top-down initiatives have their own priorities and ideas about the kinds of trade-offs that are desirable in the search for closer economic connectivity (Hill, 2015).

Clearly, different combinations of formal institutions and geopolitical contexts in the region will also affect potential shifts in the agency of riparian states. Historically, India has been the largest and strongest country in South Asia, both politically and economically. Its hegemonic situation vis-à-vis smaller neighbouring countries has affected water cooperation. In particular, Bangladesh, as a lower riparian and smaller nation to India, has always has difficulty in transboundary water cooperation with its upstream neighbour, but broader developments within and beyond the basin have arguably shifted this situation (BA2, 2016; BA3, 2016; BA4, 2016; BA6, 2016; BA7, 2016; Swain, 2010). China's recent development upstream of the Brahmaputra River means that India is now more definitively a middle riparian, concerned about the actions of its upstream neighbour (The Hindu, 2016), a shift that arguably provides an incentive for India to collaborate multilaterally over the Brahmaputra River. Similarly, China's emerging interest in the use of a port in Bangladesh as a gateway to the sea potentially raises

Bangladesh's position in negotiating with India over transit trade, a point that has long been difficult.

Cross-border collaboration between North-East India and Bangladesh, which could enhance benefit sharing, is backed by both formal and customary institutions, including, for some respondents, a sense of shared culture (BA5, 2016). There is certainly a significant precedent for drawing on shared culture to facilitate enhanced cooperation over water in the region, particularly in the case of the role that then chief minister of West Bengal, Jyoti Basu, played in fostering the 1996 Ganges Water-Sharing agreement between India and Bangladesh. Many commenters have written that Basu was important in this negotiation because he was able to draw on the common culture of Bengalis on both sides of the border. Conversely, the obstructionist attitude of the current chief minister of West Bengal, Mamata Banerjee, in delaying the signing of an agreement between India and Bangladesh over the Teesta River demonstrates that domestic sub-national political priorities can override these shared cultures and histories. Perhaps more troubling in this sense is the long history of anti-Bangladeshi-migration rhetoric in some political parties in North-East India's most populous state, Assam, which is evidence of long-standing sentiments against closer connectivity, at least in terms of population movements (Hazarika, 2000). Nevertheless, there are some positive signs to indicate that cross-sectoral collaboration is occurring in different parts of the region. For example, from 2016 onwards the Indian state of Tripura has provided 100 MW of electricity to Bangladesh in exchange for 10 gigabits-per-second internet bandwidth from Bangladesh (Express News Service, 2016).

Another formal institution that enhances cross-border collaboration is the collaboration between Bhutan, Bangladesh and India in hydropower development in Bhutan. Bangladesh had long been interested in investing in hydropower in Bhutan, but the electricity transmission needed to cross Indian territory. Reports at the end of 2017 indicate that a tripartite agreement on this investment might soon be realized, which opens new directions in multilateral cooperation over the use of the river (ANI, 2017; Tribune Desk, 2017).

And it is not only in the economic realm that potential basin-wide cooperation might occur. Indeed, if riparian countries adopt some of their existing formal institutions to the context of transboundary waters, there is a potential for basin-wide cooperation with ecological considerations. Such initiatives are extremely important because despite some of the ongoing physical alterations to the river, the Brahmaputra is relatively pristine compared to other large rivers in the world, allowing room to manage the river while ensuring ecological integrity.[14] While Bhutan has long been heralded for the space it gives in policy to environmental concerns,[15] other countries in the region have historically given this sphere less emphasis. An important step towards this is that China has embraced the concept of environmental flow in how it conceives of the management of its rivers, while more broadly adopting a policy on Ecological Civilization, which aims to ensure conservation and sustainable economic and social development (UNEP, 2016). Through this policy, the current Chinese leader, President Xi, has pointed out that 'green is gold' (CH16, 2016; UNEP, 2016). In its domestic policy, China has adopted its own version of payment for ecosystem services in a scheme called Eco-Compensation, and conducted pilot work applying the scheme to watersheds and river basins (Bennett, 2009; Zhang, Lin, Bennett, & Jin, 2010). Such changes are important in light of the comments of several of our Chinese respondents, who pointed to China's interest in being a normative actor in the international context. This desire can arguably act as a significant catalyst for cooperation over the Brahmaputra and particularly application of these progressive domestic environmental policies in the context of transboundary waters (CH13, 2016; CH14, 2016; CH20, 2016).

Potential future cooperation: the ZOPEC for the Brahmaputra River basin

Given these emerging factors allowing potential regional cooperation, shifts in actors' agency and potential for cross-sectoral benefit sharing, it is reasonable to propose that the ZOPEC for the Brahmaputra is basin-wide cooperation among all the riparian countries in conjunction with economic cooperation, allowing cross-sectoral coopera- tion and benefit sharing. Such cooperation integrates all sectors, involving water, ecology and economy in its scope, and could provide mutual gains for the riparian countries and bring solutions for sustainably managing the river basin. Any type of development in the basin, including infrastructure (in particular for hydropower, flood control, irrigation and navigation), needs to take a whole-of-basin approach. For example, the development and the level of flow and sediments needs to be coordinated jointly to maintain the ecology of the river system, as well as to ensure navigation. Benefits derived from this infrastructure can be shared fairly among riparian states.

Taking a cross-sectoral approach to water cooperation can open up space for sharing benefits from different sectors. For example, downstream countries (e.g., Bangladesh and India) can benefit from upstream hydropower generation by offering trade routes (navigation, road and rail) and access to port facilities in return for energy from the hydropower-generating country (e.g., China or Bhutan). At the same time, there are inevitably trade-offs between sectors, and these need to be discussed and negotiated in an open and transparent manner if inclusive development and environmental sustain- ability are to be achieved.

Concrete outputs suggested in the stakeholder workshop discussing the ZOPEC included joint research among riparian countries; improved data and information sharing, particularly through establishment of a knowledge platform; economic coop- eration, with mechanisms to facilitate cooperation through the development of plat- forms to support multi-track diplomacy; and as a River Commission for the Brahmaputra basin (Yasuda et al., 2017).

Discussion

Our analysis of the three cooperation action situations outlined above indicates the shifting political economy of the Brahmaputra Basin, which can organically affect relationships and the dynamics of how cooperation takes place. The emergence of Chinese activities upstream on the Brahmaputra River, coupled with its interests in expanding its trade routes using waterways, could change Bangladesh's relationship to its upper-riparian states. Similarly, analysis of Bhutanese relationships with other riparian countries identified the democratization of Bhutan and its shift away from dependence on Delhi for its foreign policy as enabling Bhutan to draw closer to China, which would contribute to broader changes in the relationships among the riparian states.

The analysis also confirmed that all the analytical factors of the Multi-Track Water Diplomacy Framework – basin-wide context, formal institutions, and customary institutions and actors – played roles in influencing the status of cooperation. Clearly, the interaction of these analytical factors influenced outcomes. For example, the contextual factors of Chinese upstream development are shaped by its formal institutions of energy development policy, which affects the positioning of India vis-à-vis its neighbouring states, including China and Bangladesh, in its water diplomacy efforts. In the analysis of the ZOPEC, the emergence of new formal institutions shifted actors' positions and agency. These interactions and influences between institutions and actors/agency prove the relevance and validity of the structure-agency approach in the analysis and the framework.

The application of the Multi-Track Water Diplomacy Framework in different tracks of water diplomacy (Tracks I, II and III), as well as different temporal scales (current and future), proved the robustness of the framework in analyzing factors affecting various forms of water diplomacy. The framework allowed analysis of institutions and actors separately, as well as their interactions, in understanding the influences and relationships among them in water cooperation. Analysis of the key factors affecting cooperation, including legal, political and economic, also enabled identification of the ZOPEC, which integrates different sectorial interests, potentially leading to a whole-of-basin approach to managing water that is conscious of the various trade-offs involved (Ringler, Bhaduri, & Lawford, 2013).

Analysis of existing and potential initiatives for water diplomacy, as well as inclusion of emerging basin context, allows us to identify 'windows of opportunity' that could facilitate a shift in thinking about mutual benefits or introduce a new dimension in transboundary water cooperation. Windows are 'particular moments in time (for instance an election or disaster) that offer opportunities for policy entrepreneurs to launch and gain support for new policy proposals' (Huitema & Meijerink, 2010, p. 30). As water diplomacy takes place within a dynamically changing basin-wide context with shifting power dynamics (Barua, Vij, & Zulfiqur Rahman, 2017), observing the right timing and identifying the window of opportunity for policy shifts is critical. However, most existing scholarship on water diplomacy and transboundary water cooperation does not specifically take into account this concept of windows of opportunity but focuses on either substantive aspects of cooperation or the shifting context of conflicts (Sadoff & Grey, 2002; Tarlock, 2015; Zeitoun & Mirumachi, 2008). Identification of the ZOPEC using the Multi-Track Water Diplomacy Framework, which takes into account factors affecting cooperation from different tracks of water diplomacy, and consideration of the temporal aspect by analyzing emerging basin context, allows transboundary water cooperation research to identify windows of opportunity.

The emergence of a civil-society-led process such as the Brahmaputra Dialogue provides an example of a process that could be the vehicle to facilitate multiple tracks of water diplomacy. The flexible nature of the civil society process allowed the dialogue to expand to basin-wide cooperation, including all the countries. Engagement of all the basin countries, including multiple layers of stakeholders, will be an important step in future basin-wide cooperation and development with integrity. While there are certainly grounds for optimism around these processes, it is also clear that they are constrained in what they can achieve. Thus, the leadership of civil society in starting dialogues can foster relationships that government actors would find more difficult to initiate with other riparian countries. On the other hand, scaling up lower-track dialogues so that they begin to influence the policies of governments is far from guaranteed, although there are examples of multi-stakeholder dialogues in the Rhine, Mekong, and Ganga-Brahmaputhra-Meghna river basins which have informed and shaped more formal negotiation and decision-making processes (Huntjens et al. 2017). Nevertheless, the trust engendered by the Brahmaputra Dialogue among various participants in the basin must be recognized as an important step forward.

Conclusion

Transboundary water cooperation over the Brahmaputra River faces important junctures. Since one of the goals of water diplomacy is to strengthen the viewpoint of mutually shared benefits in bilateral and basin-wide contexts (Keskinen, Inkinen, Hakanen, Rautavaara, & Niinioja, 2014), our results confirm that water diplomacy cannot and does not focus only on water. Our argument in this article is that each action situation holds the possibility for the renegotiation of the terms of water cooperation, and thus there is considerable potential for more appropriate institutional arrangements to emerge. Therefore, having identified the key factors of each action situation, according to our framework, and how they influence each other, we are a step closer to a holistic understanding of three aspects of water cooperation. First, it supports existing claims that broader-perspective water diplomacy can result in broader diplomatic cooperation, much as cooperation (or non-cooperation) in other sectors can be manoeuvered towards diplomatic solutions to regional challenges (Keskinen et al., 2014; Sadoff & Grey, 2002). Second, the impact of multi-track diplomacy approaches, especially the initiatives of Track III and II dialogues, become effective and popular across non-state and, to an extent, state actors. Finally, shifting political relation-ships and agencies among basin riparian states, coupled with trends towards more regional economic cooperation, supported by formal and customary institutions, provide an oppor-tune moment to move water diplomacy in the basin to the next level.

These geopolitical and economic trends also give pressure to the way development occurs in the river basin, and if appropriate measures are not taken sooner or later, the Brahmaputra could risk having uncoordinated development. The emergence of civil-society-led water diplomacy can bring various stakeholders closer to each other in the basin, where various actors' interests coexist. Identifying the ZOPEC that can bring mutual gains to different parties and exploring options for expanding the 'pie' that can be shared among basin countries are crucial in finding solutions to transboundary water conflicts. Analysis of the various factors that affect water diplomacy supports identifica-tion of solutions that exist outside the 'water box'.

Notes

1. This river has different names in different sections. It is called the Yarlung-Tsangpo in China, the Brahmaputra in India, and the Jamuna in Bangladesh. The Manas River in Bhutan flows into the Brahmaputra in the Indian section of the river. While recognizing this variety of names, this article refers to the river as the Brahmaputra River.

2. A representative selection of the kinds of articles found using 'Brahmaputra River' as keywords includes James M. Coleman, 'Brahmaputra River: Channel Processes and Sedimentation', *Sedimentary Geology* 3, 2–3 (1969); M. M. Sarin et al., 'Major Ion Chemistry of the Ganga-Brahmaputra River System: Weathering Processes and Fluxes to the Bay of Bengal', *Geochimica et cosmochimica acta* 53, no. 5 (1989); and Colin R. Thorne, Andrew P.G. Russell, and Muhammad K. Alam, 'Platform Pattern and Channel Evolution of the Brahmaputra River, Bangladesh', *Geological Society, London, Special Publications* 75, no. 1 (1993).

3. The following reports analyse the political economy perspective of the Brahmaputra River, with a limited geographic scope: Samaranayake et al. (); Prasai and Surie, 2013; Price and Mittra, (2016); Bandyopadhyay and Ghosh (2016); Hill (2015). A more recent article that includes analysis with a wider scope within the basin is Barua et al. (2017).

4. More info at http://www.thehagueinstituteforglobaljustice.org/projects/water-diplomacy-making-water-cooperation-work/.

5. The broader findings of this research can be found in Yasuda et al. (2017).

6. The literature included academic articles, reports, websites, government documents, laws, policies, newspaper and media articles, maps, and other published and unpublished data on water ecosystems and biodiversity. The literature review was conducted throughout the research, which took place from October 2015 until December 2017.

7. A total of 61 interviews were conducted in four riparian countries of the Brahmaputra River, as some of the interviewees were seen twice. Interviews were recorded or interview notes taken, with permission from the interviewee. Due to the sensitivity of the subjects discussed, all interviews are cited anonymously. Interviews cited are abbreviated in the following ways: Interviews conducted in China (CH), India (IN), Bhutan (BT) and Bangladesh (BA).

BT4. (2016): Interview with BT4 on 1 September 2016.
BA10. (2016): Interview with BA10 on 28 October 2016.
BA2. (2016): Interview with BA2 on 28th March 2016.
BA3. (2016): Interview with BA3 on 29th March 2016.
BA4. (2016): Interview with BA4 on 30th March 2016.
BA5. (2016): Interview with BA5 on 30th March 2016.
BA6. (2016): Interview with BA6 on 31st March 2016.
BA7. (2016): Interview with BA7 on 31st Mach 2016.
BA8. (2016): Interview with BA8 on 2nd April 2016.
CH1. (2016): Interview with CH1 on 3 April 2016.
CH10. (2016): Interview with CH10 on 26 April 2016.
CH11. (2016): Interview with CH11 on 26 April 2016.
CH13. (2016): Interview with CH13 on 27 April 2016.
CH14. (2016): Interview with CH14 on 27 April 2016.
CH15. (2016): Interview with CH15 on 29 April 2016.
CH16. (2016): Interview with CH16 on 2 May 2016.
CH17. (2016): Interview with CH17 on 4 May 2016.
CH18. (2016): Interview with CH18 on 5 May 2016.
CH20. (2016): Interview with CH20 on 5 May 2016.
CH3. (2016): Interview with CH3 on 5 April 2016.
CH4. (2016): Interview with CH4 on 5 April 2016.
CH6. (2016): Interview with CH6 on 6 April 2016.
CH7. (2016): Interview with CH7 on April 2016.
CH8. (2016): Interview with CH8 on 12 April 2016.
IN11. (2016): Interview with IN11 on 16 May 2016.
IN17. (2016): Interview with IN17 on 9 November 2016.
IN5. (2016): Interview with IN5 on 2 September 2016.
IN8. (2016): Interview with IN8 on 11 May 2016.

8. There were three participants from Bangladesh, five from Bhutan, five from China, five from India, two from regional organizations and seven from the workshop organizer.
9. Key literature consulted included Creighton et al. (1998), Grey et al. (2010), Priscoli and Wolf (2009), Sadoff and Grey (2002), Swain (2004), Wouters et al. (2005), Zawahri (2008), Zeitoun and Mirumachi (2008), Huntjens (2011) and Huntjens et al. (2011, 2012).
10. For details of these indicators, see Huntjens et al. (2016).
11. Most significant here are the bilateral treaties between India and Bangladesh and between India and Bhutan. Between India and Bangladesh, the key treaties are the Statute of the Indo-Bangladesh Joint Rivers Commission (1972), which established the Joint Rivers Commission, and the Ganges Treaty (1996). Between India and Bhutan, there are a Treaty of Friendship (signed in 1949 and revised in 2007) and a cooperation agreement on hydroelectric power (2006).
12. Analysis of cooperation between other riparian states is available in Yasuda et al. (2017).
13. Informal conversation with one of the participants in the Brahmaputra Dialogue Workshop, Singapore 2016. Permission to cite given by the participant.
14. Chinese experts from the Ministry of Water Resources collaborated with international experts in 2013 to publish a report on basin water allocation plan that integrates the concept of environmental flow (Speed, Yuanyuan, Quesne, Guy, & Zhiwei, 2013). A Chinese respondent indicated that China has recently adopted the concept of environmental flow in developing the master plan for the Yangtze River, which suggests that such approaches might become more widespread throughout that country (CH10, 2016).
15. Bhutan has a pro-environmental constitution that requires a minimum of 60% of the land to be covered with forest, providing important watershed for rivers (Constitution of the Kingdom of Bhutan, 2008).

Acknowledgments

The authors thank all interviewees and workshop participants for their kind contributions. The research team appreciates the contribution of IUCN Asia for the research work in Bangladesh, India and Bhutan, as well as the multi-stakeholder workshop conducted for this project. Many thanks to Archana Chatterjee and Brian Furze for their kind support of this research project. The authors sincerely thank Marian J. Neal for taking time to provide constructive feedback to the overall research leading to this article, and Rens de Man for his coordination of the multi-stakeholder workshop. We are grateful to Josh Weinberg for his input to the fieldwork in China and for written feedback on an earlier draft of this article. The authors also thank Bjørn-Oliver Magsig and Thom Almeida for their input to the legal analysis.

Disclosure statement

No potential conflict of interest was reported by the authors.

Funding

The research leading to this article was funded by the Hague Institute for Global Justice and the Stockholm International Water Institute.

References

Aggestam, K., & Sundell-Eklund, A. (2014). Situating water in peacebuilding: Revisiting the Middle East peace process. *Water International*, 39(1), 10–22.

Ali-Khan, F., & Mulvihill, P. R. (2008). Exploring collaborative environmental governance: Perspectives on bridging and actor agency. *Geography Compass, 2*(6), 1974–1994.

Amano, K. (2015). Analysis of conflict and cooperation between China and India on the Brahmaputra River Basin water resources. *Asian Studies, 61*(2), 55–68.

Aneja, A. (2015, June 26). China, India fast-track BCIM economic corridor project. *The Hindu.* Retrieved from http://www.thehindu.com/news/national/china-india-fasttrack-bcim-economic-corridor-project/article7355496.ece

ANI. (2017, July 4). Bhutan to supply hydropower to Bangladesh through India soon. *ET Energyworld.* Retrieved from https://energy.economictimes.indiatimes.com/news/power/bhutan-to-supply-hydropower-to-bangladesh-through-india-soon/59434785

Bandyopadhyay, J., & Ghosh, N. (2016). Hydro-political dynamics and environmental security in the Ganges-Brahmaputra-Meghna basin. *Social Sciences, 24,* 1–25.

Banerjee, P., Salehin, M., & Rames, V. (2014). *Water management practices and policies along the Brahmaputra River Basin: India and Bangladesh: Status report 2014.* India: SaciWATERs.

Barua, A., Vij, S., & Zulfiqur Rahman, M. (2017). Powering or sharing water in the Brahmaputra River basin. *International Journal of Water Resources Development,* 1–15. doi:10.1080/07900627.2017.1403892

Bennett, M. T. (2009). *Markets for ecosystem services in China: An exploration of China's "eco-compensation" and other market-based environmental policies.* Retrieved from http://www.forest-trends.org/documents/files/doc_2317.pdf

Bhat, V. (2017, December 13). Images show China may be using a secret tunnel to divert Brahmaputra water into desert. *ThePrint.* Retrieved from https://theprint.in/2017/12/13/first-evidence-of-complete-brahmaputra-by-china/

BIMSTEC. (2015). *Organization structure.* Retrieved from http://www.bimstec.org/index.php?page=working-structure.

BT4. (2016). [Interview with BT4 on 1 September 2016].

Calhoun, C. (2002). *Dictionary of the social sciences.* Oxford University Press on Demand. Retrieved from http://www.oxfordreference.com/view/10.1093/acref/9780195123715.001.0001/acref-9780195123715

Central Water Commission of India and Bureau of Hydrology and Water Resources of China. (2014). *Implementation plan between the central water commission, ministry of water resources, river development and Ganga Rejuvenation, the Republic of India and the bureau of hydrology and water resources, tibet autonomous region, the People's Republic of China upon provision of hydrological information of the Yarlung Zanbgu/Brahmaputra River in flood season by China to India.* Retrieved from http://indianembassybeijing.in/implementation-plan.php

Chatham House. (n.d.). *Chatham house rule.* Retrieved from https://www.chathamhouse.org/about/chatham-house-rule

Constitution of India. (1949). Retrieved from https://www.india.gov.in/my-government/constitution-india/constitution-india-full-text

Constitution of the Kingdom of Bhutan. (2008). Retrieved from http://gov.bt/wp-content/uploads/2017/08/Constitution-of-Bhutan-Eng-2008.pdf

Creighton, J. L., Priscoli, J. D., Dunning, C. M., & Ayres, D. B. (1998). *Public involvement and dispute resolution - volume 2: A reader on the second decade of experience at the institute for water resources.* Unpublished report. The Hague Institute for Global Justice. Alexandria, VA: US Army Corps of Engineers. Water Resources Support Centre. Institute of Water Resources. Retrieved from file:///C:/Users/yumya701/OneDrive%20for%20Business/Literature/Effective%20water%20cooperation/Deliscolli%201998.pdf

Dasguptal, S. (2017, October 31). China mulls 1,000 km tunnel to divert Brahmaputra waters to Xianjiang region. *The Times of India.* Retrieved from https://timesofindia.indiatimes.com/world/china/china-plans-to-divert-brahmaputra-waters-to-its-xinjiang-region-through-1000-km-long-tunnel/articleshow/61346091.cms

Economic Times. (2015, January 31). India explores scope for power trade with Bangladesh, Bhutan, Nepal. *Economic Times.* Retrieved from http://articles.economictimes.indiatimes.com/2015-01-31/news/58650561_1_power-trade-the-jwg-inter-grid-connectivity

Energy Bangla. (2015). *Power & transit cooperation between BBIN*. Retrieved from http://energy bangla.com/power-transit-cooperation-bbin/

Express News Service. (2016, March 24). Take 100 MWs of power, give 10Gbps internet speed: India-Bangladesh deal. *The Indian EXPRESS*. Retrieved from http://indianexpress.com/article/ india/india-news-india/take-100-mws-of-power-give-10-gbps-internet-speed-indo-bangla- new-deal/

Furze, B. (2016, November 8-9). Making water cooperation work in the Brahmaputra Basin. Draft report of the dialogue organized in Bangkok by the Hague Institute of Global Justice.

Giddens, A. (1984). *The constitutions of society: Outline of the theory of structuration*. Berkeley and Los Angeles: University of California Press.

Grech-Madin, C., Döring, S., Kim, K., & Swain, A. (2018). Negotiating water across levels: A peace and conflict "Toolbox" for water diplomacy. *Journal of Hydrology, 559*, 100–109.

Grey, D., Andersen, I., Abrams, L., Alam, U., Barnett, T., Kjellén, B., … Wolf, A. T. (2010). *Sharing water, sharing benefits: Working towards effective transboundary water resources management*. Retrieved from http://unesdoc.unesco.org/images/0018/001893/189394e.pdf

Hazarika, S. (2000). *Rites of passage: Border crossings, imagined homelands, India's east and Bangladesh*. Penguin Books India.

Hill, D. P. (2015). Where Hawks Dwell on water and bankers build power poles: Transboundary waters, environmental security and the frontiers of neo-liberalism. *Strategic Analysis, 39*(6), 729–743.

Hill, D. P. (2017). The discursive politics of water management in India: Desecuritising Himalayan river basins. *South Asia: Journal of South Asian Studies, 40*(4), 827–843. doi: 10.1080/00856401.2017.1380576

Huitema, D., & Meijerink, S. (2010). Realizing water transitions: The role of policy entrepreneurs in water policy change. *Ecology and Society, 15*(2). doi:10.5751/ES-03488-150226

Huntjens, P., Yasuda, Y., Swain, A., De Man, R., Magsig, B.-O., & Islam, S. (2016). *The multi- track water diplomacy framework: A legal and political economy analysis for advancing cooperation over shared waters*. The Hague: The Hague Institute for Global Justice.

Huntjens, P. (2011). *Water management and water governance in a changing climate – Experiences and insights on climate change adaptation in Europe, Africa, Asia, and Australia. Delft*. The Netherlands: Eburon Academic Publishers.

Huntjens, P., C. Pahl-Wostl, Z. Flachner, R. Neto, R. Koskova, M. Schlueter, I. NabideKiti and C. Dickens, (2011). Adaptive water management and policy learning in a changing climate. A formal comparative analysis of eight water management regimes in Europe, Asia, and Africa. *Environmental Policy and Governance, 21*(3): 145–163.

Huntjens, P., L. Lebel, C. Pahl-Wostl, R. Schulze, J. Camkin, and N. Kranz, (2012). Institutional design propositions for the governance of adaptation to climate change in the water sector. *Global Environmental Change, 22*, 67–81.

Huntjens, P, Lebel, L, & Furze, C, (2017). The effectiveness of multi-stakeholder dialogues on water: reflections on experiences in the Rhine, Mekong, and Ganga-Brahmaputhra-Meghna river basins. *International Journal of Water Governance, 5*(3), 39–60. doi:10.7564/15-IJWG98

Johnston, B. R. (1998). Culture, power and the hydrological cycle: Creating and responding to water scarcity on St. Thomas, Virgin Islands. In J. M. Donahue & B. R. Johnston (Eds.), *Water, culture, and power: Local struggles in a global context*. Washington, DC: Island Press.

Keskinen, M., Inkinen, A., Hakanen, U., Rautavaara, A., & Niinioja, M. (2014). Water diplomacy: Bringing diplomacy into water cooperation and water into diplomacy. In G. Pangare (Ed.), *Hydro-diplomacy: Sharing water across borders* (pp. 35–40). New Delhi: Academic Foundation.

Lai, H. H. (2002). China's western development program: Its rationale, implementation, and prospects. *Modern China, 28*(4), 432–466.

Lidarev, I. (2012). *History's hostage: China, India and the war of 1962*. Retrieved 2017, from http://thediplomat.com/2012/08/historys-hostage-china-india-and-the-war-of-1962/

Ministry of Development of North Eastern Region, Government of India. (n.d.). *Background*. Retrieved from http://mdoner.gov.in/content/background-1

Ministry of External Affairs, Government of India. (2013). *Memorandum of understanding between the Ministry of Water Resources, the Republic of India and the Ministry of Water Resources, the People's Republic of China on Strengthening Cooperation on Trans-border Rivers.*

Ostrom, E. (2005). *Understanding institutional diversity.* Princeton: Princeton University Press.

Prasai, S, & Surie, M. D. (2013). *Political economy analysis of the teesta river basin.* New Delhi: Asia Foundation.

Price, G., & Mittra, S. (2016). *Water, ecosystems and energy in South Asia: Making Cross-Border Collaboration Work. Research paper.* London: Chatham House.

Priscoli, J. D., & Wolf, A. T. (2009). *Managing and transforming water conflicts.* Cambridge: Cambridge University Press.

Ringler, C., Bhaduri, A., & Lawford, R. (2013). The nexus across water, energy, land and food (WELF): Potential for improved resource use efficiency? *Current Opinion in Environmental Sustainability, 5*(6), 617–624. Retrieved from http://www.sciencedirect.com/science/article/pii/S1877343513001504

Sachin, C. (2016). *East Asian regional development models: Lessons and way forward for South Asia.* Delhi: Observer Research Foundation.

SaciWATERS. (2015). *Consolidated report - Brahmaputra dialogue phase II.* India: SaciWATERS.

SaciWATERS. (2016a, October 27). *Regional level workshop: Transboundary policy dialogue for improved water governance in Yarlung Tsangpo-Brahmaputra-Jamuna River Basin.* Singapore: Nanyang Executive Centre (NEC), NTU.

SaciWATERS. (2016b, October 27). Transnational policy dialogue on improved water governance of Yarlong Zangpo-Brahmaputra-Jamuna River Basin. *Presentation at the Brahmaputra Dialogue.* Singapore.

Sadoff, C. W., & Grey, D. (2002). Beyond the river: The benefits of cooperation on international rivers. *Water Policy, 4*(5), 389–403. Retrieved from http://www.transboundarywaters.orst.edu/publications/publications/Sadoff%20%26%20Grey%20Beyond%20the%20River%2002.pdf

Sajjanhar, A. (2016). *Understanding the BCIM economic corridor and India's response.* Retrieved from http://www.orfonline.org/wp-content/uploads/2016/06/ORF_IssueBrief_147.pdf

Samaranayake, N., Limaye, S., & Wuthnow, J. (2016). *Water resource competition in the Brahmaputra River Basin: China, India, and Bangladesh.* Arlington, TX: CNA.

Sharma, R., Gorsi, M., & Paithankar, Y. (2016). *Brahmaputra.* Retrieved from http://www.india-wris.nrsc.gov.in/wrpinfo/index.php?title=Brahmaputra

Speed, R., Li, Y., Quesne, T. L., Guy, P., & Zhou, Z. (2013). *Basin water allocation planning: Principles, procedures and approaches for Basin allocation planning.* Paris: UNESCO.

State Council. (2013). *12th five-year plan for energy development.*

Swain, A. (2004). Diffusion of environmental peace? International rivers and bilateral relations in South Asia. In R. Thakur & O. Wiggen (Eds.), *South Asia in the world: Problem solving perspectives on security, sustainable development, and good governance.* Tokyo: United Nations University Press.

Swain, A. (2010). Environment and conflict in South Asia: Water-sharing between Bangladesh and India. *South Asian Journal: Quarterly Magazine of South Asian Journalists & Scholars, 28,* 27–34.

Tarlock, D. (2015). *Promoting effective water management cooperation among riparian nations.* TEC Background Papers No. 21. Stockholm: Global Water Partnership Technical Committee.

The Hindu. (2016, October 1). China blocks tributary of Brahmaputra to build dam. *The Hindu.* Retrieved from http://www.thehindu.com/news/international/China-blocks-tributary-of-Brahmaputra-to-build-dam/article15421066.ece

Tribune Desk. (2017, December 18). Bhutan, Bangladesh, India hydropower treaty soon. *Dhaka Tribune.* Retrieved from http://www.dhakatribune.com/world/south-asia/2017/12/18/bhutan-bangladesh-india-hydropower-treaty-soon/

UNEP. (2016). *Green is gold: The strategy and actions of China's ecological civilization.* Nairobi: United Nations Environment Program.

Verlare, J., & van der Putten, F. P. (2015). *'One belt one road' an opportunity for the EU's security strategy*. Retrieved from https://www.clingendael.nl/sites/default/files/One_belt_one_road_vdPutten_Verlare_Clingendael_policy_brief_2015.pdf

Wendt, A. (1987). The agent-structure problem in international relations theory. *International Organization, 41*(3), 335–350.

Wigfield, A., & Eccles, J. S. (2000). Expectancy–Value theory of achievement motivation. *Contemporary Educational Psychology, 25*(1), 68–81.

Wouters, P., Vinogrado, S., Allan, A., Jones, P., & Rieu-Clarke, A. (2005). *Sharing transboundary waters: An integrated assessment of equitable entitlement: The legal assessment model*. Retrieved from http://unesdoc.unesco.org/images/0013/001397/139794e.pdf

WWF. (2017). *Brahmaputra*. Retrieved from http://wwf.panda.org/about_our_earth/about_fresh water/rivers/brahmaputra/

Yasuda, Y., Aich, D., Hill, D., Huntjens, P., & Swain, A. (2017). *Transboundary water cooperation over the Brahmaputra River: Legal political economy analysis of current and future potential cooperation*. The Hague: The Hague Institute for Global Justice.

Zawahri, N. (2008). Capturing the nature of cooperation, unstable cooperation and conflict over international rivers: The story of the Indus, Yarmouk, Euphrates and Tigris Rivers. *International Journal of Global Environmental Issues, 8*(3), 286–310.

Zeitoun, M., & Mirumachi, N. (2008). Transboundary water interaction I: Reconsidering conflict and cooperation. *International Environmental Agreements: Politics, Law and Economics, 8*(4), 297–316. Retrieved from https://ueaeprints.uea.ac.uk/18988/1/ZeitounMirumachi_-_TBW_-_I_(2008).pdf

Zeitoun, M., & Warner, J. (2006). Hydro-hegemony-a framework for analysis of trans-boundary water conflicts. *Water Policy, 8*(5), 435–460. Retrieved from https://www.uea.ac.uk/polopoly_fs/1.147026!ZeitounWarner_HydroHegemony.pdf

Zhang, Q., Lin, T., Bennett, M. T., & Jin, L. (2010). *An eco-compensation policy framework for the People's Republic of China: Challenges and opportunities*. Retrieved from https://www.adb.org/sites/default/files/publication/28010/eco-compensation-prc.pdf

Infrastructure development and the economics of cooperation in the Eastern Nile

Marc Jeuland ⓘ , Xun Wu and Dale Whittington

ABSTRACT

This article employs a hydro-economic optimization model to analyze the effects of the Grand Ethiopian Renaissance Dam on the distribution and magnitude of benefits in the Eastern Nile. Scenarios are considered based on plausible institutional arrangements that span varying levels of cooperation, as well as changes in hydrological conditions (water availability). The results show that the dam can increase Ethiopia's economic benefits by a factor of 5–6, without significantly affecting or compromising irrigation and hydropower production downstream. However, increasing GERD water storage during a drought could lead to high costs not only for Egypt and Sudan, but also for Ethiopia.

Introduction

Large infrastructure projects can alter the dynamics of cooperation between riparian countries in transboundary river basins because they change capabilities for water flow control and may facilitate or impede opportunities for consumptive water use. There are numerous examples of water infrastructure projects contributing or giving rise to tensions between riparians in transboundary river basins, for example in the Tigris-Euphrates, Aral Sea and Mekong basins (Bagis, 1997; Bekchanov, Ringler, Bhaduri, & Jeuland, 2015; Sneddon & Fox, 2006). Yet river basin cooperation can also facilitate infrastructure development by creating new possibilities for the sharing of benefits from enhanced water control (Alam, Dione, & Jeffrey, 2009; Sadoff & Grey, 2002). Given these complex dynamics, the relationship between infrastructure density and the extent of river basin conflict and cooperation is ambiguous (Wolf, Yoffe, & Giordano, 2003).

The Eastern Nile presents an interesting case for considering the political economy of infrastructure development. On the one hand, numerous scholars have argued that the highlands of Ethiopia, and especially the steep canyons of the Blue Nile, offer some of the best undeveloped sites for hydropower development globally (Blackmore & Whittington, 2008). From a purely hydrological perspective, these projects would do little to alter the water security of downstream riparians. This is because there are limited opportunities for consumptive water use near potential Blue Nile dam sites, and because Ethiopia's financial incentives would generally favour regular and stable releases of water from its reservoirs, to maximize hydropower generation (Whittington, Wu, & Sadoff, 2005; Wu & Whittington, 2006). Yet the opposition of one downstream riparian (Egypt) to these projects has been consistent and vocal, and Ethiopia's commitment to construct the Grand Ethiopian Renaissance Dam (GERD) has been met with responses ranging from scepticism to open hostility and threats of military action (Stack, 2013; Whittington, Waterbury, & Jeuland, 2014; Yahia, 2013).

This article presents an analysis using a hydro-economic model that elucidates the nature of the current controversy over the GERD. We consider the extent to which the GERD changes steady-state (post-reservoir-filling) opportunities for upstream consumptive uses of water, as well as the distribution of hydropower and the overall economic benefits derived from the water resources of the Eastern Nile. A number of scholars have argued that the GERD has greatly altered the dynamics of cooperation in the basin (Cascão & Nicol, 2016; Tawfik, 2016). To better account for the effects of political uncertainty in the basin (Pahl-Wostl, 2002), we consider the distribution of the benefits produced under institutional regimes with different levels of cooperation. For the case of non-cooperation, we assume the Nile riparians depart from the 1959 Nile Waters Agreement between Egypt and Sudan, the only existing agreement on water allocations in this basin. The purpose of this analysis is not to suggest that this is likely, but rather to more clearly illustrate the consequences of non-cooperation, assuming Ethiopia seeks to maximize its own financial benefits.

This analysis uses the annual Nile Economic Optimization Model (NEOM) (Whittington et al., 2005), which means that the management of multi-year hydrological sequences is not considered. Additional work using simulation or dynamic optimization techniques is required to more fully understand the potential negative consequences of reservoir operation decisions on downstream riparians (Block & Strzepek, 2010; Jeuland & Whittington, 2014). Nonetheless, in our analyses of hydrological conditions with low water availability, we are able to approximate the short-term effect that droughts would have on both upstream and downstream countries. Reduced flows could also result if, for example, Ethiopia were to limit reservoir releases for the purpose of filling the GERD or for other political reasons. Importantly, we find that if Ethiopia increased water storage in the GERD during a drought, and thus released little water to downstream riparians, this would lead to high costs not only for Egypt and Sudan, but also for Ethiopia itself.

In what follows, the current situation in the basin is explained and some of the key features of the GERD project are discussed. The methodology is then described for analyzing how this project could alter the distribution of benefits to the Eastern Nile riparians – Ethiopia, Sudan and Egypt. The results of the analysis are then presented, and the article closes with a discussion of the implications and limitations of this assessment.

Background: the Nile basin today and the status of the GERD

Measured at 6700 km and covering territory in 11 countries (Egypt, Sudan, South Sudan, Ethiopia, Uganda, Kenya, Tanzania, Burundi, Rwanda, Democratic Republic of

Congo, and Eritrea), the Nile is the longest river in the world. Few other major river systems traverse so many and such diverse countries, and the basin is famous for the varied interests of its riparians and the salience they ascribe to it (Waterbury, 2002). Yet water disputes in the Nile basin have characteristics in common with those of other international rivers. For example, there is a large gap between the quantity of water available in the basin and the amount of water sought by individual riparian countries for irrigation projects. Nearly all basin governments have plans to irrigate large new tracts of land with Nile water (Jeuland & Whittington, 2014; Knott & Hewett, 1994) (Table 1). Indeed, if all the plans on the drawing board are enacted, the irrigated area will more than double, and annual water deficits in the Nile basin could exceed 50 km^3.

In addition, many of the proposed new irrigation projects – especially in Sudan and Ethiopia – would require significant new complementary investments in storage and water control, and upstream countries are increasingly willing to implement such projects, with ensuing evaporation losses (Whittington et al., 2014). Besides supplying water to new irrigation systems, large dams in upstream reaches of the Nile would hold substantial amounts of water for use in the generation of low-cost hydropower to meet growing energy demands (Jeuland & Whittington, 2014); this potential is summarized by country in Table 1. Egyptians often react to the idea of such projects with anxiety and suspicion because they fear a loss of predictability in the historical hydrology they know, or suspect that water control could be used to achieve political objectives. The experience so far with the planning and construction of the GERD exemplifies these issues because it is primarily a hydropower project that could facilitate irrigation expansion in Sudan, and therefore is viewed with intense distrust in Egypt (Hussein, 2014; UPI, 2014). Projected to cost more than US$ 5 billion to construct, the GERD is expected to generate more than 14,000 GWh of electricity annually for Ethiopia, or about three times the total annual electricity production of the country in 2014 (Whittington et al., 2014).

Some factors underlying the current tensions over the GERD are unique to the Nile basin and to this infrastructure project. Few countries in the world are as dependent on one river as Egypt is. Egypt contributes essentially nothing to the flow of the Nile, but depends upon the Nile for 97% of its water supply, and currently accounts for more

Table 1. Irrigation and hydropower potential of Nile basin countries.

Country	Irrigation potential (1000 ha)	Irrigation area (1000 ha)	Hydropower potential (MW)	Installed capacity (MW)
Burundi	80	0.05	20	0
Congo	10	0.08	78	0
Egypt	4,420	2,923	2,902	2,862
Eritrea	150	5.8	NA	NA
Ethiopia	2,220	32.1	17,355	1,946
Kenya	180	9.8	216	25
Rwanda	150	3.3	47	27
South Sudan	NA	NA	2,570	0
Sudan	4,843	1,946	4,873	1,593
Tanzania	30	14.1	280	0
Uganda	202	25.1	4,723	380
Total	12,285	4,959	33,064	6,833

The totals shown here include only potential and actual projects in the Nile basin. NA = no data available. Sources: NBI (2012), Appelgren et al. (2000).

than 80% of all water withdrawals from the river. Meanwhile, the High Aswan Dam (HAD) provides Egypt the water security to maintain a stable supply even in most sequences of drought years because the reservoir behind the HAD can store more than the annual flow of the Nile. Ethiopia meanwhile contributes 85% of the Nile flow (measured at Aswan) yet has little storage and withdraws only about 3 km^3 of water annually (or 3–4% of the total) from the river. This skewed water use pattern in the basin has arisen from a mix of climatic realities (i.e. the necessity of irrigation for agriculture in Egypt, and to a lesser extent in Sudan) and historical development factors. In particular, in the past, Egypt has been the dominant political, economic and military power in the Nile basin despite its geographic position at the downstream end of the system. This fact is largely responsible for the existing distribution of infrastructure, which has been instrumental in shaping Egypt's development trajectory (Collins, 2002).

The operation of the GERD for hydropower generation can lead to a shift in the timing of transboundary water flows (Whittington et al., 2014).[1] In fact, with the addition of the GERD, the Nile will become the only example of an international river having two man-made, over-year storage dams in different countries (International Non-partisan Eastern Nile Working Group, 2015). From the Egyptian perspective, the GERD creates redundant water storage in the Eastern Nile system because two such large storage facilities are unnecessary for smoothing the temporal variations of multi-year flows. From the perspective of the other riparians and especially Sudan, however, the GERD may enable the capture of significant additional benefits from flow control, since other existing infrastructure in the Eastern Nile provides only seasonal storage (Table 2). The GERD could increase flows to Sudan during both the low-flow summer period and long-term droughts, and this could tempt Sudanese farmers to increase irrigation abstractions well beyond current use, and beyond those specified in the 1959 Nile Waters Agreement (Whittington et al., 2014).[2]

Despite these socio-political tensions, many Nile scholars and planners have argued that hydropower projects in the Eastern Nile (and particularly the Blue Nile) offer significant opportunities for water resource development with limited negative consequences for downstream users, if they were managed to avoid such harmful impacts

Table 2. Comparison of the Grand Ethiopian Renaissance Dam (GERD) with other major dams in the Eastern Nile (adapted from Whittington et al., 2014).

Description	Gebel el Aulia	GERD	Roseires (after raising)	Sennar	Tekeze	Khasm el Girba	Merowe	High Aswan Dam
Nile tributary	White Nile	Blue Nile	Blue Nile	Blue Nile	Tekeze	Atbara	Main Nile	Main Nile
Country	Sudan	Ethiopia	Sudan	Sudan	Ethiopia	Sudan	Sudan	Egypt
Year completed	1937	2017 (expected)	1996; 2011 (raised)	1925	2009	1964	2009	1970
Historical flow at location (km^3/y)	29	48	51	51	3.1	11	78	67
Total storage (km^3)	3.3	68	5.5	1.1	9.3	1.3	12	163
Initial live storage (km^3)	2.8	31	5.4	0.9	5.3	1.2	5.7	137
Installed capacity (MW)	17	5250	400	65	300	10	1250	2100

(Blackmore & Whittington, 2008; Block, Strzepek, & Rajagopalan, 2007; USBR, 1964). For example, hydropower development at the GERD site creates an incentive for Ethiopia to continue a non-consumptive, hydropower-based water development path because abstractions upstream of the border will reduce hydropower production. Also, much Nile water is presently lost to evaporation and seepage as it flows north towards the Mediterranean, and evaporation losses from the HAD reservoir could be reduced if storage levels were lowered (Blackmore & Whittington, 2008; Jeuland & Whittington, 2014). The Main Nile flows through a harsh desert climatic zone, where net evaporation and seepage losses are high relative to those in the tropical and highland reaches of the river. If water currently held in the HAD reservoir could be shifted upstream to a reservoir in Ethiopia such as the GERD, a significant amount of additional water would become available if system water withdrawals were increased and the total amount of water held in storage was not increased. Making effective use of such evaporative gains is, however, challenging because it also increases the risks of water deficits during periods of drought; managing this risk would require cooperation and trust between the riparians of the Eastern Nile. This would necessitate a rethinking of the role of the HAD in Egypt's water security.

Methodology

Hydrological and economic modelling

This article uses a basin-wide hydro-economic model previously developed specifically for the Nile, the NEOM (Whittington et al., 2005; Wu & Whittington, 2006). Hydro-economic models help analysts better capture the complexities and interrelations that relate potential hydrological changes, water resources infrastructure development and water management regimes to economic outcomes (Bekchanov, Sood, & Jeuland, 2015; Harou et al., 2009). The NEOM is a particular type of hydro-economic model that maximizes the value of the annual production of hydropower and irrigation in Egypt, Sudan and Ethiopia. The NEOM user can specify different conditions – which in this application govern institutional regimes and water availability conditions with and without the GERD – to compare estimates of the optimal economic benefits derived from these two sectors. For the work described in this article, the model was updated to incorporate recent changes in the Nile basin (Figure 1). As shown, the full model includes 16 reservoir nodes (of which 7 are existing and 2 are under construction), 16 irrigation nodes and 2 urban/industrial nodes. Our analysis only includes the seven existing dams and the two currently under construction.

To solve for maximum economic benefits from irrigation and hydropower, the NEOM uses a nonlinear optimization routine that maximizes the benefits from these two sectors, which are the two major uses of water in the Nile basin (Whittington et al., 2005):

$$\text{Max} \sum_c \left[\sum_{i,c} p_w^{i,c} \sum_t Q_t^{i,c} + \sum_{i,c} p_e^{i,c} \sum_t E_t^{i,c} \right] \tag{1}$$

In Equation (1), the capitalized terms ($Q_t^{i,c}$ and $E_t^{i,c}$) represent the water quantity used for irrigation (in m^3) and the amount of energy production (in kWh), respectively, during month t and at node i in country c. The economic value of one unit of irrigation

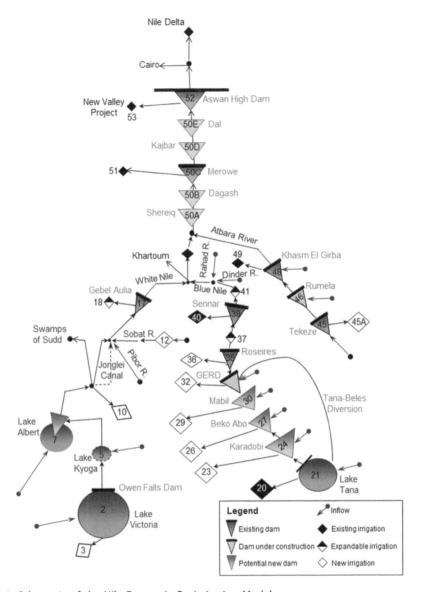

Figure 1. Schematic of the Nile Economic Optimization Model.

water (in US$/m³) is represented by the parameter $p_w^{i,c}$, and the economic value of one unit of hydropower (in US$/kWh) is indicated by the parameter $p_e^{i,c}$.[3] Importantly, Equation (1) can be maximized for the basin as a whole, or for specific countries, by specifying c, the set of countries over which the summation applies. Also, the NEOM does not include other important benefits that are likely to stem from infrastructure investment in the Blue Nile, such as sediment and flood control in Sudan. Nor does it include the costs of infrastructure development (including electricity transmission infrastructure). It is assumed that the investment costs have already been incurred.

Key model constraints then specify in more detail the physical and economic production relationships in the system. These constraints, presented in detail in Whittington et al. (2005), include:

- flow continuity and reservoir storage-elevation relationships, which also allow the calculation of evaporative and seepage losses
- requirements for satisfying municipal demands at Khartoum and in Egypt prior to allocating water to irrigation (no new municipal demands are associated with the GERD)
- reservoir and power generation capacity constraints
- hydropower generation equations, which depend on reservoir releases and storage levels
- annual storage constraints, which require the storage at the end of the year ($t = 12$) to return to the initial level of storage (at $t = 0$) to prevent depletion of water storage[4]
- mathematical representations of the pattern of crop demands at different irrigation sites across the year
- imposing a 50% net reduction in flow in the White Nile as it traverses the Sudd swamps, based on the findings of more detailed hydrological studies (Jeuland, 2009; Sutcliffe & Parks, 1999)
- rules that ensure that specific political obligations are respected (e.g. the water allocations specified in the 1959 Nile Waters agreement).

The model uses a monthly time step and is solved using a nonlinear optimization technique in the General Algebraic Modeling System. It determines the values of decision variables – storage levels in reservoirs, releases from dams, average hydropower head, water allocations to irrigated agriculture, and electricity generation – across all 12 months ($t = 1, ..., 12$) that maximize economic benefits. Accordingly, system-wide changes that occur with the addition of new infrastructure are included (e.g. hydropower uplift at downstream dams and increased water availability for irrigated areas).

Model scenarios

The model scenarios combine variation along three main dimensions: institutional regimes; assumptions about power trade; and hydrological conditions. We describe our assumptions about each of these dimensions in more detail below, and provide a summary of the scenarios we consider – each of which represents a unique combination along these three dimensions – in Table 3.

Institutional regimes
The updated NEOM is used to consider the production and distribution of economic benefits in Ethiopia, Sudan and Egypt across a set of four plausible institutional regimes with varying levels of cooperation among riparian countries, with and without the GERD (Table 4). At one extreme (institutional regime 1, or IR1), upstream riparians maximize their benefits with almost no concern for downstream users, and in particular, with unilateral increases in upstream water abstractions in both Ethiopia and Sudan. For Sudan, we allow these total abstractions to exceed those specified in the 1959

Table 3. Scenarios in the analysis.

Level	Description
1. Institutional regimes	Four assumed regimes spanning from non-cooperation to full cooperation: IR1: Non-cooperation IR2: 1959 agreement IR3: "No significant harm" IR4: System optimization
2. Power trade	Two assumed values for hydropower produced by the GERD: Low: US$ 0.07/kWh High (with power trade): US$ 0.10/kWh
3. Hydrological conditions	Two assumptions about hydrological flows: Average: Corresponding to mean runoff Low: −15% runoff relative to mean runoff
Total number of scenarios	**24 model scenarios** With GERD: $4 \times 2 \times 2 = 16$ scenarios (includes power trade) Without GERD: $4 \times 1 \times 2 = 8$ scenarios (no power trade)

Table 4. The institutional levels of cooperation considered in this analysis.

Institutional regime	Description
1. Non-cooperation	Upstream riparians maximize their benefits with only modest concern for downstream users. In this scenario, Sudan is allowed to consume more than 18.5 km^3/y (up to 22 km^3/y).
2. Non-cooperation constrained by 1959 agreement	Ethiopia maximizes its benefits without concern for downstream users. In this scenario, Sudan abides by the 18.5 km^3/y limit, and any deficits relative to the 1959 allocations are shared equally between Sudan and Egypt.
3. "No significant harm" to existing uses	Ethiopia limits its consumptive uses to privilege downstream allocations of 55.5 km^3/y (Egypt) and 18.5 km^3/y (Sudan). Any remaining deficits are shared equally between Egypt and Sudan.
4. System optimization	Objective is to maximize system benefits. This is a comparative scenario that demonstrates the magnitude of the system-wide economic benefits if the system is operated without concern over the initial distribution of benefits.

Note. Institution regimes 1–4 are listed in order of increasing cooperation. For all four, we also consider (independently) the gains from power trade for scenarios with the GERD, in which the benefits are split equally between Sudan and Ethiopia. The system optimization corresponds to the optimal allocation of water; how these benefits would be shared would depend on the nature of the transfer international agreement(s) between affected parties.

Nile Waters Agreement by 3.5 km^3/y, accounting for evaporative losses from Sudanese reservoirs (for a total of 22 km^3/y of consumptive use in Sudan). For Ethiopia, in IR1 we allow for a modest increase in abstractions for irrigation at sites throughout the basin that would not require large storage investments (2 km^3/y of additional irrigation in Ethiopia).

At the other extreme (IR4), full cooperation can be represented by a regime that maximizes system benefits. This comparative scenario demonstrates the magnitude of the system-wide economic benefits if the system is operated without concern for the initial distribution of benefits. This scenario assumes that this initial distribution could then be adjusted via transfers between riparians.

Various other institutional regimes can be formulated along the spectrum between these two extremes, based on several considerations. The first consideration is whether Sudan and Egypt abide by the 1959 agreement. Based on this agreement, Sudan's total abstractions cannot exceed 18.5 km^3/y measured at Aswan, and any deficits relative to the 1959 allocations are shared equally between Sudan and Egypt. IR2 assumes that the 1959 agreement continues to govern the relative water allocations in Egypt and Sudan, but that Ethiopia continues to operate outside it, increasing abstractions unilaterally by 2 km^3/y (as specified above under IR1).[5]

The second consideration, which is consistent with Ethiopian leaders' repeated pledges that the GERD would impose 'no significant harm' to existing users of the water resources of the Nile, is that both upstream countries would limit their unilateral development so that prior uses in Egypt and Sudan supported by the 1959 agreement would not be negatively affected. For 'no harm' to truly apply, additional water withdrawals throughout the basin would have to come only after these established rights were satisfied or otherwise compensated. The precise definition of 'no significant harm', however, is an issue to be negotiated among the riparians. In our model, this 'no significant harm' regime is considered by imposing a constraint that first requires any demand deficits to be borne by Ethiopia, which does not have water allocated under the 1959 agreement. If additional deficits still remain, these are shared equally by Egypt and Sudan, as specified in the agreement. This scenario (IR3) shows what would happen in the institutional regime that most strongly favours existing negotiated uses. Other conceptions of 'no significant harm' are clearly possible, and we do not take a position on which definition is correct or most defensible.

In modelling these various scenarios, and particularly the non-cooperative ones, we do not mean to suggest that either Sudan or Ethiopia is likely to increase abstractions in these ways or by the precise amounts we consider. Rather, we make specific assumptions about additional uses to explore the extent to which such actions would reduce downstream benefits in the absence of some form of international cooperation (i.e. if the current international cooperation were to break down in the future).

Assumptions about power trade

One issue which is somewhat orthogonal to these considerations about institutional regimes concerns the presence or absence of a power trade agreement between Sudan and Ethiopia following construction of the GERD. Currently there is only minor power trade between Sudan and Ethiopia (EDF, 2007a). With the addition of the GERD, greater power trade would become possible, although this would also require significant new investments in transmission infrastructure between the two countries. Nonetheless, in the near term, power trade would seem to be financially attractive because it might enable Sudan to obtain power more cheaply than the alternative power sources, while allowing Ethiopia to sell power that domestic markets are not able to immediately absorb.[6]

Based on the existing electricity tariffs and the alternative cost of firm power generation in Ethiopia (Foster & Morella, 2011), we assume that in the absence of a power trade agreement the economic value of firm hydropower is US$ 0.07/kWh. This value assumes that regional electricity markets other than Sudan and domestic markets in Ethiopia would eventually be able to absorb the more than 14,000 GWh of electricity produced by the GERD each year. In the medium term, if hydropower from the GERD must be transmitted to more distant markets than Sudan, with higher transmission costs and a need for additional conveyance infrastructure, it will be less valuable than in Sudan (Fichtner, 2009).[7] Our characterization of the no-power-trade scenario reflects a state of the world in the medium term when transmission lines are built, but electricity is worth less in Ethiopia and regional markets other than in Sudan. In the longer term, the Ethiopian grid should be able to utilize the GERD's hydropower, and there is no reason to assume that Ethiopia will remain a low-value user. However, in the short term, in the absence of a power trade with Sudan, hydropower from the GERD will be worth much less than US$ 0.07/kWh – perhaps nothing at all, if transmission lines are not built and water cannot be released through hydropower turbines. The value of US$ 0.07/kWh is likely to be conservative, however, if significant amounts of power are used to meet peak daily demand, or if the majority of power is used to meet demand in Sudan, where the cost of alternative generation options is higher (Jeuland & Whittington, 2014). This latter possibility motivates the assumption of additional benefits of US$ 0.03/kWh from the GERD in the model scenarios that include power trade (US$ 0.10/kWh in total). We emphasize, however, that in the long term power trade may not be necessary as development and demand in Ethiopia increase.

Hydrological conditions

Using an annual model such as the NEOM to consider the consequences of interannual variability and transient effects from reservoir filling is not appropriate, as discussed elsewhere (Block & Strzepek, 2010; Wu, Jeuland, & Whittington, 2016). For this reason, we focus on the sensitivity of economic outcomes to two illustrative water availability conditions. The first condition, historical average, uses the average of a series of stochastic flows generated to maintain the long-term spatiotemporal pattern of historical runoff (Jeuland, 2009). The second condition, low flow (−15% inflows), can be interpreted as representing what would happen if average runoff for the basin as a whole decreased according to some of the more pessimistic predictions from rainfall-runoff models of climate change (Jeuland & Whittington, 2014). An alternative interpretation of this low-flow condition is that it represents what would happen to economic benefits if Ethiopia were to reduce releases from the GERD and raise storage levels by ~9 km^3 (or 25% of the active storage in the GERD and 20% of the annual flow at the GERD) over the course of a year.

Other important assumptions and data

The data used to parameterize the NEOM for this analysis were primarily obtained from Jeuland and Whittington (2014), who developed a hydro-economic simulation model that includes all of the existing and potential new dams and irrigation projects included in the updated NEOM model. These authors obtained such information from several recent feasibility studies and master plans prepared by basin riparians (BCEOM, BRGM, & ISL, 1999; EDF, 2007a, 2007b; Norplan, Norconsult, & Shebelle Consulting Engineers, 2007; Norplan-Norconsult, 2006). The model also allows for varying hydrological conditions, and thus provides average and reduced monthly runoff values at different locations in the basin.

The major assumptions of our analysis with regard to water demands are summarized in Table 5. For IR2–4, we assume that Sudan abides by the allocations specified in the 1959 Nile Waters Agreement, which limits its consumptive use to 18.5 km^3/y (measured at Aswan). We assume that in Sudan 2.5 and 1 km^3/y are lost to evaporation and allocated to meet urban demands, respectively, leaving 15 km^3/y of water use in irrigation. Ethiopia meanwhile would demand total use of ~3–4 km^3 of water for irrigation, mostly around Lake Tana, the GERD, and the other micro-dams scattered

Table 5. Summary of water use assumptions across institutional regimes.

Water use/target	Assumptions for different institutional regimes	
	Regime 1	Regimes 2–4
Municipal/industrial use (km³/y)	Ethiopia: 0	Ethiopia: 0
	Sudan: 1.0	Sudan: 1.0
	Egypt: 4.3	Egypt: 4.3
Irrigation water use target (km³/y)[1]	Ethiopia: 4	Ethiopia: 4
	Sudan: 18	Sudan: 15
	Egypt: 51.2	Egypt: 51.2
Evaporative losses from dams (km³/y)[2]	Ethiopia: 2.5	Ethiopia: 2.5
	Sudan: 3.5	Sudan: 3.5
	Egypt: 10–12	Egypt: 10–12
Total use target (km³/y)	Ethiopia: 6.5	Ethiopia: 6.5
	Sudan: 21.5	Sudan: 18.5
	Egypt: the balance of flow	Egypt: the balance of flow

[1]For Ethiopia, this includes ~2 km³/y of current irrigation water use in the vicinity of Lake Tana. Regimes 2 and 3 assume that Sudan adheres to the allocation specified in the 1959 Nile Water Agreement; i.e. Sudan uses a total of 18.5 km³/y (for municipal/industrial + irrigation + evaporation from new dams), and Egypt maintains a target water demand of 55.5 km³/y (for municipal/industrial + irrigation) but also uses excess (remaining) water. Regime 1 assumes that Sudan increases water use to 21.5 km³/y overall.

[2]For Sudan, this does not include evaporative losses from Gebel el Aulia, Sennar, or the original Roseires, because these dams preceded or were accepted as part of the 1959 Nile Waters Agreement. For Ethiopia, only man-made dams are included; losses from Lake Tana are not.

throughout the upstream catchments of the Blue Nile. Under normal flow conditions, this balance of upstream water uses allows Egypt to withdraw 55.5 km³/y. In the most extreme scenario of non-cooperation (IR1), we allow Sudan to increase its irrigation withdrawals from 15 km³/y to 18 km³/y. Egypt then uses the balance of the water that is available downstream. Egypt's model allocation thus decreases in the non-cooperation scenarios, and in the low-water-availability scenario, since the model does not allow Egypt to supplement water use by drawing down storage in Lake Nasser. Importantly, this may overstate the reduction in benefits for Egypt under these scenarios, since Egypt would almost certainly reduce storage levels and revise HAD operations according to these new hydrologic conditions, perhaps sacrificing hydropower production in order to maintain higher water deliveries to downstream users.

Results

Comparison of economic results across institutional regimes

We begin our discussion with a comparison of the results of the optimization modelling under average hydrological conditions with and without the GERD. This section first summarizes the results without the GERD across all institutional regimes, and then describes how these results change when the GERD is added to the system. Without this new dam, IR2 and IR3 yield identical solutions because there is sufficient water in the system to meet the 1959 Nile Waters Agreement allocations or to do 'no significant harm' to downstream riparians, given the assumed irrigation targets in Ethiopia (Table 6). The non-cooperative case (IR1), with Sudan taking more than 18.5 km³/y, yields somewhat lower benefits overall because the benefits of additional water use in Sudan lead to a reduction of irrigation in Egypt and a loss of hydropower production from the dams along the Main Nile (the HAD and Merowe). Ethiopia is unaffected because it is upstream of the additional Sudanese withdrawals. It is perhaps surprising

Table 6. Comparison of economic benefits with and without the Grand Ethiopian Renaissance Dam (US$ millions/y), with normal water availability, under different institutional regimes (as numbered).

	Ethiopia	Sudan	Egypt	Total
Without the GERD				
1: Non-cooperation	253	1691	2999	**4943**
2: 1959 agreement	253	1570	3168	**4991**
3. No significant harm to existing uses	253	1570	3168	**4991**
4: System optimization	192	1363	3579	**5134**
With the GERD				
1: Non-cooperation	1317	1668	2891	**5875**
1+PT: Non-cooperation + power trade	1556	1907	2891	**6353**
2: 1959 agreement	1317	1507	3092	**5917**
2+PT: 1959 agreement + power trade	1556	1746	3092	**6395**
3: No significant harm to existing uses	1217	1577	3164	**5958**
3+PT: No significant harm to existing uses + power trade	1434	1794	3164	**6393**
4: System optimization	1240	1370	3570	**6180**
4+PT: System optimization + power trade	1465	1595	3570	**6629**

that Sudan would be able to withdraw this much additional water in the absence of the GERD, since the lack of storage in Sudan has historically constrained irrigation development there. However, this problem has largely been solved with two recent developments on the Blue Nile. First, Roseires has been raised and can now hold 5 km^3 of water in storage. Second, the Tana-Beles project diverts a regular flow of water from Lake Tana into the Blue Nile throughout the year.

Finally, the system optimization (IR4) results without the GERD are notable mainly because additional water (in excess of the allocations specified in the 1959 Nile Waters Agreement) is allocated to Egypt at the expense of irrigation in both Ethiopia and Sudan. This result stems from the fact that any water not consumed in Ethiopia and Sudan passes through energy-generating turbines at multiple dams along the Blue and Main Nile. Thus, because we assume that the value of water in irrigation is at least as high in Egypt as elsewhere in the basin (which is certainly true at this time but may not always be), the 'system value' of water is maximized by taking full advantage of all hydropower facilities on the river before water is withdrawn for irrigation (Sadoff, Whittington, & Grey, 2002), even after accounting for evaporation and seepage losses as water flows from Sudan to Egypt.

The addition of the GERD changes these results somewhat, but not dramatically. Ignoring power trade for the moment, we note that the system optimization solution (IR4) yields nearly identical economic benefits in Egypt with and without the GERD (the GERD reduces benefits by just 0.2% compared to IR4 without the GERD). In fact, this solution allocates an identical amount of irrigation water to Egypt as the optimal pre-GERD solution, the only difference being that the HAD is operated at slightly lower levels in order to reduce evaporative losses and maintain these water allocations, and thus hydropower production at the HAD is reduced.

For Sudan, IR4 with the GERD implies somewhat lower irrigation benefits but higher hydropower benefits, and a very small net gain of 0.5% over the pre-GERD IR4 situation. This reallocation follows from a nearly one-to-one shift in the internal trade-off between hydropower production (mostly at Merowe, but also at Roseires and Sennar) and irrigation in Sudan, where the former gains value as a consequence of the new flow regulation provided by the GERD. Finally, for Ethiopia, economic benefits increase by a factor of nearly 6.5 due to the large amounts of new hydropower produced at the GERD. For Ethiopia, the optimal solution does not allocate any water to irrigation in Ethiopia beyond the 2 km^3 at Lake Tana because such upstream irrigation would lead to the loss of comparatively more valuable hydropower throughout the downstream system, as well as loss of irrigation in Egypt.

Turning to the results for the other institutional regimes (IR1, IR2 and IR3), we observe that outcomes with the GERD are again equivalent for Ethiopia under non-cooperation (IR1) and the 1959 agreement (IR2), as would be expected, since Ethiopia's objectives are the same in these two cases. Ethiopia's benefits increase by a factor of more than 5 relative to the situation without the GERD. Compared to the system optimization solution, we observe modest irrigation expansion in Ethiopia (an additional ~1 km^3 of irrigation), and slightly higher (+2%) hydropower production since the GERD and Tana-Beles releases are uncoordinated with the schedule of downstream demands. Meanwhile, in the non-cooperation institutional regime (IR1) with the GERD, Sudan sacrifices some energy production (~3–4%) for a 26% increase in irrigation water use and benefits relative to IR2 (1959 agreement), such that its benefits are about 9% higher than they would be under the water allocations from the 1959 agreement with the GERD. Despite evaporative losses from the GERD and the effects of this on water flow downstream, the hydropower and irrigation benefits under IR1 and IR2 with the GERD are only slightly lower (by 1.5–4%) for Sudan than without the GERD because the GERD also provides more regular flow and hydropower uplift in Sudan.

In Egypt, the GERD leads to somewhat lower benefits in IR1 and IR2 compared to IR1 and IR2 without the GERD (by 2–3%). This is because Egypt bears most of the cost of lower water availability (less in IR2 since deficits are shared with Sudan) and releases less water from the HAD. These lower releases translate to reductions in irrigation water use and hydropower generation.

Finally, the GERD coupled with the 'no significant harm' institutional regime (IR3) is least favourable for Ethiopia (reducing benefits by 8% or US$ 100 million/y) compared to IR1 and IR2 with the GERD because IR3 places greater weight on downstream water uses relative to new upstream uses and the timing of reservoir releases. Even so, benefits still increase by a factor of almost 5 for Ethiopia with the GERD compared to the pre-GERD situation in IR3. The benefits for Egypt under the 'no significant harm' institutional regime (IR3) are essentially unchanged with and without the GERD (they drop by about 0.1% with the GERD due to marginally lower operating levels at the HAD), and also for Sudan (they increase by 0.4% as a result of hydropower uplift). Relative to IR1 with the GERD (which favours Sudan because irrigation is increased beyond the allocations specified in the 1959 agreement), the benefits of IR3 with the GERD are 5% (US$ 90 million/y) lower for Sudan and 9% (US$ 270 million/y) higher for Egypt with the GERD.

If we add in power trade and assume an equal sharing of the benefits of power trade between Sudan and Ethiopia, the benefits to each of these two countries increase by US$ 220–240 million/y, depending on the institutional regime. With power trade structured and hydropower benefits shared in this way, the GERD improves outcomes across all institutional regimes for both Sudan and Ethiopia. Benefits to Egypt are essentially unchanged under the 'no significant harm' (IR3) and system optimization (IR4) regimes, and are reduced by 4% with non-cooperation (IR1) and the 1959 agreement (IR2). It is thus evident that sharing even a modest amount of the power benefits from the GERD with Sudan through trade would increase the attractiveness of the GERD from Sudan's perspective. In the short term, a power trade deal is necessary for Ethiopia to utilize the hydropower generation from the GERD and to receive a financial return on its investment, an issue that we return to below in the discussion.

Sensitivity of results to reduced water availability

We next consider the results with and without the GERD when downstream flows are reduced, noting that such a reduction could arise from climate change and multi-year droughts (i.e. a decrease of Nile runoff by 15%). The total benefits with and without the GERD for the four institutional regimes and distribution under these reduced flow conditions are shown in Table 7, and the changes compared to normal flow conditions are disaggregated by country and sector in Figure 2. This analysis illustrates four important points about the economic consequences of flow reductions. First, the reduction in total benefits due to flow reductions is clearly larger following construction of the GERD than without the GERD. This is logical because lower water availability affects hydropower production from the GERD as well as the remainder of the system.

Second, the impacts of lower water availability on Ethiopia are minor in the absence of the GERD, but become more important after construction and filling of the GERD are completed, especially if the GERD is managed with a constraint to avoid down-stream harm. Third, holding infrastructure constant, the overall reduction in benefits across the four institutional regimes is similar when water availability changes, with the decline in benefits being modestly larger under IR1 and IR2 compared to the optimal regime, IR4. Nonetheless, the distribution of the reduction in benefits from decreased water availability varies considerably. Specifically, Sudan and Egypt share the harm to irrigation if the former adheres to the 1959 agreement, while Egypt bears nearly all of this cost under non-cooperation (Sudan only suffering from some reduction in hydro-power production).

Table 7. Comparison of economic benefits with and without the Grand Ethiopian Renaissance Dam (in US$ millions/y); low water availability.

	Ethiopia	Sudan	Egypt	Total
Without the GERD				
1: Non-cooperation	158	1595	2209	3962
2: 1959 agreement	158	1164	2695	4017
3. No significant harm to existing uses	158	1164	2695	4017
4: System optimization	107	844	3335	4287
With the GERD				
1: Non-cooperation	1020	1558	2046	4624
1+PT: Non-cooperation + power trade	1205	1743	2046	4995
2: 1959 agreement	1020	1125	2533	4678
2+PT: 1959 agreement + power trade	1205	1310	2533	5049
3: No significant harm to existing uses	713	1150	2693	4556
3+PT: No significant harm to existing uses + power trade	832	1270	2693	4795
4: System optimization	944	940	3152	5036
4+PT: System optimization + power trade	1123	1120	3152	5395

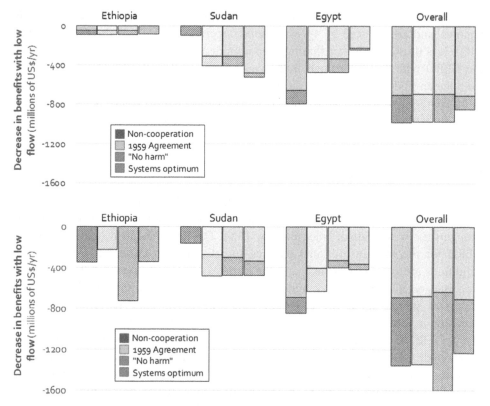

Figure 2. The effects of lower flows on the generation of economic benefits in the Eastern Nile without (top) and with (bottom) the Grand Ethiopian Renaissance Dam. Changes in benefits are relative to average flow conditions; stippling indicates changes in irrigation benefits, and hatching indicates changes in hydropower.

Fourth, under the 'no significant harm' institutional regime (IR3) with the GERD, system-wide benefits are lower than for all other regimes. This reflects the fact that the GERD is then operated in a way that avoids downstream shortfalls relative to the existing withdrawal levels in Sudan and Egypt, at the cost of larger declines in hydropower production in Ethiopia. When water is more scarce in the basin, the difference in benefits between the optimal regime (IR4) and sub-optimal regimes (IR1, IR2 and IR3) tends to increase.

Summary of results

All three riparians stand to benefit from system-wide optimization if agreement can be obtained on an equitable sharing of aggregate benefits. However, if the system optimization solution is politically infeasible (note that the upstream riparians in the basin would have to limit their irrigation use to well below the current use), the preferred institutional regimes with the GERD, based on total benefits to each party, are IR1 or IR2 for Ethiopia, IR1 for Sudan, and IR3 for Egypt (Table 8). In addition, Sudan and Ethiopia together stand to gain at least US$ 400–500 million/y from power trade,

Table 8. Preferred institutional regime for each riparian, and for producing overall benefits.

	Ethiopia	Sudan	Egypt	Total benefits
Normal flow				
Pre-GERD	Indifferent	IR1	IR2 or IR3	IR2 or IR3
Post-GERD	IR1 or IR2	IR1	IR3	IR2 (barely)[a]
Restricted flow				
Pre-GERD	Indifferent	IR1	IR2 or IR3	IR2 or IR3
Post-GERD	IR1 or IR2	IR1	IR3	IR2

Note. IR = institutional regime (see Table 4). GERD = Grand Ethiopian Renaissance Dam. The system optimization solution is not shown, since it entails large (and likely unpalatable) reductions in water use in Sudan.
[a] The difference from the IR3 solution is negligible in this case. The IR2 solution is only 2 million US$/y (or 0.03%) better.

depending on the institutional regime.[8] The precise distribution of these gains would depend on the negotiated agreement between the countries.

In the absence of the GERD and if the optimal solution (IR4) is politically infeasible, Ethiopia is indifferent between the three other institutional regimes. One might posit that the GERD would push Ethiopia towards non-cooperation as it allows Ethiopia to pursue limited irrigation while operating the dam for maximum power generation. Yet the analysis shows that the gains for Ethiopia from this strategy are surprisingly small relative to the substantial benefits that accrue to it after completion of the GERD. Furthermore, even if Ethiopia acts non-cooperatively, seeking to maximize its own benefits, the potential losses for Egypt would be smaller than many expect.

Discussion

Large infrastructure projects can alter the dynamics of cooperation between riparian countries in transboundary river basins because they change capabilities for water flow control and may facilitate or impede opportunities for additional water withdrawals (Wu et al., 2016). This article has presented an analysis of changes in the distribution and magnitude of benefits for the three riparians of the Eastern Nile following construction of a new, large and controversial dam project, the GERD, near the border between Ethiopia and Sudan. To more fully understand the implications of this infrastructure, its effects were analyzed across varying degrees of cooperation among riparians, from non-cooperation to various increasing levels of cooperation, and with or without power trade between Ethiopia and Sudan. We also considered the interaction of these changes with a reduction in water availability, to determine the extent to which flow reductions, whether natural or directly related to dam operations, would differentially affect the riparians in this system.

The analysis produces a number of important insights concerning the general effects of large infrastructure projects on the hydro-political dynamics in river basins, as well as the specific consequences of the GERD. With regard to specific impacts, the GERD dramatically alters the distribution of the economic benefits of water resource management in the Eastern Nile in favour of Ethiopia (increasing its benefits by a factor of 5–6), without significantly affecting or compromising irrigation and hydropower production in downstream countries. Across different institutional regimes, in a normal hydrological year, the GERD changes the magnitude of hydropower and irrigation benefits to Sudan and Egypt by only 0–4%, and both Sudan and Ethiopia clearly benefit if power trade can be arranged. The GERD is essentially a non-consumptive water project, and the higher evaporation from its reservoir can be largely offset by modifications in the operating rules for other downstream dams. Our calculations also do not include a number of other positive impacts for Sudan, including reduced siltation at Roseires, flood and sediment control along the Blue Nile, and enhanced navigation. These results are therefore probably conservative, and understate the potential value of this infrastructure project to the Nile system.[9]

In addition, reductions in flows (or in releases from the GERD) reduce economic benefits in all three countries, but they result in a relatively larger reduction of the benefits accruing to Ethiopia than to Egypt and Sudan. In this respect, the downstream concerns about loss of control over water flows due to the GERD appear to be exaggerated, assuming that Ethiopia does not act to intentionally harm downstream riparians. Here it is worth emphasizing that the 'non-cooperation' case analyzed in this article represents a situation where Ethiopia, and then Sudan, each seeks to maximize its own financial benefits, and not one in which these upstream riparians try to operate water storage facilities to achieve other (e.g. political or military) objectives.

One notable finding from the analysis is that there are significant differences in the preferred allocation regimes across countries. This has of course long been recognized: the 1959 Nile Waters Agreement after all represented a compromise between Sudan and Egypt over these two countries' conflicting water resource development objectives (Waterbury, 2002). At the time of that agreement, which allowed for the construction of the HAD, Egypt traded a more favourable water allocation for one that was more equitable but also more secure. That agreement and Egypt's asymmetric power advantage in the basin provided it unprecedented water security for nearly half a century (Whittington et al., 2014).

The GERD represents an important shift towards the upstream riparians of the Nile in the balance of power in the basin, and adds a layer of complexity to negotiations among riparians. When existing institutions do not include all major riparians, an interconnected transboundary river basin system can be destabilized by shifts in countries' relative power that may have little to do with water resources. Ethiopia has never accepted the 1959 Nile Waters Agreement as legitimate, so Egypt's water security has been (and continues to be) contingent on Ethiopia's inability to do much that would affect the prior treaty allocations. In the absence of the GERD (or some similar project), our analysis shows that the institutional regime matters little to Ethiopia. This may be why the current regime, representing a compromise between Egypt's and Sudan's interests, has been stable for so long. With the GERD, however, Ethiopia gains a clear preference for institutional arrangements that allow more upstream control over water allocations, even if its own withdrawals remain modest. Given the limited water withdrawals in Ethiopia, the critical water sharing agreement will remain the one between Egypt and Sudan. Seen in this light, one of Egypt's main concerns about the construction of the GERD should be a possible destabilization of the water sharing compromise established under the 1959 Nile Waters Agreement.

Importantly, system-wide optimization with equitable sharing of benefits could improve outcomes for all three riparians. The GERD will provide significant new benefits to Ethiopia and Sudan, and restricting releases from this dam in the short term (beyond the need to fill the reservoir) for political purposes might be harmful to

the economies of all three countries: Ethiopia would suffer delayed hydropower production and benefits; Sudan might also experience delays in such benefits as provided
through power trade; and Egypt will probably have to cope with greater short-term
variability in its water supply. Without a power trade deal in place, there is no obvious
market for the large amount of energy that will soon be produced at the GERD, and sale
of this power is essential to make this project financially beneficial to Ethiopia. The
filling of the large reservoir behind the GERD, meanwhile, could impose significant
short-term costs on the downstream riparians if that process is not managed cooperatively (Wheeler et al., 2016). Nonetheless, to date there has been only limited discussion
of how the operation of the GERD should be coordinated with that of other control
infrastructure in the system; there are no established water-sharing rules between
Ethiopia and the other Eastern Nile riparians; and there does not yet appear to be a
power trade deal between Ethiopia and Sudan. If mutually beneficial outcomes are to be
achieved, these operational and water-sharing issues require prompt and careful attention from the riparians.

Notes

1. Whittington et al. (2014) estimate that 80% of the economic benefits of the GERD are from
   the dam's hydropower generation, and that the remaining benefits are largely from hydropower uplift, flood and sediment control, and irrigation water supply enhancement in Sudan.
2. This argument was somewhat weakened by the raising of Roseires in 2011. Prior to this
   heightening, Sudan suffered from perpetual summer water shortages owing to the limited
   live storage of this dam, and the fact that it was largely silted up. In fact, as our analysis will
   make clear, Sudan can expand irrigation significantly even without the GERD project.
3. Due to lack of data and because we do not want to assume that the economic value of water
   upstream in the basin will continue to be lower than that in Egypt, we follow Whittington
   et al. (2005) in setting the value of irrigation water to US$ 0.05/m^3 throughout the basin. The
   assumptions about the value of hydropower are explained further below.
4. This constraint has important implications for interpretation of our analyses of reduced
   flows, because it does not allow riparians to draw down storage to mitigate the consequences
   of reduced flows or severe droughts. As such, those reduced flows would have to be
   sustained over a long period of time to have the effects we ascribe to them; short-term
   deficits would be managed very differently.
5. Ethiopia has never accepted the legitimacy of the 1959 agreement.
6. These gains for Sudan are due to the difference between tariffs in Ethiopia and the alternative cost of production in Sudan. Our analysis assumes for illustrative purposes that
   Ethiopia and Sudan would split these benefits equally.
7. For example, research in a range of sub-Saharan countries suggests that the economic value
   of rural electrification is fairly low at this time (Peters & Sievert, 2016), although this may
   change in the long term.
8. The amount is at least USD 400–500 million/y assuming that all power generated at the
   GERD could otherwise be consumed in Ethiopia (albeit at lower value). This seems unlikely,
   given that the demand for energy in the country is currently well below the installed capacity
   that will exist in the country once the GERD is complete.
9. There are of course some negative impacts as well that we have not included, particularly to those
   engaged in flood-recession agriculture, and for those whose livelihoods will be displaced by the
   new reservoir. A full benefit–cost analysis of the dam would include these various costs and
   benefits, as well as the capital and O&M costs of the new dam.

Acknowledgments

We are most grateful to Selina Ho and Huang Jing of the Centre on Asia and Globalization of the National University of Singapore for organizing the workshop during which this work was first presented and discussed. We thank the participants in that workshop, who provided many useful comments, and particularly the discussant for our article, Dr John Waterbury. We also thank the anonymous reviewers for their feedback and suggestions.

ORCID

Marc Jeuland ⓘ http://orcid.org/0000-0001-8325-2622

References

Alam, U., Dione, O., & Jeffrey, P. (2009). The benefit-sharing principle: Implementing sovereignty bargains on water. *Political Geography*, *28*(2), 90–100. doi:10.1016/j.polgeo.2008.12.006

Appelgren, B., Klohn, W., & Alam, U. (2000). *Water and agriculture in the Nile Basin*. Rome: FAO.

Bagis, A. I. (1997). Turkey's Hydropolitics of the Euphrates-TigrisBasin. *International Journal of Water Resources Development*, *13*(4), 567–582. doi:10.1080/07900629749647

BCEOM, BRGM, & ISL. (1999). *Main report abbay river basin integrated development master plan project*. Addis Ababa, Ethiopia: Ministry of Water Resources, Federal Democratic Republic of Ethiopia.

Bekchanov, M., Ringler, C., Bhaduri, A., & Jeuland, M. (2015). How would the Rogun Dam affect water and energy scarcity in Central Asia? *Water International*, *40* (5–6), 856–876. doi:10.1080/02508060.2015.1051788

Bekchanov, M., Sood, A., & Jeuland, M. (2015). Hydro-economic models to address river basin management problems: Structure, applications, and research gaps. In IWMI (Ed.), *International water management institute working paper*. Colombo, Sri Lanka.

Blackmore, D., & Whittington, D. (2008). *Opportunities for cooperative water resources development on the eastern Nile: Risks and rewards*. Independent report of the scoping study team to the Eastern Nile Council of Ministers. Washington: The World Bank.

Block, P., & Strzepek, K. (2010). Economic analysis of large-scale upstream river basin development on the Blue Nile in Ethiopia considering transient conditions, climate variability, and climate change. *Journal of Water Resources Planning and Management*, *136*(2), 156–166. doi:10.1061/(ASCE)WR.1943-5452.0000022

Block, P. J., Strzepek, K., & Rajagopalan, B. (2007). *Integrated management of the Blue Nile Basin in Ethiopia*. IFPRI Discussion Paper 00700. Washington, DC: International Food Policy Research Institute.

Cascão, A. E., & Nicol, A. (2016). GERD: New norms of cooperation in the Nile Basin? *Water International*, *41*(4): 550–573. doi:10.1080/02508060.2016.1180763

Collins, R. O. (2002). *The Nile*. New Haven, CT: Yale University Press.

EDF. (2007a). *Pre-feasibility study of border hydropower project, ethiopia: Draft final report (eastern nile power trade program study)*. Addis Ababa: Eastern Nile Technical Regional Office.

EDF. (2007b). *Pre-feasibility study of mandaya hydropower project, ethiopia: Final report*. Addis Ababa: Eastern Nile Technical Regional Office.

Fichtner. (2009). *Ethiopia-Kenya power systems interconnection project: Final feasibility study report*. Addis Ababa: Nile Basin Initiative Regional Power Trade Project.

Foster, V., & Morella, E. (2011). Ethiopia's infrastructure: A continental perspective. *World Bank Policy Research Working Paper Series*, 5595. Washington, DC: World Bank.

Harou, J. J., Pulido, M. A., Rosenberg, D. E., Medellín-Azuara, J., Lund, J. R., & Howitt, R. E. (2009). Hydro-economic models: Concepts, design, applications, and future prospects. *Journal of Hydrology*, *375*, 627–643. doi:10.1016/j.jhydrol.2009.06.037

Hussein, H. (2014, February 6). *Egypt and Ethiopia spar over the Nile* (Al Jazeera America). Retrieved July 31, 2015 from http://america.aljazeera.com/opinions/2014/2/egypt-disputes-ethiopiarenaissancedam.html

International Non-partisan Eastern Nile Working Group. (2015). *The Grand Ethiopian Renaissance Dam: An opportunity for collaboration and shared benefits in the Eastern Nile Basin*. Cambridge, MA: Abdul Latif Jameel World Water and Food Security Lab, Massachusetts Institute of Technology.

Jeuland, M. (2009). *Planning water resources development in an uncertain climate future: A hydro-economic simulation framework applied to the case of the Blue Nile* (PhD). University of North Carolina at Chapel Hill, Chapel Hill.

Jeuland, M., & Whittington, D. (2014). Water resources planning under climate change: Assessing the robustness of real options for the Blue Nile. *Water Resources Research*, *50*(3), 2086–2107. doi:10.1002/2013WR013705

Knott, D., & Hewett, R. (1994). Water resources planning in the Sudan. In P. P. Howell & J. A. Allan (Eds.), *The Nile: Sharing a scarce resource: A historical and technical review of water management and of economical and legal issues* (pp. 205–216). Cambridge, UK: Cambridge University Press.

NBI. (2012). *The state of the river nile basin*. Entebbe, Uganda: Nile Basin Initiative.

Norplan, Norconsult, & Shebelle Consulting Engineers. (2007). *Beko-abo multipurpose project reconnaissance study report*. Addis Ababa: Eastern Nile Technical Regional Office.

Norplan-Norconsult. (2006). *Karadobi multipurpose project pre-feasibility study: Draft final report (Vol. 1)*. Addis Ababa: Ministry of Water Resources, Federal Democratic Republic of Ethiopia.

Pahl-Wostl, C. (2002). Towards sustainability in the water sector–The importance of human actors and processes of social learning. *Aquatic Sciences*, *64*(4), 394–411. doi:10.1007/PL00012594

Peters, J., & Sievert, M. (2016). Impacts of rural electrification revisited–The African context. *Journal of Development Effectiveness*, *8*(3), 327–345. doi:10.1080/19439342.2016.1178320

Sadoff, C. W., & Grey, D. (2002). Beyond the river: The benefits of cooperation on international rivers. *Water Policy*, *4*(5), 389–403. doi:10.1016/S1366-7017(02)00035-1

Sadoff, C. W., Whittington, D., & Grey, D. (2002). *Africa's international rivers: An economic perspective. Direction in Development Series*. Washington DC: The World Bank.

Sneddon, C., & Fox, C. (2006). Rethinking transboundary waters: A critical hydropolitics of the Mekong basin. *Political Geography*, *25*(2), 181–202. doi:10.1016/j.polgeo.2005.11.002

Stack, L. (2013, June 6). With cameras rolling, Egyptian politicians threaten Ethiopia over dam. *New York Times*. Retrieved from http://thelede.blogs.nytimes.com/2013/06/06/with-cameras-rolling-egyptian-politicians-threaten-ethiopia-over-dam/

Sutcliffe, J. V., & Parks, Y. P. (1999). *The hydrology of the Nile*. Oxfordshire, UK: International Association of Hydrological Sciences.

Tawfik, R. (2016). The grand Ethiopian renaissance dam: A benefit-sharing project in the Eastern Nile? *Water International*, *41*(4), 574–592. doi:10.1080/02508060.2016.1170397.

UPI. (2014, February 27). Egypt plans dam-busting diplomatic offensive against Ethiopia. Retrieved July 31, 2015, from http://www.upi.com/Business_News/Energy-Resources/2014/02/27/Egypt-plans-dam-busting-diplomatic-offensive-against-Ethiopia/UPI-13631393533111/

USBR. (1964). *Land and water resources of the blue nile basin: Main report and appendices I–V*. Washington, DC: United States Bureau of Reclamation.

Waterbury, J. (2002). *The Nile Basin: National determinants of collective action*. New Haven, CT: Yale University.

Wheeler, K. G., Basheer, M., Mekonnen, Z. T., Eltoum, S. O., Mersha, A., Abdo, G. M., …
    Dadson, S. J. (2016). Cooperative filling approaches for the grand ethiopian renaissance dam.
    Water International, 41(4): 611–634. doi:10.1080/02508060.2016.1177698

Whittington, D., Waterbury, J., & Jeuland, M. (2014). The Grand Renaissance Dam and
    prospects for cooperation on the Eastern Nile. *Water Policy, 16*(4), 595–608.

Whittington, D., Wu, X., & Sadoff, C. (2005). Water resources management in the Nile basin:
    The economic value of cooperation. *Water Policy, 7*(3), 227–252.

Wolf, A. T., Yoffe, S. B., & Giordano, M. (2003). International waters: Identifying basins at risk.
    Water Policy, 5(1), 29–60.

Wu, X., Jeuland, M., & Whittington, D. (2016). Does hydropolitical ambiguity affect water
    resources development? The case of the Eastern Nile. *Policy and Society, 35* (2), 151–163.

Wu, X., & Whittington, D. (2006). Incentive compatibility and conflict resolution in interna-
    tional river basins: A case study of the Nile Basin. *Water Resources Research, 42*(2), 15.
    doi:10.1029/2005WR004238

Yahia, M. (2013). Leaked report sparks disagreement between Egypt and Ethiopia over dam.
    Nature Middle East. doi:10.1038/nmiddleeast.2013.99

The remarkable restoration of the Rhine: plural rationalities in regional water politics

Marco Verweij

ABSTRACT

The restoration of the Rhine basin is widely viewed as an exemplary case of international water protection. The river's clean-up has been characterized by a number of puzzling developments. These include chemical companies reducing their toxic effluents by more than legally required, and riparian governments quarrelling internationally over environmental measures that each of them were undertaking domestically. It is argued that the plural rationality (or cultural) theory pioneered by Dame Mary Douglas offers an empirically valid explanation of these remarkable processes.

Introduction

During the last 50 years, the Rhine watershed has been transformed from the "open sewer of Europe" (as the river was commonly known in the early 1970s) to "the cleanest river in Europe" (as *Le Monde* proclaimed on 17 October 1996). During this time, the release of chemical pollutants and heavy metals by corporations and municipalities into the watershed has been greatly reduced (Villamayor-Tomas, Fleischman, Perez Ibarra, Thiel, & van Laerhoven, 2014), the dumping of salt into the river has been largely discontinued (Dieperink, 2011), flood protection systems have been improved (International Commission for the Protection of the Rhine [ICPR], 2012), a comprehensive warning system against accidental spills has been put in place (Frijters & Leentvaar, 2003), the runoff of nitrogen and other harmful substances from agricultural sources has been somewhat diminished (Farmer & Braun, 2002), monitoring of the Rhine's water quality has been improved (ICPR, 2007), passageways for migratory fish have been built (Bölscher, van Slobbe, van Vliet, & Werners, 2013), and ecosystems have partly been restored (Schmitt et al., 2012). As a result, oxygen levels in the waters of the Rhine have rebounded to healthy levels, while many species of fish (including salmon) and other animals have begun to return to the river (Molls & Nemitz, 2006). Today, the governance of the Rhine is routinely held up as an example for other watersheds around the globe (Chase, 2011; da Silveira & Richard, 2013; Myint, 2005).

None of these successes has been a foregone conclusion. In the Rhine basin, massive interests and values clash, and diverse countries are involved (Uehlinger, Wantzen, Leuven, & Arndt, 2009). The Rhine begins in the Swiss and Austrian Alps, where a number of small brooks flow together in the Bodensee. From this lake, the waters of the Rhine start their 1250 km trip through Switzerland, France, Germany and the Netherlands, where the river drains into the North Sea. Other countries included in the Rhine catchment area are Belgium, Luxembourg, Liechtenstein, Austria and Italy, which are connected to the river by tributaries. In 1992, the Rhine–Main–Danube Canal was completed, linking the North Sea with the Black Sea. For centuries, transport of bulk goods over the Rhine has been of paramount importance to the economies of Western Europe. About 50% of all inland navigation within the European Community currently takes place on the Rhine, with about 311 million tonnes of goods and 700 ships crossing the border between the Netherlands and Germany each day. The watershed hosts some of the largest chemical companies in the world, including Hoechst, Bayer, BASF, Novartis and Shell Chemicals. Today, the river receives the wastewater of about 58 million people and a multitude of chemical and other companies, as well as agricultural sources. The basin is also dotted with more than 2000 hydroelectric, and 10 nuclear, power plants. In addition, it is used as a source of drinking water for over 25 million people.

As the governance of the Rhine watershed has often been touted as a model for other transboundary water basins, it is vital that valid theoretical conclusions and policy implications are derived from the case. This is a conceptual challenge, as the Rhine's restoration has involved processes that appear surprising from the viewpoint of standard theories of international environmental cooperation. Below, I first describe how the remarkable restoration of the Rhine has come about. Thereafter, I assert that to explain the puzzles that characterize the river's clean-up, it is necessary to look beyond the standard, interest-based frameworks that abound in environmental studies. Specifically, I argue that the plural rationality theory pioneered by anthropologist Dame Mary Douglas offers a convincing explanation of the remarkable restoration of the Rhine. This approach posits that resilient environmental governance depends on the creative interplay between adherents of a limited set of alternative ways of defining and resolving the issues at hand. If decision makers facilitate, and make use of, the interplay between these opposing viewpoints, then widely acceptable and sustainable solutions to environmental issues can emerge. In contrast, if decision makers insist on addressing the issues in a more monolithic manner, then policy failure will ensue. After setting out Douglas's approach, I show that these hypotheses explain the paradoxical restoration of the Rhine. Finally, on the basis of plural rationality theory, I conclude by suggesting that policy makers in the Rhine catchment area appear to have drawn the wrong lessons from the river's restoration, putting at risk the continued improvement of the water basin.

The restoration of the Rhine

International cooperation on the protection of the Rhine started in 1946, when the Dutch government raised the issue with the governments of the other riparian states. Four years later, Switzerland, France, Germany, Luxembourg and the Netherlands formed the International Commission for the Protection of the Rhine against Pollution (ICPR), which acquired official status in 1963. Ever since, it has been tasked to: (1) report on the state of the Rhine's environment; (2) propose and coordinate policy solutions to the river's ecological problems; (3) organize regular international consultations; and (4) monitor and implement any intergovernmental agreements that have been reached. The European Economic Community (now the European Union) joined the commission in 1976. The efforts to restore the Rhine basin, coordinated by the ICPR, can be usefully separated into three distinct phases.

1963–1986: intergovernmental strife and effective domestic policies

The efforts to stop the environmental degradation of the Rhine between 1963 and 1986 were paradoxical. The attempts to do so with the help of international agreements were ineffective at best, and counterproductive at worst (Kiss, 1985). Yet, simultaneously, the domestic efforts to clean up the Rhine were quite effective in each riparian country.

In 1976, after years of tense negotiations, both the Convention on the Protection of the Rhine against Chlorides and the Convention on the Protection of the Rhine against Chemical Pollution were signed. The first of these was not ratified until 1983, and never implemented to any significant degree (Bernauer, 1995). This non-compliance even resulted in the recall of the Dutch ambassador from France in 1979. The Chemicals Convention was equally unsuccessful. It called for the establishment of 'black' and 'grey' lists of toxic substances. The black list was to contain the most dangerous chemicals, whose reduction needed priority. Grey-listed substances were considered somewhat less toxic, but still in need of regulation. Between 1976 and 1986, only three chemical substances found their way onto the ICPR black list. The insignificance of this number becomes apparent when one realizes that the European Commission had drawn up a list of some 1500 chemical substances suspected of being toxic.

Meanwhile, the same ministries that were fighting in the international arena were setting up elaborate water protection programmes at home. Between 1970 and 1987 the levels of many toxic substances in the Rhine were reduced by 60–80% (Beurskens, Winkels, de Wolf, & Dekker, 1994). Part of this achievement has to be attributed to the domestic water protection policies developed in the Rhine countries. In each of these countries, a command-and-control system for water protection was set up. Point-source dischargers (municipalities and large firms) were required to obtain permits for their discharges of wastewater into rivers and lakes. This made it necessary for them to build sewage treatment plants from the late 1960s onwards. In addition, riparian governments levied a water pollution tax on companies and cities. Nevertheless, these domestic protection policies cannot receive the sole credit for the clean-up of the Rhine. Remarkably, the large chemical firms along the river often took measures that went beyond the required legal standards (Bernauer & Moser, 1996). Non-governmental organizations (NGOs) also played their part by organizing large-scale protests, suing salt mines, and, in 1983, holding a much-publicized International Water Tribunal (Myint, 2005).

The restoration of the Rhine watershed from 1963 until 1986 was therefore characterized by a paradox: within each riparian country, national and local governments, as well as large corporations, had begun to significantly reduce their release of toxic effluents into the basin, while at the intergovernmental level discord about how to reduce the pollution of the Rhine abounded. Several processes were involved in the making of this puzzle. First, until the end of the 1980s, the domestic water protection programmes in all Rhine countries took the form of command-and-control systems. These approaches worked reasonably well *within* the Rhine countries. Yet these approaches all differed from each other – for instance, some relied on effluent limits, others on water quality standards. As each national delegation to the ICPR tried to impose its own method of organizing water pollution control on the other delegations, it proved difficult for them to agree on a single command-and-control approach at the international level. A second factor that hampered the intergovernmental deliberations until 1987 was the insistence on following the formal procedures of international public law. This was a slow process with many veto points, as it required unanimous agreement among the governments of the riparian countries, and ratification by their parliaments. Moreover, the governments appeared weary of committing themselves to agreements that are legally binding under international public law (Wieriks & Schulte-Wülwer-Leidig, 1997).

1987–1999: implementation of effective international programmes

The international cooperation on the protection of the Rhine drastically changed in November 1986. From then on, the international policies have led, rather than lagged behind, domestic efforts. This was the month in which the Sandoz accident took place (Giger, 2009). During the dousing of a fire in a warehouse of Sandoz AG in Basel, Switzerland, about 15,000 m^3 of water, mixed with highly dangerous chemicals, flowed into the Rhine, forming a 70 km red trail that moved slowly downstream from Switzerland, through France, Germany, and the Netherlands. Hundreds of thousands of dead fish and waterfowl washed up along the banks of the river. All the processing plants of the drinking water companies using Rhine water shut down.

The Sandoz incident was widely perceived as an indictment of the reigning international approach to the protection of the Rhine, and shook up the intergovernmental negotiations on this issue (Plum & Schulte-Wülwer-Leidig, 2014). Environmental groups staged novel demonstrations, the media zoomed in, and ministers became more involved. A sense of alarm swept over the delegations to the ICPR. The Dutch minister of transport, Neelie Kroes, cleverly exploited this to strengthen international cooperation on the protection of the Rhine. Minister Kroes took a step that is unusual in international relations: to broker international agreement on the clean-up of the Rhine she relied on private initiative. She hired a team of consultants from McKinsey Amsterdam to outline a comprehensive international agreement on the restoration of the Rhine basin, and to build up intergovernmental support for this plan. The final report of the McKinsey team likewise contained several innovative proposals. First, it recommended keeping intergovernmental agreements concerning the Rhine non-binding. Second, it gave responsibility for the implementation of any international agreements to the lowest possible government levels: the Swiss *cantons*, the French *agences de l'eau*, the German *Länder* and the Dutch *waterschappen*. Last, it kept intergovernmental regulation to a minimum. The report outlined a small number of environmental goals to be achieved: (1) the governments should strive to eliminate a restricted list of the most toxic chemicals from the Rhine watershed; and (2) the governments should ensure the return of salmon and other indigenous species to the

Rhine. How these goals were to be reached was left up to the riparian countries themselves.

The McKinsey report was endorsed at the 1987 Ministerial Rhine Conference and adopted as the Rhine Action Programme (RAP). This programme was hugely successful. It aimed to reduce the discharge of a number of highly toxic chemicals into the river by 50% in 1995 (as compared to 1985). Moreover, its goal to allow the salmon to return introduced an ecosystem approach, as it required the redevelopment of upstream spawning grounds that had disappeared because of industrial, agricultural or city development, or had become unattainable because of dams and weirs in the waterway. The anti-pollution measures, as well as the ecosystem approach, envisioned by the RAP went beyond existing national legislation and initiated new domestic protection laws and policies. The RAP was also costly. The German Association of Chemical Firms (VCI) has estimated that its members along the Rhine alone spent DM 6.6 billion on sewage treatment during 1987–1991. The costs incurred by municipalities are generally thought to be higher.[1] In addition, a total cost of DM 110 million was foreseen for the construction of fish ladders and fish sluices (ICPR, 1991).

Nevertheless, the RAP was quickly implemented. Originally, all of the goals of the RAP were meant to be achieved by the year 2000. By the end of 1994, most had already been reached. Discharges of a large number of the most toxic substances had been reduced by 70–90%, instead of the targeted 50% (ICPR, 2003). Salmon and other species had returned to the river, for the first time in 40 years (Cazemier, 1994). A sophisticated warning system had been put in place to react better to accidental spills (Malle, 1994). The RAP was indeed one of the most successful international programmes for the restoration of a major water basin in the Northern Hemisphere (Dieperink, 2000; Huisman, de Jong, & Wieriks, 2000). A large part of its success can be attributed to its unique approach to international coordination: (1) non-binding; (2) limited to goal-setting, instead of aiming at agreement on both goals and means of environmental protection; (3) leaving implementation up to the lowest possible government level. This pragmatic approach to international governance allowed the government officials involved to reduce their preoccupations with legal formalities, sovereignty, relative gains, and defence of their own regulatory approach, and allowed the officials to focus on how to clean up the Rhine.

The massive floods of the Rhine and Meuse in late 1993 and early 1995 served to strengthen the ecosystem approach that the RAP had brought about (van Stokkom, Smits, & Leuven, 2005). In January 1995, in the Netherlands, Germany and Belgium, around 250,000 citizens had to be evacuated. The total economic damage of the 1995 events has been estimated at around €1 billion. The immediate cause of the floods was a combination of adverse weather events in both basins. The underlying reason, however, was the centuries-old human encroachment upon the rivers for the purposes of urban expansion, better navigation, greater electricity supply and increase of farmland. As a result of this encroachment, riverbeds had become eroded, rainwater runoff had become more rapid (resulting in higher water discharge peaks) and natural floodplains had disappeared. In response, the Action Plan on Floods for the Rhine was adopted by the riparian countries in 1998. This plan calls for floodplain expansion, habitat restoration, the lowering of groynes, moving dykes further inland, dredging the riverbed where sedimentation occurs, increasing citizens' awareness of the risk of flooding, and the installation of a public warning system. Thus, the Action Plan builds upon, and expands, the ecosystem approach initiated by the RAP. The plan leaves it up to national and local authorities to decide upon specific measures, and calls for public participation in the making of these decisions. It was swiftly implemented by the governments in the catchment area in a coordinated, but decentralized, manner – to the tune of €10 billion (ICPR, 2012).

Habitat restoration had been a long-standing priority of the environmental NGOs in the Rhine watershed. The support that the environmental groups in the watershed showed for the RAP and the Action Plan on Floods helped Dutch Minister of Transport, Public Works and Water Management Annemarie Jorritsma overcome initial resistance from the French, German and Swiss delegations to include NGOs more in the governance of the Rhine watershed. From 1997 onwards, NGOs have not only had observer status at the Ministerial Rhine Conferences, but have also been involved in the deliberation on restoration activities.

Between 1987 and 1999, the international policies to clean up the Rhine were therefore highly effective, and led (rather than trailed) domestic restoration efforts. The only exception was the agricultural sector: runoff from diffuse sources (especially nitrogen) did decrease during those years, but not by much (de Vries, Boers, Heinis, Bruning, & Sweerts, 1998).

2000–present: the EU Water Framework Directive

Despite the many steps taken to restore the Rhine catchment area, a number of environmental problems remain or have recently emerged. Apart from continued pollution from diffuse sources, these include: contaminated sediments as a result of past effluents (Uehlinger et al., 2009); the possible effects of climate change (van Slobbe, Werners, Riquelme-Solar, Bölscher, & van Vliet, 2014); invasion of non-native species (Leuven et al., 2009); and the inability to achieve fully self-sustaining populations of salmon and other species due to hard-to-reverse human changes to the watershed (bij de Vaate, Breukel, & van der Velde, 2006). Since 2000, several initiatives for overcoming these obstacles have been launched. The most important of these is the EU's Water Framework Directive (WFD), which came into force in December 2000 (Mostert, 2009). The WFD has rapidly become the focal point of the protection of the Rhine. The initial objective of the directive was to achieve, by 2015, good ecological quality for all EU waters with the help of river basin management. Under the directive, all 27 EU member states are obliged to identify their river basins and set up river basin districts. For each district, a six-yearly River Basin Management Plan has to be written. The directive also specifies which biological, physical and chemical parameters should be measured, how deviations from reference conditions should be enumerated, how monitoring programmes should be designed, in what manner results should be presented, and which timetables for the completion of different tasks should be followed. Furthermore, member states have to organize three consultation rounds in each basin. Last, the directive asks member states in each river basin to calculate, and agree on, the environmental and resource costs of using water, and then incorporate these costs into the pricing of water services (Gawel, 2014).

The WFD represents a major divergence from the RAP. Unlike the RAP, the WFD legally commits the governments of the Netherlands, Germany, Luxembourg and France. With its binding targets, the WFD resembles a supersized version of the formal and ineffective approach that characterized the international cooperation on the clean-up of the Rhine before 1987 more than the informal, successful international regime that came into being after the Sandoz incident. The directive has come in for criticisms from politicians, water managers and scientists (Hering et al., 2010). This may not be surprising given that, in 2015, "No Member State reported having completed all measures [of the WFD] and only one Member State (AT) reports that the status of water bodies is improving" (European Commission, 2015, p. 81). Discussion of these criticisms is reserved for the concluding section of the article.

An uncommon explanation

It is difficult enough to describe the ecological recovery of the Rhine in detail, but harder still to explain it. This is because the Rhine's environmental restoration has involved a number of surprising developments. First, before the Sandoz spill in late 1986, domestic and local measures by governments, municipalities and firms in all riparian countries had greatly reduced discharges of chemicals into the river, while at the same time the governments of these countries were quarrelling over such reductions in international fora. Second, during the same period, the large chemical companies in the watershed reduced the toxicity of their effluents by more than they were obliged to do under national (and international) water policies. Last, in the course of just a few months in late 1986 and early 1987, the international decision making regarding the clean-up of the Rhine turned from acrimonious and ineffective to harmonious and cutting-edge.

It is clear that interest-based theories of environmental cooperation (e.g. Sprinz & Vaahtoranta, 1994) do not offer much leverage in the case of the Rhine. Such approaches aim to explain the actions of stakeholders in terms of their (financial or other) self-interests. But in the Rhine basin, upstream users helped downstream users, companies went beyond legal standards, and the effectiveness of international cooperation changed nearly overnight. None of these processes can be explained by interest-based approaches without overstretching their central concepts. Even Elinor Ostrom's (2010) common-pool resources theory offers relatively limited purchase. This framework sets out 20 characteristics of resource systems that are presumed to increase the chances of sustainable resource use. Yet according to Villamayor-Tomas et al. (2014), nearly one-third of these characteristics have not been present in the Rhine catchment area. In addition, Villamayor-Tomas et al. conclude that various factors not mentioned in common-pool resources theory have been instrumental in bringing about the Rhine's restoration. To fully explain the remarkable clean-up of the Rhine, and extract policy lessons for other cases, I therefore believe that it is helpful to look for other approaches. Below, I argue that the plural rationality theory initiated by Douglas offers a more convincing explanation.

Plural rationality theory

After anthropologist Dame Mary Douglas's (1978, 1982) pioneering research, plural rationality (or cultural) theory[2] was developed into a full-fledged social and political theory (Rayner, 1992; Thompson, 2008; Thompson, Ellis, & Wildavsky, 1990). The theory distinguishes between four primary ways of organizing, perceiving and justifying social relations (called 'ways of life' or 'rationalities'): *egalitarianism, hierarchy, individualism* and *fatalism*. The theory postulates that these four ways of life emerge in contradistinction to each other in every conceivable domain of social life. Most such domains (say the way a school operates, or how an international regime functions) will consist of some dynamic combination of these pure forms. Because many social domains can be distinguished within and between societies (and because many societies can be distinguished around the world), the theory allows one to perceive a wide and ever-changing cultural and social variety – while still enabling the formulation of general propositions about social and political life. These propositions include possible ways in which people perceive and attempt to stave off a threat such as the degradation of a transboundary watershed. In order to explain this, I have to set out plural rationality theory in some detail.

Each of the theory's four ways of life consists of a pattern of social relations as well as a supporting cast of perceptions, values and interests. The typology is derived from two dimensions of sociality that Douglas called 'grid' and 'group'. Grid measures the extent to which ranking and stratification constrain the behaviour of individuals. Group, by contrast, measures the extent to which an overriding commitment to a social unit constrains the thought and action of individuals. Assigning two values (high and low) to each of the two dimensions gives the four ways of organizing social relations. Egalitarianism is associated with a low grid score (little stratification) and a high group score (strong group boundaries and solidarity). The combination of a high score on the grid dimension (lots of stratification) with a high score on the group dimension (much solidarity) gives hierarchy. The third way of life, individualism, is associated with low scores on both the grid and group scales. Last, fatalism is characterized by a high grid score and a low group score.

According to plural rationality theory, each of these four ways of organizing tends to induce, and be supported by, a particular way of perceiving nature, human nature, time, space, risk, technology, justice and governance. Since it was first formulated, this classification has helped illuminate the paradoxical and sometimes contradictory ways people approach a welter of contemporary public issues (Hartmann, 2011; Levin-Keitel, 2014; Swedlow, 2014; Verweij, 2011). Moreover, plural rationality theory posits that social domains and policy discourses are forever in flux due to the never-ending waxing and waning, splitting and merging of its four ways of life. Adherents to a particular way of life constantly compare its truth claims (regarding nature, human nature, risk, technology, etc.) with perceived reality. When this distance becomes too large, they will start to adjust their views (and social relations).

The theory's classification can be usefully applied to the protection of international watersheds, such as the Rhine.[3] In the hierarchical perspective, preservation of transboundary water basins requires the provision of public goods that are under-produced by the market or the anarchical international system. In order to remedy this, watersheds need to brought under the formal control of a single authority that is responsible for regulating, in a coherent and integrated manner, all aspects of the ecosystem involved. This authority, preferably enshrined in domestic and international law, should keep the activities of stakeholders (such as companies, ships, municipalities, farmers and fishers) within precise targets. The limits themselves should be determined by experts, if possible with the help of such techniques as objective risk analysis, long-term planning, assessment of best available technology, and computer-based future scenarios. They should be imposed through legally binding standards, as well as prices that are appropriately adjusted to incorporate externalities with the help of taxes, fees and/or subsidies. Public participation can be allowed, but will mostly serve as input into the decision makers' calculations, and to inform citizens of what needs to be done. In the Rhine watershed, the ICPR, European Commission, and national governmental delegations, as well as many scientists, have frequently adhered to such a hierarchical perspective. These views were somewhat submerged during the implementation of the RAP, but resurfaced when that programme had run its course in 2000.

The egalitarian take on how to restore transboundary watersheds is to consider these as highly vulnerable common-pool resources that are imminently threatened by greedy corporations, aloof bureaucrats, abstruse technocrats and rampant consumerism. Rather than top-down regulation or market solutions, what is needed is a moral revival, resulting in a significant reduction of human interference with watersheds. Ecocentrism and animal rights are called for, not egocentrism and speciesism. This desired state of affairs can be brought about by adhering to a very strict interpretation of the precautionary principle (according to which companies have to prove that a particular chemical is harmless, in any doses or combinations, before mass-producing it) and by letting waterways meander freely again. Decisions regarding the catchment area should ideally be made on the basis of consensus among all citizens affected by them. If this cannot be done, then citizens and citizens' groups should nevertheless be involved in the basin's governance as much as possible. In the Rhine watershed, many (though not all) environmental groups have espoused such ideals. These groups have included Greenpeace, Reinwater, Stichting Natuur en Milieu, Bundesverband Bürgerinitiativen Umweltschutz, Naturschutzbund Deutschland, Alsace Nature and Schweizerischer Bund für Naturschutz.

The individualistic perspective on how to protect international watersheds emphasizes the need to retain the opportunity to freely produce, and consume, private goods. In this view, water basins are highly resilient, and the extent of their degradation is frequently exaggerated. The use of chemicals should not be regulated unless abundant evidence of their toxicity has come to light. Even then, the costs of their regulation should be carefully weighed against the benefits of their use. Moreover, voluntary covenants between companies and governments are preferable to rigid, top-down imposition of environmental rules. The latter would risk preventing the technological improvements that could usher in a brighter environmental and economic future. Governments should therefore restrict themselves to declarations of intent, and leave the achievement of these goals to the polluters themselves. In the Rhine basin, the large chemical companies have often endorsed such views.

The fatalistic stance on watershed restoration is that it cannot be done, or at least not planned for. This is because all actors (be they companies, states or farmers) are too preoccupied with improving their own, short-term fortunes to care much about the environment or each other. In this view, upstream actors will not take into consideration the interests and wishes of downstream users unless they are coerced by superior force to do so. Upstream actors merely exploit the club goods that they can eke out of the watershed. In any case, the ecosystems involved are too complicated and unpredictable to allow rational analysis and planning. So why bother? Quite a few farmers in the Rhine catchment area appear to have adhered to such a view, at least more than other actors.

Hence, plural rationality theory's fourfold classification captures many of the protagonists involved in the governance of the Rhine watershed rather well. Of course, not all stakeholders involved have adhered to a single rationality. One hybrid actor has been the influential representative of the 120 water supply companies in the basin, called the Internationale Arbeitsgemeinschaft der Wasserwerke in Rheineinzugsgebiet (IAWR). These companies have had to make sure that their drinking water, taken from the Rhine, complies with very strict national and European quality standards. They therefore have had a strong financial incentive to induce other companies to reduce their pollution of the river. Moreover, the water supply companies have long assembled detailed information on the effluents of companies and cities. Until 2006, they maintained the only fine-grained measurement system of water quality in the entire Rhine basin. Backed up by this knowledge, the IAWR has used a series of sticks (in the form of lawsuits, lobbying for stricter laws, and threats to name and shame) and carrots (in the shape of a yearly award for good environmental practice) to motivate the chemical industry and municipalities to reduce their pollution of the Rhine. In their quest, the drinking water companies have combined rationalities. On the one hand, they have shared with the more egalitarian NGOs a limited tolerance for chemicals in the watershed. On the other hand, being companies themselves, they have not employed a radical anti-capitalist, anti-industry rhetoric.

Plural rationality theory has also important normative and policy implications. These flow from the theory's premise that each of its four ways of life is not only different from (and in competition with) but also dependent on all the others. That is to say, each way of life can survive only with the help of the other ways of life. From this premise, the following hypothesis can be derived: *Attempts to resolve pressing social and environmental problems that flexibly combine all ways of defining and resolving the issues at hand tend to be more successful than attempts that rely on fewer ways of life.* The latter will not only fail according to the goals, norms and values prioritized in the excluded ways of life, but they will also fail on their own terms – because each way of organizing and perceiving is complementary to, and co-dependent on, the other three. These efforts have often been labelled "clumsy solutions" (Verweij & Thompson, 2006), although the terms "polyrational" (Davy, 2012) and "robust" (Offermans & Valkering, 2016) have been used as well.

The constructive interplay among opposing ways of life that plural rationality theory advocates helps explain the various phases of the Rhine's restoration. As noted above, before the adoption of the RAP in 1987, two puzzling processes occurred: (1) chemical corporations reduced their toxic effluents beyond what they were legally required to do in the 1970s and 1980s; and (2) while companies and municipalities in all riparian countries were treating their wastewater, intergovernmental negotiations over precisely this topic broke down in acrimony. Plural rationality theory offers the following explanation for these puzzles.

The national and local authorities in the Rhine catchment area tended to follow a hierarchical perspective on water protection. That is, from the 1960s onwards, they

implemented domestic water protection policies and laws, often in the form of pollution fees and command-and-control systems rooted in 'best available technology'. This
compelled large chemical companies, as well as municipalities, to start building wastewater treatment plants. As a result, the chemical pollution of the Rhine began to decline
from the early 1970s onwards. However, the chemical multinationals reduced the
toxicity of their effluents to a larger degree than was legally mandated. In part, this
had to do with the strong pressure exerted, especially on the chemical industry, by
egalitarian organizations such as Greenpeace, as well as by the IAWR. Still, it could be
argued that, in the face of this pressure, chemical companies along the Rhine could have
limited themselves to what was legally required. Hence, something else must also have
been going on. The chemical firms have been the more individualistic actors in the
Rhine basin. For instance, they have tended to view the Rhine as a more resilient
ecosystem than other actors have (Disco, 2007). Before the 1970s, "Public waterways
were largely seen as a 'free good' by [German] industry. Industrial pollution was also
seen as essentially 'harmless', and the great rivers like the Rhine were ascribed great
natural cleaning powers" (Rüdig & Kraemer, 1994, p. 59).

Nevertheless, due to several accidents, it became unfeasible to maintain belief in the
"great natural cleaning powers" of the Rhine. In June 1969, Hoechst accidentally dumped
large amounts of the insecticide Endosulfan into the Main (a major tributary), killing large
numbers of fish (Greve & Wit, 1971). And during several days in June 1971, more than
100 km of the Rhine was left without oxygen, killing off all fish in that part of the
watershed. Even the chemical corporations had to admit that the Rhine was in need of
saving in the early 1970s. When faced with this incontrovertible evidence, the chemical
corporations reacted in an individualistic fashion: where others saw an environmental
problem, they saw a market opportunity. Instead of relying on 'the best available technology', several companies invented new, more efficient wastewater treatment technology
(such as Hoechst's Biohoch Reaktor and Bayer's 'tower technology'). They not only
installed the novel technologies at their own plants, but also proceeded to sell these to
companies and municipalities around the world. Thus, the chemical companies on the
Rhine were able to branch out into providing ecosystem services, including wastewater
treatment. For example, in 2014, Bayer Technology Services had around 2300 employees
and an output of €480 million, while Infraserv Hoechst had some 2700 employees and an
output of €1.1 billion.[4] In this manner, between 1970 and 1986, the Rhine chemical
corporations reduced their effluents by more than the hierarchical government agencies
would have thought technically possible, or the egalitarian NGOs would have believed
morally feasible – while building up lucrative subsidiaries. This is clearly a "clumsy
solution" that resulted from the often fractious, but ultimately virtuous, interactions
among egalitarian environmental groups, hierarchical governmental agencies, individualistic corporations and the culturally hybrid IAWR.

Plural rationality theory also offers an explanation for why, during the same period,
international cooperation on the Rhine came to naught – and for why this later
changed. Before the Sandoz incident, only hierarchical principles informed intergovernmental cooperation. Non-state actors, and their different perspectives, were barred from
the international halls of power. By themselves, national governments attempted to
construct a legally binding, transboundary control-and-command system out of the
divergent domestic control-and-command systems already in place. Moreover, they

attempted to reach such an agreement in a highly formalistic and time-consuming manner. This typically hierarchical approach to the international governance of the Rhine watershed only led to deadlock.

In contrast, the shock of the Sandoz spill facilitated the infusion of other principles and practices into the Rhine cooperation. The approach that Minister Kroes and the McKinsey team got accepted included a number of individualistic elements: the informal and non-binding character of the RAP; its small set of targets; and the decision to leave the responsibility for reaching these aims to the lowest possible government levels. Indeed, Kroes's initial decision to let consultants outline an international agreement on the restoration of the Rhine basin, and to build up the necessary intergovernmental support for this plan, can itself be seen as individualistic, as it replaces governmental duties with private enterprise. With the subsequent decisions to restore the natural habitat of the Rhine, and to consult much more frequently with NGOs, egalitarian principles also became part of the international governance of the watershed. Only with this much clumsier arrangement in place did the international cooperation on the protection of the Rhine start to function well.

Conclusion: a foreboding

In this article, I have argued that the plural rationality (or cultural) theory pioneered by Dame Mary Douglas offers a plausible explanation of both the successes and failures of the ecological protection of the Rhine. This approach has begun to be applied by Dutch policy makers and scientists in their efforts to increase the resilience of the Rhine delta (Middelkoop et al., 2004; Offermans, Valkering, Vreugdenhil, Wijermans, & Haasnoot, 2013).

But plural rationality theory has certainly not informed the WFD. Although it has frequently been claimed that this directive was influenced by the successful RAP (e.g. Plum & Schulte-Wülwer-Leidig, 2014), from the viewpoint of plural rationality theory the WFD appears very different from the RAP. The latter was a polyrational mix of egalitarian, hierarchical and individualistic ideas and practices. The former – with its strict timetables, comprehensive lists of priority substances, prescribed price adjustments, binding biological, hydromorphological and physical-chemical targets, limited public participation, complicated measurements methods and legalistic nature – is exclusively hierarchical. From the perspective of plural rationality theory, it is therefore not surprising that the directive has received "major criticism from politicians, water managers and scientists" (Hering et al., 2010, p. 4008), while the European Environmental Bureau (2010), which represents more than 140 environmental citizens' groups from across the EU, has called the directive a "toothless tiger". Indeed, the European Commission recently admitted that all member states were "well short of meeting the WFD targets required by 2015", while "differences and tensions have emerged across Europe about the future of EU environmental targets under the Water Framework Directive" (Waterbriefing, 2015). Only a less monolithic approach to water management will be able to sustain the restoration of the Rhine and other river basins.

Notes

1. Uehlinger et al. (2009, p. 25) estimate that around €40 billion was spent on building and improving sewage treatment facilities in the Rhine basin between 1970 and 1990.
2. Previously, this approach was often called 'cultural theory'. To avoid confusion with post-essentialist cultural theories, the term 'plural rationality theory' is increasingly favoured.
3. This section is based on semi-structured interviews held in 1996 and 1997 with 54 stake-holders in the Rhine basin (Verweij, 2001), analysis of primary documents, and close reading of secondary literature. Recent corroboration for the Dutch part of the basin comes from Offermans and Cörvers (2012).
4. Data retrieved from http://www.bayertechnology.com/unternehmen/zahlen-fakten.html (11 October 2015) and http://www.infraserv.com/de/unternehmen/daten_und_fakten_inkl_standorte/ (11 October 2015).

References

Bernauer, T. (1995). The international financing of environmental protection: Lessons from efforts to protect the river Rhine against chloride pollution. *Environmental Politics, 4*, 369–390. doi:10.1080/09644019508414212

Bernauer, T., & Moser, P. (1996). Reducing pollution of the river Rhine: The influence of international cooperation. *The Journal of Environment & Development, 5*, 389–415. doi:10.1177/107049659600500402

Beurskens, J. E. M., Winkels, H. J., de Wolf, J., & Dekker, C. G. C. (1994). Trends of priority pollutants in the Rhine during the last fifty years. *Water Science Technology, 29*, 77–85.

Bij de Vaate, A., Breukel, R., & van der Velde, G. (2006). Long-term developments in ecological rehabilitation of the main distributaries in the Rhine delta: Fish and macroinvertebrates. *Hydrobiologia, 565*, 229–242. doi:10.1007/s10750-005-1916-4

Bölscher, T., van Slobbe, E., van Vliet, M. T. H., & Werners, S. E. (2013). Adaptation turning points in river restoration? The Rhine salmon case. *Sustainability, 5*, 2288–2304. doi:10.3390/su5062288

Cazemier, W. G. (1994). Present status of the *salmondis atlantic* and sea-trout in the Dutch part of the river Rhine. *Water Science Technology, 29*, 37–41.

Chase, S. K. (2011). There must be something in the water: An exploration of the Rhine and Mississippi rivers' governing differences and an argument for change. *Wisconsin International Law Review, 29*, 609–641.

Da Silveira, A. R., & Richard, K. S. (2013). The link between polycentrism and adaptive capacity in river basin governance systems: Insights from the river Rhine and the Zhujiang (Pearl River) basin. *Annals of the Association of American Geographers, 103*, 319–329. doi:10.1080/00045608.2013.754687

Davy, B. (2012). *Land policy: Planning and the spatial consequences of property*. Cheltenham: Edward Elgar.

De Vries, I., Boers, P. C. M., Heinis, F., Bruning, C., & Sweerts, J. P. R. A. (1998). *Targets for nitrogen in the river Rhine: Nitrogen as a steering factor in marine and freshwater ecosystems* (Internal Report RIKX/OS 98.129X). The Hague: Institute for Coast and Sea.

Dieperink, C. (2000). Successful international cooperation in the Rhine catchment area. *Water International, 25*(3), 347–355. doi:10.1080/02508060008686842

Dieperink, C. (2011). International water negotiations under asymmetry, Lessons from the Rhine chlorides dispute settlement (1931–2004). *International Environmental Agreements: Politics, Law and Economics, 11*(2), 139–157. doi:10.1007/s10784-010-9129-3

Disco, C. (2007). Accepting father Rhine? Technological fixes, vigilance, and transnational lobbies as 'European' strategies of Dutch municipal water supplies 1900–1975. *Environment and History, 13*(4), 381–411. doi:10.3197/096734007X243140

Douglas, M. (1978). *Cultural bias* (Occasional Paper No. 35). London: Royal Anthropological Institute.

Douglas, M. (Ed.). (1982). *Essays in the sociology of perception*. London: Routledge.

European Commission. (2015). *Report on the progress in implementation of the water framework directive programme of measures* (SWD (2015) 50 final). Brussels: Author.

European Environmental Bureau. (2010). *10 years of the water framework directive: A toothless tiger?* Brussels: Author.

Farmer, A., & Braun, M. (2002). Fifty years of the Rhine commission: A success story in nutrient reduction. *Scope Newsletter, 47*, 1–15.

Frijters, I. D., & Leentvaar, J. (2003). *Rhine case study* (Technical Documents in Hydrology No. 17). Paris: UNESCO International Hydrological Programme.

Gawel, E. (2014). Article 9 of the EU water framework directive: Do we really need to calculate environmental and resource costs? *Journal for European Environmental & Planning Law, 11*, 249–271. doi:10.1163/18760104-01103004

Giger, W. (2009). The Rhine red, the fish dead—The 1986 Schweizerhalle disaster, a retrospect and long-term impact assessment. *Environmental Science and Pollution Research, 16*, 98–111. doi:10.1007/s11356-009-0156-y

Greve, P. A., & Wit, S. L. (1971). Endosulfan in the Rhine river. *Journal (Water Pollution Control Federation), 43*(12), 2338–2348.

Hartmann, T. (2011). *Clumsy floodplains: Responsive land policy for extreme floods*. Farnham: Ashgate.

Hering, D., Borja, A., Carstensen, J., Carvalho, L., Elliott, M., Feld, C. K., & van de Bund, W. (2010). The European water framework directive at the age of 10: A critical review of the achievements with recommendations for the future. *Science of the Total Environment, 408*, 4007–4019. doi:10.1016/j.scitotenv.2010.05.031

Huisman, P., de Jong, J., & Wieriks, K. (2000). Transboundary cooperation in shared river basins: Experiences from the Rhine, Meuse and North Sea. *Water Policy, 2*(2), 83–97. doi:10.1016/S1366-7017(99)00023-9

International Commission for the Protection of the Rhine. (1991). *Ecological master plan for the Rhine: Salmon 2000*. Koblenz: Author.

International Commission for the Protection of the Rhine. (2003). *Outcome of the Rhine action programme*. Koblenz: Author.

International Commission for the Protection of the Rhine. (2007). *Bericht über die Koordinierung der Überblicksüberwachungsprogramme gem. Artikel 8 und Artikel 15 Abs. 2 WRRL in der internationalen Flussgebietseinheit Rhein*. Koblenz: Author.

International Commission for the Protection of the Rhine. (2012). *Aktionsplan Hochwasser 1995–2010: Handlungsziele, Umsetzung und Ergebebnisse*. Koblenz: Author.

Kiss, A. (1985). The protection of the Rhine against pollution. *Natural Resources Journal, 25*, 613–637.

Leuven, R. S. E. W., van der Velde, G., Baijens, I., Snijders, J., van der Zwart, C., Lenders, H. J., & bij de Vaate, A. (2009). The river Rhine: A global highway for dispersal of aquatic invasive species. *Biological Invasions, 11*(9), 1989–2008. doi:10.1007/s10530-009-9491-7

Levin-Keitel, M. (2014). Managing urban riverscapes: Towards a cultural perspective of land and water governance. *Water International, 39*(6), 842–857. doi:10.1080/02508060.2014.957797

Malle, K. G. (1994). Accidental spills: Frequency, importance, control, and countermeasures. *Water Science & Technology, 29*(3), 149–163.

Middelkoop, H., van Asselt, M. B. A., van 't Klooster, S. A., van Deursen, W. P. A., Kwadijk, J. P. C., & Buiteveld, H. (2004). Perspectives on flood management in the Rhine and Meuse rivers. *River Research and Applications, 20*, 327–342. doi:10.1002/rra.782

Molls, F., & Nemitz, A. (2006). Restoration of Atlantic salmon and other diadromous fishes in the Rhine river system. *American Fisheries Society Symposium, 49*, 587–603.

Mostert, E. (2009). International co-operation on Rhine water quality 1945–2008: An example to follow? *Physics and Chemistry of the Earth, Parts A/B/C, 34*(3), 142–149. doi:10.1016/j.pce.2008.06.007

Myint, T. (2005). *Strength of 'weak' forces in multilayer environmental governance: Cases from the Mekong and Rhine river basins* (Unpublished doctoral dissertation). Department of Political Science, Indiana University, Bloomington.

Offermans, A., & Cörvers, R. (2012). Learning from the past; changing perspectives on river management in the Netherlands. *Environmental Science & Policy, 15*(1), 13–22. doi:10.1016/j.envsci.2011.10.003

Offermans, A., & Valkering, P. (2016). Socially robust river management. *Journal of Water Resources Planning and Management, 142*(2), 1–9. doi:10.1061/(ASCE)WR.1943-5452.0000615#sthash.A7K6cv8I.dpuf

Offermans, A., Valkering, P., Vreugdenhil, H., Wijermans, N., & Haasnoot, M. (2013). The Dutch dominant perspective on water; risks and opportunities involved. *Journal of Environmental Science and Health, Part A, 48*(10), 1164–1177. doi:10.1080/10934529.2013.776438

Ostrom, E. (2010). Beyond markets and states: Polycentric governance of complex economic systems. *American Economic Review, 100,* 641–672. doi:10.1257/aer.100.3.641

Plum, N., & Schulte-Wülwer-Leidig, A. (2014). From a sewer into a living river: The Rhine between Sandoz and salmon. *Hydrobiologia, 729*(1), 95–106. doi:10.1007/s10750-012-1433-1

Rayner, S. (1992). Cultural theory and risk analysis. In S. Krimsky (Ed.), *Social theories of risk* (pp. 83–115). Westport, CN: Praeger.

Rüdig, W., & Kraemer, R. A. (1994). Networks of cooperation: Water policy in Germany. *Environmental Politics, 3*(4), 52–79. doi:10.1080/09644019408414167

Schmitt, L., Roy, D., Trémolières, M., Blum, C., Dister, E., Pfarr, U., & Späth, V. (2012). 30 years of restoration works on the two sides of the Upper Rhine River. In *Final proceedings of the first international conference on "Integrative sciences and sustainable development of rivers"* (pp. 1–3), June 26–28, 2010. Lyon: I.S. Rivers.

Sprinz, D., & Vaahtoranta, T. (1994). The interest-based explanation of international environmental policy. *International Organization, 48*(1), 77–105. doi:10.1017/S0020818300000825

Swedlow, B. (Ed.). (2014). *Advancing policy theory with cultural theory.* Special issue of *Policy Studies Journal* (Vol. 42(4), pp. 465–697). doi:10.1111/psj.12079

Thompson, M. (2008). *Organising and disorganising: A dynamic and non-linear theory of institutional emergence and its implications.* Axminster: Triarchy.

Thompson, M., Ellis, R. J., & Wildavsky, A. (1990). *Cultural theory.* Boulder, CO: Westview.

Uehlinger, U., Wantzen, K. M., Leuven, R. S. E. W., & Arndt, H. (2009). The Rhine river basin. In K. Tockner, U. Uehlinger, & C. T. Robinson (Eds.), *Rivers of Europe* (pp. 199–245). London: Academic Press.

Van Slobbe, E., Werners, S. E., Riquelme-Solar, M., Bölscher, T., & van Vliet, M. T. H. (2014). The future of the Rhine: Stranded ships and no more salmon? *Regional Environmental Change, 14,* 1–11. doi:10.1007/s10113-014-0683-z

Van Stokkom, H. T. C., Smits, A. J. M., & Leuven, R. S. E. W. (2005). Flood defense in the Netherlands: A new era, a new approach. *Water International, 30*(1), 76–87. doi:10.1080/02508060508691839

Verweij, M. (2001). *Transboundary environmental problems and cultural theory: The protection of the Rhine and the Great Lakes.* New York, NY: Palgrave.

Verweij, M. (2011). *Clumsy solutions for a wicked world.* New York, NY: Palgrave.

Verweij, M., & Thompson, M. (Eds.). (2006). *Clumsy solutions for a complex world.* New York, NY: Palgrave Macmillan.

Villamayor-Tomas, S., Fleischman, F. D., Perez Ibarra, I., Thiel, A., & van Laerhoven, F. (2014). From Sandoz to salmon: Conceptualizing resource and institutional dynamics in the Rhine watershed through the SES framework. *International Journal of the Commons, 8,* 361–395. doi:10.18352/ijc.411

Waterbriefing. (2015, April 13). Tensions emerge in Europe over future of water framework directive. Retrieved from http://www.waterbriefing.org/home/regulation-and-legislation/item/10623-tensions-emerge-in-europe-over-future-of-water-framework-directive

Wieriks, K., & Schulte-Wülwer-Leidig, A. (1997). Integrated water management for the Rhine river basin, from pollution prevention to ecosystem improvement. *Natural Resources Forum, 21,* 147–156. doi:10.1111/j.1477-8947.1997.tb00686.x

The dilemma of autonomy: decentralization and water politics at the subnational level

Scott M. Moore

ABSTRACT
This article develops a framework for understanding the role of subnational states in water politics in decentralized federal systems. First, that role has increased worldwide as a result of decentralization. Second, the quest for autonomy sometimes leads subnational officials to prefer weak forms of cooperation. Third, the interaction of subnational states, central governments and non-governmental actors largely explains interjurisdictional conflict and cooperation in shared river basins. This framework is applied to the case of the Colorado River basin to help explain a long-term shift towards more cooperative relationships between the riparian states.

Introduction: exploring subnational hydro-politics

Traditionally, scholars of water politics have emphasized the role of the central state in shaping the dynamics of conflict and cooperation at both subnational and international levels. But in many river basins subnational units like states and provinces play an important role in water resource management, and their interests often vary considerably from those of their parent nation-states. The role of these subnational states has begun to receive some attention from scholars of water politics. Several detailed accounts exist of the role of subnational states and provinces in basins as diverse as the Ganges, the Missouri and the Murray-Darling (Franda, 1968; Garrick, Anderson, Connell, & Pittock, 2014; Thorson, 1994). At the same time, a small but intriguing literature examines the links between water-related conflict and disputes at the subnational level and tensions in international transboundary basins, particularly in South Asia (Alam, Dione, & Jeffrey, 2009; Giordano, Giordano, & Wolf, 2002; Wirsing, 2007). Finally, a subset of the collective-action literature examines how different subnational jurisdictions cooperate to manage estuaries and other shared water bodies (Berardo & Scholz, 2010; Sabatier et al., 2005; Weber, 2003). However, the existing literature leaves several important questions unanswered. First, exactly what role do subnational states play in hydro-politics relative to the central states and non-governmental actors that have been the primary subjects of past scholarly attention? Second, under what circumstances do subnational states engage in conflict versus cooperation over shared water resources?

This article attempts to present a tentative answer to these questions, and in the process to develop a framework for understanding the role of subnational jurisdictions in water politics. It does so by examining the role of riparian states, the federal government and environmental advocacy organizations in the case of the Colorado River basin. The Colorado case is well suited to the theory-building objectives of this article for two reasons. First, the American federal system defines a constitutional role for subnational states in water resource management. Second, nearly a century of conflict and cooperation in the basin between the states, the federal government and NGOs evinces several shifts in relationships between these actors. This within-case variation presents an opportunity to explain the role of subnational states, and to identify the factors that induce them to engage in conflict versus cooperation (King, Keohane, & Verba, 1994).

Accordingly, I rely on a wide range of documentary resources, including news articles, personal papers and official records, to construct a historical case study of Colorado River management. While this case-study approach cannot validate a generalizable theory, it is suited to the theory-building objective I have set out, namely to conceptualize the role of subnational states in water politics. I identify several distinct shifts in the degree of cooperation between the federal government, the riparian states and civil society actors in the Colorado basin. These shifts illustrate that the behaviour of the riparian states was informed by a desire to maintain their autonomy relative to the central government, which led them to prefer weak forms of cooperation with their neighbours. In this context, third-party pressure from NGOs has been critical to establishing more robust forms of cooperation, which have in turn promoted more sustainable approaches to Colorado River governance.

The increasingly prominent role of subnational states in water resource management stems from the global trend towards the transfer of administrative responsibilities, fiscal resources and political power from central to subnational levels of government. While countries have implemented decentralization in different ways and to varying degrees, the overall effect has been to greatly enlarge the power of subnational levels of government relative to that of the central government (Treisman, 2007). The effect of decentralization has been felt to some extent by nearly every country in the world, as even highly centralized states such as Denmark, Indonesia and Uganda have devolved decision making and fiscal powers in many policy areas from central to local governments (Hooghe & Marks, 2003; Montero, 2001). As a result of this global trend towards decentralization, subnational states have become important players in water resource management. They have always been important actors in the water politics of federal countries, where constitutional responsibility for water resources is usually divided between central and state governments (Forum on Federations, 2009). But under decentralization, states, provinces and municipalities have become increasingly important players in natural and water resource management across the globe (Andersson, Gibson, & Lehoucq, 2006; Ascher, 2007; Mullin, 2009).

While decentralization has played out very differently across countries, its most important and universal effect is to create an acute dilemma of autonomy. The essence of decentralization is to permit subnational jurisdictions to act with some independence from the central government. In some countries, subnational governments have narrowly defined and pedestrian responsibilities like trash collection, while in others they possess powers in virtually every area except defence and foreign affairs, and maintain independent judiciaries and legal systems (Watts, 2008). Even in more centralized countries, however, the division of powers and responsibilities between central and subnational levels of government creates ambiguity in who is responsible for certain issues, as well as inherent coordination problems between levels of government. Given the inherently intersectoral nature of water resource management, these challenges often hinder effective responses to interjurisdictional issues like water resource allocation, pollution control and groundwater regulation (UNFAO, 2010). The division of power is in particular frequently faulted for failures of interjurisdictional collective action in shared river basins. A study of flood control in the Missouri River basin concludes, for example, that "the Constitution provides few opportunities for resolving natural resource conflicts between states or within regions, and it provides even fewer formal avenues for intergovernmental cooperation in the management of shared natural resources – particularly rivers" (Thorson, 1994, p. 4).

However, the division of power under decentralization also means that subnational governments have to constantly protect their autonomy from over-reach by the central state. Decentralization is often employed by central governments as an expedient, most typically either to quiet the demands of peoples or regions for independence or to improve public services in areas where the central government has been ineffective (Hooghe & Marks, 2003). As a result, decentralization is often subject to periodic renegotiation, and even reversal. The challenge of maintaining subnational independence is especially acute in countries where the powers of subnational governments lack firm constitutional foundations. However, even federal countries often experience prolonged debates over the diminution of subnational authority. In the United States in particular, the power of the states relative to the federal government has been subject to constant flux since its foundation, and is perhaps the foundational question in American political life (Elazar, 1984). The struggle between central states and local communities over control of water has been observed in a large number of contexts, including the American West (Bakker, 2010; Worster, 1985). Accordingly, subnational officials across countries face a common dilemma: how to preserve their independence while also cooperating with the centre and with their neighbours on matters of shared concern.

For subnational officials, the solution to this dilemma is often to pursue weak forms of cooperation with their neighbours. Joseph Zimmerman (2011, p. 201) notes, for example, that in the United States, "The modus operandi of most states does not encourage extensive interstate joint ventures because states, as semiautonomous entities, naturally are reluctant to engage in such ventures due to the loss of exclusive control accompanying them." This tendency to prefer weak forms of cooperation with neighbouring jurisdictions poses a particular challenge for interjurisdictional collective-action problems in shared river basins. Most scholars agree that problems like water scarcity and pollution are most effectively addressed by institutions that can convene all relevant stakeholders at an appropriate scale of decision making, usually that of the watershed or river basin (Imperial, 2005; Lankford & Hepworth, 2010; Teclaff, 1996). While these institutions can include less formal bodies like water user associations as well as formal river basin commissions, they must have sufficient powers and resources to facilitate collaborative and participatory decision making. By necessity, assuming these powers dilutes the independence of individual subnational jurisdictions. As a result, subnational officials tend to prefer weaker forms of cooperation that enable them to veto decisions favoured by either the central government or their neighbours. As a study of US river basin commissions observed, "Most states appear to join commissions for defensive purposes.... They seek defenses against one another as well as against federal action" (Derthick, 1974, p. 151).

In the United States, this desire to protect their autonomy has led state officials to prefer weak forms of cooperation over shared waterways in the form of interstate compacts, which are the predominant instrument employed in the United States to address both water quantity and water quality externalities (Schlager & Heikkila, 2009). Though numerous, these arrangements often lack enforcement and conflict-resolution capacities (Heikkila, Schlager, & Davis, 2011). Instead, compacts are typically subsumed within wider interstate politics. As one observer has commented, "No compacts have truly succeeded in eliminating externalities because the regional governments that they establish do not coordinate well with pre-existing governments which create the compacts" (Thorson, 1994, p. 140). At the same time, the US Constitution provides states with an alternative to cooperative agreements like interstate compacts, namely by pursuing competing claims to water through judicial means (Watts, 2008). The number of such lawsuits has increased significantly in recent years; as of the time of writing four interstate water-sharing disputes were before the Supreme Court (Jacobs, 2015).

Yet while some features of decentralization and federalism may raise the cost of collective action in interstate river basins, others create opportunities for inclusive and cooperative governance of shared water resources. In general, the more decentralized the political system, the broader is what social movement scholars call the political opportunity structure, meaning that more opportunities exist for non-governmental actors, such as environmental organizations, to gain both political voice and influence (Kitschelt, 1986; Kriesi, Koopmans, Duyvendak, & Giugni, 1992; Marks & McAdam, 1996). This broader political opportunity structure promotes water resource management that is both collaborative, meaning that decisions are made based on shared values and norms of reciprocity (Lubell, 2004), and participatory, meaning that all relevant actors are included in the decision-making process. Both practices, the consensus runs, are essential to achieving environmental and social objectives, especially under changing circumstances (Blatter & Ingram, 2001; Sabatier et al., 2005; Weber, 2003). In the United States, conservation organizations effectively exploited points of access at both state and federal levels to gain support for national water quality legislation in the early twentieth century, for example (Bouleau, 2009; Paavola, 2006).

Environmental non-governmental organizations (ENGOs) and civil society groups have long been acknowledged to play a critical role in fostering positive, cooperative outcomes in water resource management (Keck & Sikkink, 1998; Wood, 2007). This is because non-governmental and civil society representation provides inputs into decision making that balance the often narrow views of governmental actors on the one hand, and of self-interested economic interests on the other. Perhaps most significantly, ENGOs and civil society organizations help confer legitimacy on decision-making processes, an imprimatur that can be critically important given the contention which surrounds many aspects of water resource management. This feature of ENGO and civil society participation in water resource governance, vividly illustrated in the case of

Brazil, is decisive in promoting cooperation and preventing conflict among the users of a shared water resource (Abers & Keck, 2006, 2009, 2013). In many ways, ENGO and civil society participation is the glue that holds the cumbersome architecture of water resource management in place, especially in large, multi-stakeholder river basins. Their influence and pressure is critical to fostering collaborative, participatory institutions for water resource management, as illustrated by the case of the Colorado River basin.

Multi-level institution building in the Colorado River basin

For the better part of a century, the US federal government and states of the Colorado River basin have attempted to construct an adequate regional institutional framework to manage the region's scarce water resources. The degree of cooperation between them has steadily increased over time (Figure 1). A key feature in this evolution has been the changing relationship between the states, the federal government and environmental advocacy organizations. In the initial phase of institution building in the Colorado basin, fear that Mexico might lay claim to large quantities of the river's flow led riparian states to forge the Colorado River Compact, which created the nucleus for continued cooperation in subsequent decades. However, the concern of states with maintaining their individual autonomy resulted in a weak and brittle compact architecture that proved ill-suited to meeting growing demands placed on the river's waters. This conflict intensified in the second stage of the controversy as states and water users jockeyed for rents from federally financed water infrastructure projects. Finally, in the third, most recent period, the growing power of environmental movements has created pressure for the development of more robust, collaborative and participatory governance institutions. The case of institution building in the Colorado River basin therefore suggests that subnational states face a dilemma of autonomy which discourages robust institution building. Instead, cooperation is most often induced by third-party advocacy organizations.

The Colorado River is located in the western United States, and is shared between seven American states as well as between the United States and Mexico (Figure 2). Large portions of Arizona, Colorado and Utah lie within the basin, while the remaining states contribute a smaller share of the river's flow. The waters of the Colorado are apportioned by a 1922 interstate compact between an Upper Basin, which consists of those parts of Wyoming, Colorado, Utah and New Mexico which lie above Lee's Ferry, Arizona, and a Lower Basin, constituting those parts of Arizona, Nevada and California which lie below; 7.5 million acre-feet of water (MAF) are apportioned to each part of the basin annually and in perpetuity by the compact. In the Upper Basin, a 1948 interstate compact allocates 51.75% of the water to Colorado, 11.25% to New Mexico, 23% to Utah and 14% to Wyoming, and 50,000 acre-feet annually to Arizona. In the Lower Basin, a 1964 Supreme Court decree apportioned 4.4 MAF to California and 2.8 MAF to Arizona, and 300,000 acre-feet annually to Nevada. A 1944 treaty guarantees Mexico a downstream flow of 1.5 MAF annually, subject to minor restrictions (US Bureau of Reclamation, 2008).

Subsequent agreements under this treaty framework provide for the control of salinity in waters delivered to Mexico, as well as for the maintenance of baseline levels of environmental flow in the Colorado River Delta (Bennett, 2000). Collectively, these compacts and agreements are known as the Law of the River. This regime has historically been hobbled by a number of structural deficiencies, particularly its failure

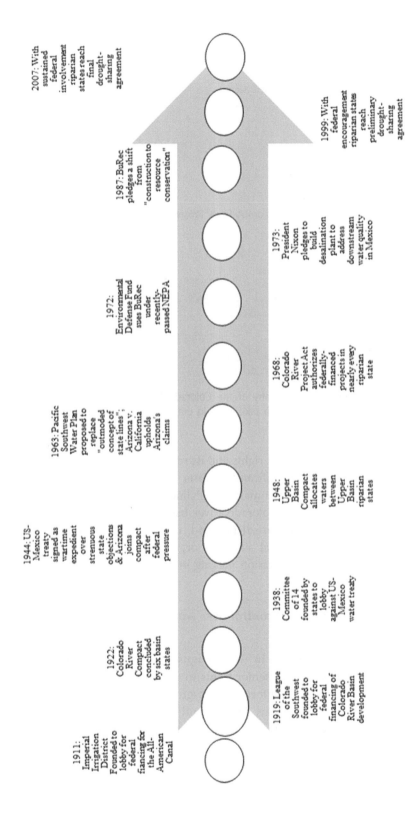

Figure 1. Timeline of institution building in the Colorado River basin. Source: Author.

Figure 2. Physical and political geography of the Colorado River basin. Note: The shaded ovals represent discrete historical events which take place in different years. Source: Author.

to recognize Native American water rights and its reliance on a baseline Colorado River flow that research suggests is significantly greater than long-term historical averages (Christensen, Wood, Voisin, Lettenmaier, & Palmer, 2004). But one of the biggest challenges to cooperation in the Colorado River basin has been competition between federal and state governments for jurisdiction and control over the river's waters, joined more recently by non-governmental advocacy organizations who have lobbied for a more participatory approach to decision making in the basin.

The struggle for autonomy and control over water: roots of interstate conflict in the Colorado River basin

The roots of interstate conflict lie in the attempts of riparian states to permanently secure their autonomy, along with senior rights to Colorado River water. Beginning in the late nineteenth century, California's fast-growing and water-scarce southern regions, particularly the cities of Los Angeles and San Diego and the irrigated lands of the Imperial and Coachella Valleys, turned to the Colorado for their water needs. Accordingly, these regions began to lobby for a share of Colorado River water. As a 1924 report noted, "California makes no contribution to the waters of the Colorado but … on account of the low lying lands within her borders can use advantageously the greater part of the water allocated to the lower basin" (Clark, 1924, p. 15). Irrigated agriculture quickly expanded throughout the early twentieth century, and in 1911 the Imperial Irrigation District (IID) was organized as the world's largest single agricultural unit. The IID quickly emerged as a powerful political force, and soon began calling for federal assistance to build the All-American Canal, which would divert water from the Colorado on the American side of the border, instead of relying on water from Mexico as had previously been done (Nadeau, 1974). At a 1919 conference, the IID allied with Los Angeles and Coachella farmers to seek congressional appropriations for the canal, in return for which they were guaranteed Colorado water. Less than two months later, IID lobbyists were in Washington, "handing out cantaloupes produced in the [Imperial] valley" (Hundley, 1975, pp. 41–42).

A similar process played out in the neighbouring state of Arizona, where constituencies emerged in support of water diversion projects to irrigate vast areas in central Arizona. In the early 1920s, an Arizona entrepreneur named George Maxwell helped start an organization advocating construction of the High-Line Canal to channel water from the main stem of the Colorado River to central Arizona. The project, Maxwell argued, was made necessary by the combined water allocations implied for both California and Mexico under the proposed terms of the Colorado River Compact, leaving little for Arizona. In particular, Maxwell argued that the compact "was a smokescreen to cover the design of turning the waters of the Colorado over to ... northern California", leaving only enough to irrigate some 282,000 acres in Arizona, as opposed to 900,000 in California and 810,000 in Mexico (*Arizona Republican*, 1922). Maxwell's charge that Arizona faced a poor bargain gained considerable political support in the Arizona legislature, and after the 1923 legislative session began, rarely was "a day allowed to pass ... without an added measure designed to stave off action on the Colorado River compact" (*Los Angeles Times*, 1923).

Arizona's hostility to the Colorado River Compact was perhaps the most extreme example of the extent to which the dilemma of autonomy led riparian states to prefer weak forms of institution building in the basin. However, the breadth of this concern is illustrated by the creation of the League of the Southwest, a lobbying group founded to support basin-wide initiatives like the compact. In its founding declaration, the league pledged to uphold state sovereignty over water: "In the arid States of the West the irrigation projects undertaken by or with the aid of the Federal Government should in every instance be based upon a full compliance with the laws of the State wherein the projects are located so far as the appropriation of water and other matters of purely State control are concerned." The basin states thus approached institution building as a counterweight to the US and Mexican governments, who they feared would assert control over Colorado River waters. A remark by Delph Carpenter, the Colorado lawyer who first proposed the Colorado River Compact, makes clear his position that "the States of the Union ... have the same power to enter into compacts with each other as do independent nations". Without guarantees to its claims to water, Carpenter (1923) proclaimed, "The upper State has but one alternative, that of using every means to retard development in the lower State until the uses within the Upper State have reached their maximum."

This suspicion of their neighbours as well as of the federal government led the states to pursue a loose form of cooperation among themselves. However, by the end of the 1930s, growing water demand and the prospect of a large Mexican water claim under the proposed US–Mexican boundary treaty persuaded Arizona's leaders that they required main-stem Colorado River water, which necessitated ratification of the compact. The Gila Project Association, which desired federal appropriations for irrigation projects in the Yuma region, stridently criticized Arizona's refusal to sign the Colorado River Compact, and the risk of remaining outside the compact was driven home when in May 1937 Upper Basin congressional representatives voted *en masse* to defeat an appropriations bill for the Gila Project (*Los Angeles Times*, 1937). Henceforth, the political pendulum in Arizona shifted towards ratification of the compact, though only with the goal of supplying its long-sought High-Line Canal, which Democratic Governor Rawghlie Stanford claimed would mean that "existing water disputes and water shortages ... on the same streams within Arizona will be ended" and would allow Arizona to "become the most prosperous State in the Union" (Smith, 1937).

Indeed, opposition to alleged federal and Mexican water claims proved to be the best stimulant to cooperation between the riparian states. In 1938, largely in response to fears regarding the amount of Colorado River water promised to Mexico under a pending boundary treaty, all seven Colorado basin states formed the Committee of Fourteen, with two representatives from each state, to persuade Washington not to surrender a drop of Colorado River water to Mexico (Nadeau, 1974). The committee petitioned President Roosevelt to abandon negotiations on the US–Mexico treaty, and instead to serve notice to the government of Mexico that "it was the policy and purpose of the United States to reserve for use within this country all waters of the Colorado River". Reflecting their distrust of Washington, however, the conference also asked the federal government to withdraw claims asserted during the Arizona Supreme Court litigation as to federal ownership and control of non-navigable streams within the Colorado basin (*Los Angeles Times*, 1938). At the same time that the committee advocated confrontation with Washington, however, it promoted comity among the basin states. As an Arizona state government report noted, "Through this Committee of Fourteen, friendly relations have been built up between all the basin states" (Colorado River Commission of Arizona, 1940, p. 27).

These improved relations did not, however, end interstate conflict in the basin. Instead, they merely pitted the other states against a new feared water hegemon in the form of California and its senior claims to Colorado River water. In 1943–1944, all five Upper Basin states rewarded Arizona for its opposition to the US–Mexico treaty by defying California and voting to approve Arizona's entry into the Colorado River Compact under favourable terms (*Los Angeles Times*, 1943). The federal government encouraged the states to resolve their differences, but in a concession to the animosity which existed between California and the other basin states, nonetheless maintained a degree of remove. In 1944, again over California's strenuous objections, Interior Secretary Harold Ickes approved a contract delivering 2.8 MAF annually to Arizona, provided that its legislature ratify both the contract and the compact, which was promptly done. Reflecting federal officials' reluctance to become embroiled in the deep-seated conflict, however, the contract "specifically provides that it does not resolve the issues between the two States and that it is without prejudice to their respective claims" (*Los Angeles Times*, 1944). After the war, enhanced federal efforts to induce cooperation among the riparian states would meet with even greater frustration.

Preserving autonomy and seeking appropriations: frustrated federal involvement in interstate disputes

After 1945, driven by rising demand for water as the economy boomed and the population of the western states increased dramatically, the US federal government made a determined attempt to resolve interstate conflict in the Colorado River basin. In the early 1960s, Interior Secretary Stewart Udall attempted to finally overcome the problem of autonomy by proposing the gigantic Pacific Southwest Water Plan (PSWP). The PSWP aimed to meet the energy and water needs of the entire western United States by developing large dams, canals and transmission lines to channel water and power from the Pacific Northwest to the Southwest, side-stepping the Colorado River basin dispute (Dominy, 1964). This regional approach, Udall proclaimed, promised to finally transcend "the outmoded concept of state lines.... Only regional planning and action will enable us to meet the growth needs of this area" (Sherman, 1963). Yet despite the stridency of Udall's support for such a regional solution, the secretary was soon forced to acknowledge the significance of the very state lines he had initially disparaged. In testimony before Congress, Udall promised that "The rights of the individual states likewise must be respected and the aspirations of the states accommodated" (US Senate Committee on Interior and Insular Affairs, 1963, p. 311).

Indeed, state political leaders and their allies in Congress quickly mobilized to thwart aspects of the PSWP which threatened state control over water. Powerful Democratic Senator Henry Jackson of Washington grew concerned about proposed water transfers from the Columbia River to the more arid parts of the Southwest, and demanded that the PSWP be reviewed by an independent National Water Commission. As Washington's governor testified before Congress, "We are a downstream state within a basin whose waters have consistently been looked upon and prejudged as a source of import supply" (US Senate Committee on Interior and Insular Affairs, 1967, p. 631). Meanwhile, national conservation organizations lobbied representatives of eastern states, such as Republican Congressman John Saylor of Pennsylvania, to oppose the PWSP because of its threatened effects on the West's dramatic natural scenery (Fleming, 1954). The combined influence of this congressional opposition outside the Colorado River basin was sufficient to dramatically reduce the scale of the PSWP, returning the congressional debate to projects favoured by the individual states.

In light of the effective state opposition to the attempts of federal agencies to settle the Colorado River basin dispute themselves, a California–Arizona lawsuit was left to determine the final partition of Lower Basin water. The matter remained as contentious as in previous phases of the dispute; a pre-trial conference so exasperated a court-appointed special master that he resigned, saying simply, "I give up" (*Los Angeles Times*, 1956). After a 26-month trial, a re-appointed special master upheld Arizona's claim to an additional million acre-feet of Colorado River water, allocating it 2.8 MAF, and reducing California's allotment by 978,000 acre-feet annually, to 4.4 MAF (*Chicago Daily Tribune*, 1960). The court itself upheld this partition in its final 1963 decision in *Arizona v. California*, and further re-affirmed the interior secretary's power to apportion water among water users in the Lower Basin states, particularly in times of drought (Blair, 1963). Federal officials, expectant that the court's decision had finally solved the dispute, appeared eager to wash their hands of the matter. Secretary Udall refused to become embroiled in continued controversy surrounding California's 4.4 MAF, telling Congress, "It is our view that this is a matter between the states, and however they want to work this out is satisfactory so far as the administration is concerned" (US Senate Committee on Interior and Insular Affairs, 1967).

The court's decision set the stage for the Colorado River Project Act in 1968, which authorized a raft of water infrastructure projects throughout the basin, finally promising states the sovereignty over water resources they had long sought. But during debates over the act, a new force emerged which would once again highlight the tension between state sovereignty over water and true basin-wide cooperation. The salience of environmental issues in Colorado River basin governance became clearly marked in 1969, when President Johnson used his last 90 minutes in office to protect several areas along the river from ranching, mining and other extractive activities. While conservation groups like the Sierra Club praised the move as "true conservation in the public interest", state officials were furious. Utah's director of natural resources expressed hope that the next administration would "undo the damage Johnson has done", while Utah Republican Senator Bob Bennett called the federal designation "the most blatant type of greed I can imagine" (Goodman, 1969). Despite this familiar state objection to perceived federal over-reach, the growing political salience of environmental issues induced Washington to take a significantly more active and constructive role in governance of the Colorado basin after the 1970s than it had in previous decades.

In part, the federal government's increased role in basin governance relative to the states during the 1970s was both the result of and enabled by its long-standing prerogatives in the field of foreign relations. Soon after taking office, President Nixon made improving ties with Mexico a priority, including solving the growing problem of salinity in the Lower Basin, which Mexican officials claimed made Colorado water essentially worthless by the time it entered Mexico, in violation of the 1944 treaty (Semple, 1970). The issue became so politically heated that a 1972 article noted that "When Mexicans discuss relations between their country and the United States, they tend to mention first the level of mineral salts in the Colorado River" (Severo, 1972). This continued attention to the issue bore fruit when in 1973 President Nixon announced that the US would commit to building what was then the world's largest desalinization facility, to ensure that water flowing into Mexico remained of sufficient quality to support irrigation (Childs, 1973). Concentrated economic interests in the United States, however, grew concerned that this commitment would reduce water availability in the southwestern states, and exerted their influence in Congress by persuading it to designate the Department of the Interior, over which they held more influence, as the lead agency on the desalinization project, rather than the State Department as the administration had proposed (McElheny, 1974). This victory proved short-lived, however, as the environmental movement changed the way that all federal bureaucracies approached institution building in the basin.

The creation of cooperation: rise of the environmental movement

The end result of the federal effort to tackle desalination was Washington's close involvement in environmental protection in the Colorado River basin, a process in

turn facilitated by the growing influence of the environmental movement. The involvement of third-party ENGOs created a critical constituency for basin-wide cooperation
that eventually helped diffuse the long-running interstate dispute. The 1970s opened
with passage of several landmark pieces of environmental legislation, most notably the
National Environmental Policy Act (NEPA) of 1970 and the Endangered Species Act of
1973, both of which transformed the federal government's approach to Colorado River
basin governance. Nonetheless, as state opposition to Johnson's conservation measures
illustrates, success was far from inevitable, and it required the skilful exploitation by
environmental organizations of new lobbying opportunities created by environmental
legislation. As early as 1972, barely two years after NEPA created the environmental
impact assessment process, organizations like the Sierra Club and Friends of the Earth
provoked a distinct change in orientation among federal agencies like the Bureau of
Reclamation (BuRec) and the US Army Corps of Engineers. Where once these agencies
had been obsessed with "quick technological fixes", by 1972 they were actively considering conservation-oriented approaches (Ripley, 1972). The same year, the
Environmental Defense Fund illustrated the growing power of environmental organizations by suing BuRec under NEPA for proposing to sell water allocated by Congress for
irrigation to coal companies, an unprecedented challenge to what had previously been
an unassailable bureaucratic prerogative (Kenworthy, 1973).

In the Upper Basin, meanwhile, environmentalists joined with residents of
Colorado's West Slope region to oppose completion of the Fryingpan-Arkansas
Project, which was strongly favoured by BuRec and Colorado congressional representatives, including powerful Democrat Wayne Aspinall. One of the most powerful
congressmen of the postwar period, Aspinall represented what was at the time the
largest single congressional district in the continental United States, and one with little
appetite for the environmentalism that was increasingly influential in other parts of the
country. A 1971 *New York Times* profile quoted a congressional colleague as saying that
"In many ways he is very, very provincial. It is fair to say that he has repeatedly put his
district above the national interest" (Fradkin, 1971). But such provincialism had by the
early 1970s ceased to be the *modus operandi* in the Colorado River basin. The changing
balance of power between environmental and rights organizations on one hand and
BuRec and state officials on the other was illustrated at the 1972 dedication of the
Fryingpan-Arkansas Project, when Interior Secretary Rogers Morton acknowledged
"diverse opinions" concerning its wisdom, and Colorado Republican Governor John
Love declared that the project represented "the end of an era" of large water infrastructure projects (*New York Times*, 1972).

At the same time, provisions of the Endangered Species Act of 1973 began to fracture
the previous alliance of convenience between federal agencies and state governments in
support of reclamation projects. Following passage of the act, elements of the federal
bureaucracy began to ally themselves with the environmental movement, instead of
supporting reclamation projects favoured by states as they had hitherto done almost
without exception. In 1981, the Colorado River Water Conservation District, which
supported additional dam construction, protested to the Interior Department that "As
viewed by the federal courts and agencies, the Endangered Species Act ... stands
supreme in the land no matter what it costs the country and its people in lost
opportunities" (Associated Press, 1981). While this objection would at one time have

reached sympathetic ears in the Department of the Interior, pressure from a national conservation organization, the National Wildlife Federation, compelled Interior Secretary James Watt to cancel a planned expansion of hydropower facilitates at Glen Canyon Dam, despite his reputation as an anti-conservationist (Shabecoff, 1981).

Within the executive branch, environmental interest groups were growing rapidly in influence, and altered agency priorities in the Colorado River basin in favour of conservation and management rather than reclamation. A watershed moment occurred in 1987, when BuRec officials announced that the agency would cease initiating major construction projects, and would instead change its "orientation from construction to resource management" (Coates, U.S. pulls plug on new dams 1987). The Environmental Protection Agency, which hitherto had played little role in western water issues, helped solidify this trend by blocking construction of a Colorado dam strongly favoured by state political leaders, citing its disruption of a number of threatened species (Coates, 1989). Although tensions between environmental, flood control and irrigation priorities persisted, especially in the Lower Basin, federal agency leaders increasingly identified with conservation objectives (Lancaster, 1990). In 1994, the Clinton administration announced a change in managing releases from Glen Canyon Dam to minimize variation in flows downstream, a move that BuRec's commissioner said heralded "the difference between the old way we have treated the river and the way of the future" (Kenworthy, 1994). Interior Secretary Bruce Babbitt, meanwhile, proclaimed a "new era for ecosystems" and "a new era for dam management" (Associated Press, 1996).

Even more significant than this change in policy, however, was a newfound will-ingness on the part of federal leaders to constructively engage in Colorado River basin governance. In 1997, Interior Secretary Babbitt announced a new federal rule designed to encourage interstate water transfers as an alternative to litigation and political re-allocation of increasingly scarce water. Babbitt, breaking with a long tradition of federal neutrality in interstate water disputes, warned California, "the time has come for me as the River Master to play a more active role" and impose limits on deliveries of Colorado River water to the state (Perry, 1997). On the strength of this unprecedented personal involvement, the seven riparian states of the Colorado River basin in 1999 reached a preliminary agreement to resolve the remaining issues concerning allocation on the Colorado. Under the agreement, the states agreed to a process under which the federal government would begin reducing water deliveries, especially to California, which for its part committed to an ambitious effort to line canals and construct underground storage to accelerate transfers from agricultural to urban areas (Purdum, 2000).

The federal government's newfound commitment to cooperation in the Colorado River basin further increased during the Bush administration. Beginning in the mid-1990s, multi-state Adaptive Management Working Groups representing federal, state and non-governmental stakeholders had been formed, and in the early 2000s these groups scored notable successes, including building support for plans to adjust stream-flow to protect endangered species (Blakeslee, 2002a, 2002b). Even more importantly, when a vote by the IID board threatened to imperil the 1999 interstate agreement, Interior Secretary Gale Norton did not back down on a threat to reduce deliveries of Colorado River water in order to compel IID to approve the sale of water to San Diego (Murphy, 2003). Despite IID's protests that this threat constituted "an example of heavy-handed and unwarranted federal interference" (Murphy, 2003c), Norton quickly gained the enthusiastic support of the other six basin states, environmental groups, and California state legislators, who were loath to surrender deliveries of Colorado River water. IID finally bowed to this combination of federal, state and non-governmental pressure by agreeing to the transfer, thus signalling a dramatic shift in the stance of the Colorado River basin's most important single economic actor (Murphy, 2003b).

This shift in IID's attitude towards cooperation in turn set the stage for another, even more productive, phase of institution building in the Colorado River basin, even as physical water scarcity pressures grew acute. In 2007, with the determined personal involvement of Norton's successor, Dick Kempthorne, a set of guidelines were con-cluded which superseded the 1999 agreement and specified how the Lower Basin states would share the burden of water use reductions during drought (Archibold, 2007). As the drought grew more acute in the mid-2010s, the Lower Basin states, along with the Bureau of Reclamation, agreed to further curtail water use by implementing conserva-tion measures to preserve water levels in Lake Mead, and to increase storage by 1.5 MAF over time (Brean, 2014). California's willingness to accept further restrictions marked a striking change from the prior history of basin competition. As one partici-pant recalled, the state's "original attitude was 'We have the priority and those look like difficult problems – good luck.'" However, the very severity of the drought led California officials to calculate that senior water rights would not be sufficient to protect the state from rationing in the event that Lake Mead water levels dropped further. As a California water official stated, "We absolutely agree we're better off having the lake higher. We can't only rely on our priority and say we have our priority and we're protected" (Davis, 2014). The dilemma of autonomy which produced weak institution building among the Colorado riparian states in the twentieth century has not disap-peared in the twenty-first, but it has been attenuated by a constructive partnership between environmental groups and the federal government.

Conclusion

This article has attempted to propose an initial framework for understanding the role of subnational states in water politics. This framework consists of three key elements: a process and structure of decentralization, which creates a role for subnational states in water resource management; a basic concern with autonomy, which leads subnational states to prefer loose forms of cooperation over shared water resources; and finally pressure from third-party NGOs and national political leaders, which creates the constituency for collaborative and participatory institution building. This framework has been applied to the case of the Colorado River basin, and I have argued that it explains the evolution of cooperation in the basin over time, from an initial phase of weak institution building to a more institutionalized and cooperative approach in recent decades. More specifically, the role of subnational states vis-à-vis the federal government and civil society explains variation in the extent of cooperation in the basin, and suggests that subnational states may be the primary determinants of insti-tutionalized collective action in shared river basins.

This observation can be formulated into a testable hypothesis that might inform further research: the greater the extent of political decentralization in a given country, the more likely it is to feature weak, less institutionalized forms of cooperation over shared water resources. However, the case of the Colorado has unfolded in the distinctive context of American federalism, and validating this hypothesis would require testing against a greater number and variety of cases. Such additional research would help clarify the impact of decentralization on interjurisdictional collective action. In the meantime, the framework I have proposed suggests the need for greater attention to the interests and incentives of subnational political actors in water politics, especially when decentralization accords them a more significant role relative to national-level actors. Indeed, while scholars concerned with water resource management often direct their focus to the national level, the case of the Colorado suggests that subnational officials frequently determine the success or failure of national water resource policy reforms. This phenomenon is potentially of great consequence, for it may help explain why, despite widespread consensus on the need for water resource management to adopt more collaborative and participatory approaches, progress towards implementation in many countries remains limited.

Disclosure statement

No potential conflict of interest was reported by the authors.

References

Abers, R., & Keck, M. (2013). *Practical authority: Agency and institutional change in Brazilian water politics*. New York, NY: Oxford University Press.

Abers, R., & Keck, M. E. (2009, June). Mobilizing the state: The erratic partner in Brazil's participatory water policy. *Politics & Society*, *37*(2), 289–314. doi:10.1177/0032329209334003

Abers, R. N., & Keck, M. E. (2006, September). Muddy waters: The political construction of deliberative river basin governance in Brazil. *International Journal of Urban and Regional Research*, *30*(3), 601–622. doi:10.1111/ijur.2006.30.issue-3

Alam, U., Dione, O., & Jeffrey, P. (2009). The benefit-sharing principle: Implementing sovereignty bargains on water. *Political Geography*, *28*, 90–100. doi:10.1016/j.polgeo.2008.12.006

Andersson, K. P., Gibson, C. C., & Lehoucq, F. (2006). Municipal politics and forest governance: Comparative analysis of decentralization in Bolivia and Guatemala. *World Development*, *34*(3), 576–595. doi:10.1016/j.worlddev.2005.08.009

Archibold, R. (2007, December 10). Western states agree to water-sharing pact. *New York Times*, p. A18.

Arizona Republican. (1922, December 17). Division of the waters of the Colorado river is Maxwell's theme. *Proquest Historical Newspapers*, December 2011.

Ascher, W. (2007). Issues and best practices in the decentralization of natural resource control in developing countries. In G. S. Cheema & D. Rondinelli (Eds.), *Decentralizing governance: Emerging concepts and practices* (pp. 292–305). Washington, DC: Brookings Institution Press.

Associated Press. (1981, December 6). Species Act Battle; law protecting endangered animals, plants faces review in Congress. *Boston Globe*, p. 1.

Associated Press. (1996, March 27). Artificial flood created to rejuvenate the Grand Canyon. *New York Times*, p. B8.

Bakker, K. (2010). *Privatizing water: Governance failure and the world's urban water crisis*. Ithaca, NY: Cornell University Press.

Bennett, L. L. (2000). The integration of water quality into transboundary allocation agreements lessons from the southwestern United States. *Agricultural Economics*, *24*, 113–125. doi:10.1111/agec.2000.24.issue-1

Berardo, R., & Scholz, J. T. (2010, July). Self-organizing policy networks: Risk, partner selection, and cooperation in estuaries. *American Journal of Political Science, 54*(3), 632–649. doi:10.1111/(ISSN)1540-5907

Blair, W. (1963, June 3). 3 Justices strongly oppose provision allowing US to apportion supplies. *New York Times*.

Blakeslee, S. (2002a, April 26). U.S. Panel backs a risky effort to save a Grand Canyon fish. *New York Times*, p. A16.

Blakeslee, S. (2002b, June 11). Restoring an ecosystem torn asunder by a dam. *New York Times*, p. F1.

Blatter, J., & Ingram, H. (2001). *Reflections on water: New approaches to transboundary conflicts and cooperation*. Cambridge, MA: MIT Press.

Bouleau, G. (2009). La contribution des pecheurs a la loi sur l'eau de 1964. *Economie rurale, 309*, 9–21.

Brean, H. (2014, December 15). Cooperation keys latest pact to protect Lake Mead. *Las Vegas Review-Journal*. Retrieved January 1, 2015, from http://www.reviewjournal.com/news/coopera tion-keys-latest-pact-protect-lake-mead

Carpenter, D. (1923). Report and supplemental report of Delph Carpenter, commissioner for Colorado on the Colorado River Commission. In Colorado River Commission (Ed.), *Pamphlets on water rights and the use of water power* (Vol. 1). Denver, CO: Colorado River Commission.

Chicago Daily Tribune. (1960, May 9). California hard hit by Colorado River ruling. *Chicago Daily Tribune*, p. 1.

Childs, M. (1973, September 11). Cleaning up the Colorado. *Washington Post*, p. A21.

Christensen, N. S., Wood, A. W., Voisin, N., Lettenmaier, D. P., & Palmer, R. N. (2004, January). The effects of climate change on the hydrology and water resources of the Colorado River Basin. *Climatic Change, 62*(1–3), 337–363. doi:10.1023/B:CLIM.0000013684.13621.1f

Clark, W. G. (1924). *The Colorado River: History, seven states compact and future development*. Phoenix, AZ.

Coates, J. (1987, November 15). U.S. pulls plug on new dams. *Chicago Tribune*, p. 25.

Coates, J. (1989, October 8). Water woes put Denver in a bind. *Chicago Tribune*, p. 3.

Colorado River Commission of Arizona. (1940). *Annual report of the Colorado River Commission of Arizona, November 1, 1939 - November 1, 1940* (Annual Report). Phoenix, AZ: Colorado River Commission of Arizona.

Davis, T. (2014, December 21). How California got on board for the water deal. *Arizona Daily Star*. Retrieved January 2, 2015, from http://tucson.com/news/blogs/desertblog/how-califor nia-got-on-board-for-the-water-deal/article_b55d3960-88e4-11e4-ae69-67118550eb7f.html

Derthick, M. (1974). *Between state and nation: Regional organizations of the United States*. Washington, DC: Brookings Institution.

Bureau of Reclamation Report; Dominy, F. (1964). *Pacific southwest water plan*. Washington, DC: US Department of the Interior.

Elazar, D. (1984). *American federalism: A view from the states*. New York, NY: Harper and Row.

Fleming, R. (1954, September 24). Eisenhower pushes upper Colorado basin project. *Christian Science Monitor*, p. 3.

Forum on Federations. (2009, January/February). Water and intergovernmental relations. *Federations*, pp. 8–27.

Fradkin, P. (1971, December 5). Rep. Aspinall, boss of the west: Coloradan backs land and water uses. *Washington Post*, p. E1.

Franda, M. (1968). *West Bengal and the federalizing process in India*. Princeton: Princeton University Press.

Garrick, D., Anderson, G., Connell, D., & Pittock, J. (2014). Federal rivers: A critical overview of water governance challenges in federal systems. In D. Garrick, G. Anderson, D. Connell, & J. Pittock (Eds.), *Federal rivers: Managing water in multi-layered political systems* (pp. 3–19). Cheltenham: Edward Elgar.

Giordano, M., Giordano, M., & Wolf, A. (2002, December). The geography of water conflict and cooperation: Internal pressures and international manifestations. *The Geographical Journal*, *168*(4), 293–312. doi:10.1111/geoj.2002.168.issue-4

Goodman, J. (1969, February 2). Protecting the Colorado's banks. *New York Times*, p. XX15.

Heikkila, T., Schlager, E., & Davis, M. W. (2011, February). The role of cross-scale institutional linkages in common pool resource management: Assessing interstate river compacts. *Policy Studies Journal*, *39*(1), 121–145. doi:10.1111/psj.2011.39.issue-1

Hooghe, L., & Marks, G. (2003). Unraveling the central state, but how? Types of multi-level governance. *American Political Science Review*, *92*(2), 233–243.

Hundley, N. (1975). *Water and the west: The Colorado River compact and the politics of water in the American west*. Berkeley: University of California Press.

Imperial, M. T. (2005). Using collaboration as a governance strategy: Lessons from six watershed management programs. *Administration & Society*, *37*(3), 281–320. doi:10.1177/0095399705276111

Jacobs, J. (2015, October 29). Rising tide of interstate battles could swamp supreme court. *Energy and Environment Publishing*. Retrieved from http://www.eenews.net/stories/1060027067

Keck, M., & Sikkink, K. (1998). *Activists beyond borders: Advocacy networks in international politics*. Ithaca, NY: Cornell University Press.

Kenworthy, E. W. (1973, October 17). Environmental Defense Group suing federal agency to prevent the use of irrigation water for industry. *New York Times*, p. 13.

Kenworthy, T. (1994, January 7). Dam limits proposed to aid grand canyon. *Washington Post*, p. A3.

King, G., Keohane, R., & Verba, S. (1994). *Designing social inquiry: Scientific inference in qualitative research*. Princeton: Princeton University Press.

Kitschelt, H. P. (1986, January). Political opportunity structures and political protest: Anti-nuclear movements in four democracies. *British Journal of Political Science*, *16*(1), 57–85. doi:10.1017/S000712340000380X

Kriesi, H., Koopmans, R., Duyvendak, J. W., & Giugni, M. G. (1992). New social movements and political opportunities in Western Europe. *European Journal of Political Research*, *22*(2), 219–244. doi:10.1111/ejpr.1992.22.issue-2

Lancaster, J. (1990, April 27). Lujan bars testimony by 2 officials. *Washington Post*, p. A17.

Lankford, B., & Hepworth, N. (2010). The Cathedral and the Bazaar: Monocentric and poly-centric river basin management. *Water Alternatives*, *3*(1), 82–101.

Los Angeles Times. (1923, January 30). Arizona swats compact again. *Proquest Historical Newspapers*, December 2011.

Los Angeles Times. (1937, May 15). River states hit Arizona. p. 1.

Los Angeles Times. (1938, June 24). States push water plans: Organization formed at peaceful session on Colorado river supply. p. 17.

Los Angeles Times. (1943, December 19). Utah approves river pact entry by Arizona. p. 13.

Los Angeles Times. (1944, February 10). Ickes approves water accord with Arizona. *Los Angeles Times*, p. 3.

Los Angeles Times. (1956, April 12). Agreement on Colorado water fails. *Los Angeles Times*, p. 25.

Lubell, M. (2004, August). Collaborative watershed management: A view from the grassroots. *Policy Studies Journal*, *32*(3), 341–361. doi:10.1111/psj.2004.32.issue-3

Marks, G., & McAdam, D. (1996). Social movements and the changing structure of political opportunity in the European Union. *West European Politics*, *19*(2), 249–278. doi:10.1080/01402389608425133

McElheny, V. (1974, June 27). Desalting plant faces year delay. *New York Times*, p. 7.

Montero, A. P. (2001, January). Decentralizing democracy: Spain and Brazil in comparative perspective. *Comparative Politics*, *33*(2), 149–169. doi:10.2307/422376

Mullin, M. (2009). *Governing the tap: Special district governance and the new local politics of water*. Cambridge, MA: MIT Press.

Murphy, D. (2003, January 1). California water users miss deadline on pact for sharing. *New York Times*, p. A11.

Murphy, D. (2003b, January 5). In a first, U.S. officials put limits on California's thirst. *New York Times*, p. L1.

Murphy, D. (2003c, January 12). County water board sues U.S. in dispute over Colorado River. *New York Times*, p. L16.

Nadeau, R. (1974). *The water seekers*. Santa Barbara, CA: Peregrine Smith.

New York Times. (1972, July 5). Water diversion tunnel is dedicated in Colorado as its opponents protest. *New York Times*, p. 78.

Paavola, J. (2006). Interstate water pollution problems and elusive federal water pollution policy in the United States, 1900-1948. *Environment and History*, *12*(4), 435–465. doi:10.3197/096734006779093659

Perry, T. (1997, December 19). Babbitt warns California district on Colorado River usage. *Washington Post*, p. A41.

Purdum, T. (2000, July 28). 7 states in West reach outline. *New York Times*, p. A12.

Ripley, A. (1972, January 17). West's thirst for water is questioned. *New York Times*, p. 1.

Sabatier, P., Focht, W., Lubell, M., Trachtenberg, Z., Vedlitz, A., & Matlock, M. (2005). *Swimming upstream: Collaborative approaches to watershed management*. Cambridge, MA: MIT Press.

Schlager, E., & Heikkila, T. (2009, August). Resolving water conflicts: A comparative analysis of interstate river compacts. *Policy Studies Journal*, *37*(3), 367–392. doi:10.1111/psj.2009.37.issue-3

Semple, R. (1970, August 22). U.S. - Mexico pact details boundary. *New York Times*, p. 1.

Severo, R. (1972, June 11). Colorado River salt annoys Mexicans. *New York Times*, p. 6.

Shabecoff, P. (1981, October 30). Watt drops a power project threatening Colorado River. *New York Times*, p. A14.

Sherman, G. (1963, July 15). Arizona fears federal politics. *Los Angeles Times*, p. 2.

Smith, R. (1937, April 25). Arizona, seeking water, continued Boulder Dam fight. *Washington Post*, p. B7.

Teclaff, L. (1996). Evolution of the River basin concept in national and international water law. *Natural Resources Journal*, *36*(359), 359–391.

Thorson, J. (1994). *River of promise, river of peril: The politics of managing the Missouri River*. Lawrence: University Press of Kansas.

Treisman, D. (2007). *The architecture of government: Rethinking political decentralization*. New York, NY: Cambridge University Press.

UNFAO. (2010). Waterlex. Retrieved January 8, 2010, from http://waterlex.fao.org/waterlex/srv/en/home

US Bureau of Reclamation. (2008, December). The Colorado River. Retrieved February 17, 2012, from http://www.usbr.gov/lc/hooverdam/faqs/riverfaq.html

US Senate Committee on Interior and Insular Affairs. (1963). *Hearings before the subcommittee on irrigation and reclamation on S.1658 a bill to authorize, construct and maintain the Central Arizona Project and for other purposes* (Senate Committee Report).Washington, DC: US Government Printing Office.

US Senate Committee on Interior and Insular Affairs. (1967). *Hearings before the subcommittee on water and power resources of the Committee on Interior and Insular Affairs- Central Arizona Project*. Washington, DC: US Government Printing Office.

Watts, R. (2008). *Comparing Federal Systems*. Montreal: McGill-Queen's University Press.

Weber, E. (2003). *Bringing society back in: Grassroots ecosystem management, accountability, and sustainable communities*. Cambridge, MA: MIT Press.

Wirsing, R. G. (2007). Hydro-politics in South Asia: The domestic roots of interstate river rivalry. *Asian Affairs: An American Review*, *34*(1), 3–22. doi:10.3200/AAFS.34.1.3-22

Wood, J. (2007). *The politics of water resource development in India: The Narmada dams controversy*. New Delhi: SAGE Publications.

Worster, D. (1985). *Rivers of empire: Water, aridity, and the growth of the American West*. New York, NY: Oxford University Press.

Zimmerman, J. (2011). *Horizontal federalism: Interstate relations*. Albany: State University of New York Press.

Index